I0605120

Lab Dog

Lab Dog

WHAT GLOBAL SCIENCE OWES AMERICAN BEAGLES

Brad Bolman

The University of Chicago Press *Chicago and London*

The University of Chicago Press, Chicago 60637
The University of Chicago Press, Ltd., London
© 2025 by The University of Chicago

All rights reserved. No part of this book may be used or reproduced in any manner whatsoever without written permission, except in the case of brief quotations in critical articles and reviews. For more information, contact the University of Chicago Press, 1427 E. 60th St., Chicago, IL 60637.
Published 2025
Printed in the United States of America

34 33 32 31 30 29 28 27 26 25   1 2 3 4 5

ISBN-13: 978-0-226-82553-3 (cloth)
ISBN-13: 978-0-226-83974-5 (paper)
ISBN-13: 978-0-226-83973-8 (e-book)
DOI: https://doi.org/10.7208/chicago/9780226839738.001.0001

Library of Congress Cataloging-in-Publication Data

Names: Bolman, Brad, author.
Title: Lab dog : what global science owes American beagles / Brad Bolman.
Description: Chicago : The University of Chicago Press, 2025. | Includes bibliographical references and index.
Identifiers: LCCN 2024042374 | ISBN 9780226825533 (cloth) | ISBN 9780226839745 (paperback) | ISBN 9780226839738 (ebook)
Subjects: LCSH: Dogs as laboratory animals. | Beagle (Dog breed)—United States. | Animal experimentation—United States. | Animal experimentation—Moral and ethical aspects—United States. | Animal rights—United States.
Classification: LCC SF407.D6 B65 2025 | DDC 616.02/73—dc23/eng/20241128
LC record available at https://lccn.loc.gov/2024042374

♾ This paper meets the requirements of ANSI/NISO Z39.48-1992 (Permanence of Paper).

For Joan and Lee
and for Katie and Laszlo

In memory of Douglas and Van Gogh

Let us go beagling, reader dear. Let us be retiring to bed in a state of gloom brought about by the way the world is using us, by the recent increase of five shillings in the pound on the School Board rate, by the never-ending slump in the stock market. Suddenly our eyes brighten, and we draw ourselves up and absolutely grin: for to-morrow is beagling day, and the slumps are all gone in a moment.

J. F. SULLIVAN, 1896

The proper subject for the study of man is man. Or as Marvin Goldman says, the dog is always right, our job is to find out why he's doing it.

ROBLEY EVANS, 1978

CONTENTS

PREFACE

In September 2019, I found myself nestled in the back corner of a Massachusetts State House hearing concerning "animals and other local and miscellaneous matters." A sizeable crowd had assembled within the small chamber, and some complained about the heat. Although a few people were there to discuss the market for illicit elephant tusks and rhino horns in Massachusetts, most had come for one particular piece of legislation, H.758, which required that cats and dogs be released for adoption, rather than euthanized, after participating in scientific research. Cats, however, were hardly mentioned. What inspired public comment on this mild Tuesday morning was the fate of dogs, specifically the thousands of beagles used annually by researchers in the United States. The legislation was known, colloquially, as "The Beagle Bill."

A vibrant procession of beagle fans crowded the room. An older man with a gray ponytail held a large, framed photograph of two dogs: "My girls," he clarified. A woman nearby wore an oversized button on her jean jacket that displayed someone very much like her hugging a beagle. Tucker, a rescue, waited with caretakers outside the State House entrance while a companion testified on his behalf. Many speakers stressed the powerful human bond with dogs, who were friends, companions, and family members. Some expressed gratitude for their sacrifices in the interests of science. Later, when the ponytailed man concluded his comments, rich with descriptions of "beagle dance parties," he offered a vision of the future for everyone present. It was his slightly corny dream, he explained, to one day see a rescue beagle at every firehouse in Massachusetts. The only ones opposed, quipped one lawmaker, would be the dalmatians.

Readers of a certain vintage may remember when beagle studies of chemical weapons and cigarette smoking made headlines in the 1970s.

For most, those controversies are ancient history. Yet beagles have returned to the public spotlight over the last few decades, garnering an outsized role in emotional debates about animal experimentation. In 2017, for instance, criticism over their possible mistreatment in studies funded by the US Department of Veterans Affairs led to bipartisan introduction of a fittingly acronymed "PUPPERS Act," which would end public funding for dog experiments. Soon after, a committee of the National Academies of Sciences, Engineering, and Medicine was tasked with analyzing how Veterans Affairs research with dogs should continue, if at all.

Some in the animal rights community expect that within one or two decades, beagles will largely vanish from laboratory research in the United States and Europe. Groups such as the Beagle Freedom Project, a nonprofit that drafts and supports legislation such as H.758, see this as an end to inexcusable torment and a possible step toward the abolition of animal experimentation. In July 2022, the release of four thousand beagles from the Virginia facilities of contract research organization Envigo, following findings of animal welfare violations, appeared to confirm their optimism. And, although it took three years to wind its way through the sclerotic Massachusetts legislature, Governor Charlie Baker signed the Beagle Bill into law in August 2022.

You have probably heard something about beagles. But how did traditional British hunting hounds come to be rhetorical centerpieces in critiques of some of America's most prominent infectious disease experts? This is a book about how beagle dogs became densely entangled with the day-to-day work of international scientific research and subject to fierce debates about the value and ethics of animal experimentation. It is neither a condemnation nor a defense of such work. Instead, in seeking to answer a deceptively simple question—Why beagles?—this book excavates the history of the laboratory dog in modern science and shows how dogs became intensely scientific beings over the last hundred years, reimagined as cognitively complex human analogues who merit moral treatment and daily pharmaceuticals.

Modern life, after all, is a resolutely dogged endeavor. The popular John Wick film franchise, which has grossed over a billion dollars globally, is symptomatic: A former hitman, portrayed by Keanu Reeves, carries out an elaborate campaign of revenge after a group of mobsters steal his muscle car and, most crucially, kill his beagle, Daisy.[1] Wick spends three subsequent films balancing escalatingly ludicrous action sequences with care for an adopted pit bull. In the United States, where 65.1 million households are thought to own a pup, advertising has particularly gone to the dogs: they are reasons to lease expensive sports cars, shop at Amazon, or just spend more on pet food.[2] Service dogs are widespread and legally protected, and emotional support animals now appear routinely in institutions. That an athlete has "got that dog in him" means he is tough and performing at a high level, while a viral inversion of the idea, featuring a painting by Alison Friend of a diminutive terrier seated on a plush chair, explains, "When I say I got that dog in me this is what I mean."[3]

We see ourselves in dogs and dogs in ourselves, whether on screens or in our homes. This makes their role in scientific research special. Yet despite countless books on dogs, their training, and their behavior, surprisingly few histories of the laboratory dog exist.[4] Why is that? The philosopher Søren Kierkegaard once distinguished a thinker's conceptual system from his everyday existence by comparing a palace to a dog kennel: intellectual work aims at the erection of a majestic palace, but life happens in the kennel.[5] The same could be said for the popular history of science, which is crowded with transformative discoveries based on animal research but has less to say about the quotidian aspects of experimental life. Ivan Pavlov is a household name, but few know how he got his dogs or why he chose them over another species. To collect gastric secretions, he and his assistants first had to find mixed-breed strays on the streets of St. Petersburg. They also needed collars and leashes in order to handle them. All of this presupposed a wider universe of knowledge about dogs, or a dog culture, that predated the first drooly ring of the bell.[6] Our understanding of science, then, needs to account for the kennel as well as the palace.[7]

On the other hand, our contemporary understanding of dogs is profoundly shaped by science. Magazine spreads and social media posts continuously extoll remarkable findings about canines: that they can recognize objects by name, for instance, or sniff out cancers. Knowing dogs, then, requires studying the history of science. In one approach to doing so, scholars sometimes draw a distinction between scientific *practice* and scientific *culture.* "A hammer, nails, and some planks of wood are not the same as the act of building a dog kennel," sociologist Andrew Pickering offers by way of analogy, "though a completed dog kennel might well function as a resource for future practice (training a dog, say)."[8] *Lab Dog* shows how real, completed kennels functioned as vital resources not only for training dogs, but for the future practice of science; how knowledge and skill from hunting, sports, and pet keeping transformed laboratory experimentation; and how those leisure and consumer practices were, in turn, reshaped by the practice and culture of science. We have learned much about ourselves by studying beagles, this book shows, but far too little about the dogs or those who studied them.

Introduction

Beagles, like any breed of dog, have a history. Once, like plastics and personal computers, these small black, white, and tan hounds with plaintively musical howls did not exist. When *Canis familiaris* first appeared in the archaeological record, perhaps eleven thousand years ago, early Pleistocene dogs bore few similarities to their future Westminster Dog Show–winning kin. What happened over the thousands of years that followed is still the subject of concerted debate among the world's canine geneticists and evolutionary biologists, but something akin to beagles began to emerge during the fifteenth century. In Britain, shifting class politics gradually supported the popularization of pet ownership and the specialization of breeders, who crafted increasingly distinct styles of dog for a variety of purposes. Over the centuries that followed, dogs called "beagles" followed their masters across the world's oceans to the Americas and far beyond.

The basic pieces of that trajectory hold for many recognized breeds, such as the Welsh corgi and cocker spaniel. But the story of beagles, unlike that of many other dogs, is also intricately interwoven with the history of modern science and the struggle to understand what it means to live a healthy human life. Efforts to determine the danger of nuclear reactors, the safety of oral contraceptives, the toxicity of cigarettes, and the causes of Alzheimer's all drew on extended studies of laboratory beagles, who were imagined to share physiology, diseases, and behavior with human beings. Between the Middle Ages and now, the noisy little hounds went from working animals chasing hare in the British countryside to specially bred "experimental" dogs on nearly every continent, hot in pursuit of that most elusive prey of all: an ideal nonhuman analogue for your body and mine.

Beagles were not the only dogs to become experimental subjects: a pioneering 1940s study of behavioral genetics at the Jackson Memorial Laboratory in Bar Harbor, Maine, for example, used an assortment, including beagles, spaniels, and basenjis. Each represented a different canine type, researchers argued, and revealed unique insights into the genetic foundations of doggy behavior.[1] Since that time, terriers, retrievers, pointers, and more have all had their uses and their evangelists. The first dog to have its genome sequenced was a boxer, and the international scientific collaboration Dog10K recently announced plans to sequence whole genomes from ten thousand widely varying dogs.[2]

Yet for many twentieth-century scientists, the humble, workmanly, and affectionate beagle was not simply one experimental creature among many, but *the* laboratory dog; not just a norm, but a legislated standard. While the modern history of dogs is often told as an accelerating division of *Canis familiaris* into ever more distinct breeds, beagles partially reversed that trend. In the parlance of modern laboratory life, "beagle" became nearly synonymous with "dog." This book traces a metamorphosis from popular hunting companions to trademarked, mass-produced, and controversial laboratory commodities, showing how a host of new ways of thinking about dogs emerged and took hold of the popular imagination.

Meet the Beagles

Bred originally for hunting foxes and hare, a "Beagle" with a capital B, to designate its status as a recognized breed, is distinguished from two of its kin, the medium-sized "Harrier" and larger "Foxhound."[3] The three derive from similar stock, but for divergent purposes: trotting beagles had exceptional scenting capacity, making them an ideal companion for ambulatory hunts amid Britain's bluebells and wood sorrel. Speedy foxhounds, on the other hand, used longer legs to keep up with fleet riders astride powerful thoroughbreds. Harriers, often confused with both, tended to have a size and a function somewhere in-between. As with foxhounds, their name followed from their occupation: harriers are those who hunt hares.

According to the National Beagle Club of America, beagles possess coats of any "true hound color," which generally means a mixture of white, black, and tan.[4] The American Kennel Club today acknowledges two varieties: one thirteen inches tall and below, the other between thirteen and fifteen inches. Get much larger, and one ceases, officially speaking, to be a beagle. But this is only the view from the present; and before the development of modern breed standards, the picture was much more confusing. In those days, *the* beagle was not so obviously one kind of creature.

In 1923, Otho Paget, known in British dog communities for his carefully bred "Thorpe Satchville" beagles, reflected on changes to the breed:

> "The Harrier and Beagle Association . . . have done much in improving the appearance of the hound devoted to the chase, and the Beagle Club have done their best in popularizing the variety in the showring. Whether the latter has done good or harm to the beagle in view of its proper vocation is a question that is open to doubt, as breeding from any sporting dog, without being able to ascertain its capabilities for work, must always be a step in the dark."[5]

The beagle's proper vocation was sport, the hunt—work threatened by an emphasis on aesthetically pleasing hounds.[6] Men who selected dogs principally for a charming appearance were not following the path of exceptional noses and single-minded dedication to the pursuit of scents, in Paget's view. The beagle's escalating popularity risked losing what made the dogs special, what made them *beagles.*

But in the centuries before Paget wrote, disagreement extended beyond beagles' vocation to basic questions of nomenclature. One of few certainties among beagle chroniclers is that no one knows quite where the word came from. Mentions date to the fifteenth century, but consistent usage of "beagle" took at least a hundred years more. In the meantime, various terms appeared interchangeably to describe the many dogges littering Britain's streets and countryside.[7] Even as "beagle" gradually gained popular currency, its obscure derivation remained controversial. Many beaglers traced the term to the Old English "begle," meaning small. That beagles were once "begles" implied a *British* origin. But linguists suggest the hard *g* in beagle would have softened, or palatalized, had its etymology been English, suggesting a painful alternate origin: the French "bégueule," meaning a gaping throat.[8] Early allusions to loud yelping beagles seem to support the notion, such as the following retort in an adaptation of Plautus's *Menaechmi:* "You, you, you: shall we fetch a kennel of beagles that may cry nothing but you, you, you."[9]

Even as their name stabilized, the dogs known as beagles differed meaningfully from those kept today.[10] Paintings, such as John Dalby's *Beagles in Full Cry* (1845), reveal a diversity in physique, size, and color that persisted into the twentieth century. Sometimes, the forms beagles were said to take stretched the limits of plausibility, including pocket-sized hounds that supposedly kept lonely duchesses company.[11]

One reason, as historian Michael Worboys and others have shown, is that "breed" became the principal way of differentiating dogs only in the middle of the nineteenth century, as Queen Victoria's love of dogs encouraged broader attention to breeding beautiful canines.[12] The modern dog, including the visually differentiable kinds we recognize today, was invented in this time and constructed through "point" systems and "conformation" standards that affirmed a coat should have so many colors, feet should look this way, and so on. Before conformation, on the other hand, dogs varied widely within types and mixed at the margins.[13] A large beagle and small harrier needed less active discrimination.

Paget acknowledged this multiform history. The Beagle Club in England had "adopted a standard and fixed on a type which," he supposed, "they consider represents the ancient beagle."[14] But finding the ancestral beagle, the "Adam of dogdom" in the words of later enthusiast George Whitney, was a dubiously speculative enterprise.[15] Efforts to recreate the original beagle missed an essential truth:

> The beagle, as we have it to-day, is an animal of many types, and it would be absurd for anyone to assert that one particular sort is correct. This difference is not the result of breeders' vagaries within the last few years, but these various sorts were in existence two hundred years ago, and have been handed down to us with little alteration since then.[16]

Paget's claim was a radical one: *the* beagle was never a singular entity persisting through time, but a fluid multiplicity.[17] There could be no perfect ancestor, no pure origin. The hound, at its best, was "a composite production."[18]

Despite the monolithic singularity of its capitalized name, "the Beagle" was always a work in progress. This was true in Paget's time and remains so today, even as national clubs and laboratory breeders stress uniformity. Dogs are not stable beings but living organisms locked, like us, in a process of becoming, "whose meaning is never so clearly revealed to our understanding as when it disconcerts it," as Georges Canguilhem once wrote.[19] Some seek to pin down this continuous process of change by identifying and stabilizing points of similarity, but heterogeneity reigns. For these reasons, I write throughout of "beagles" rather than "the Beagle," unless referring to official designations of the breed.[20] The composite was constantly shifting, so the important question is less what a beagle *is* than what they *have been* and what those who kept them *thought* they were.

Conformations of Life

Those who raised beagles shared ideas about dogs, but they also shared a broader cosmology. Sportsmen, their wives, and intrepid single women met on weekends in organized hunts.[21] They paid dues to clubs, which maintained conformation standards, and traded tips about proper attire: sturdy boots were a must, dresses to be avoided. Club-sponsored uniforms emerged; photographs were shared and point charts exchanged. Techniques for keeping dogs fit were hotly debated, and sometimes clubs raised funds to support research into canine health. "Beaglers," as such people called themselves, even catalogued appropriate names for each other's dogs.[22] They did all this in a vocabulary so occasionally arcane that it inspired printed glossaries. "The more occult and sacred forms of Beaglerese are used by the hierarchy alone," jested writer J. F. Sullivan in 1896.[23] To be a beagler was to be a specific sort of person, one who might imagine his or her breeding work as part of the hound's perpetual advancement before joining the "Happy Hunting Grounds" in the sky.[24]

All this activity constituted a shared world of practice and meaning: a form of life, a conformation between dog and human.[25] The particularities appear strange to those beyond the fold, but beaglers were equally perplexed by unschooled outsiders. "I have a town friend," wrote Sullivan, "who on all other known subjects is a brilliantly well-informed man; and yet, the first time I spoke to him of beagling, his gross mediæval ignorance filled me with a positive nausea."[26] Scientists working with beagles, too, shared a form of life: they produced papers, competed for grants, used lingo that left the uninitiated behind, and wore purpose-made clothing of their own. Sometimes, these forms of life converged, as with the recently defunct Atomic Beagle Club, founded near the Oak Ridge National Laboratory, in Tennessee, to entertain scientists and their nuclear families.[27]

I think of these shared practices and cultures as *conformations*—in the sense of forming together, rather than a standard applied unyieldingly from above—because they shaped the lives and forms of dogs and people.[28] Following centuries of mixed breeding, beagles emerged as sporting dogs who lived and hunted in packs. Their bodies were molded to make them dedicated seekers of the country hare, a fate very different from that of beagle-y strays living on crowded city streets. Purpose-bred beagles were biotechnologies, whether for sport or for leisure, and they were modified to suit different conditions.[29] When imported English beagles arrived in Sweden in the nineteenth century, for instance, the dogs were retrained and bred to work alone, a better fit for the Scandinavian style of hunting.[30] Over time, the ways of being a beagle expanded: after "Hound" became a valid competition category, dogs increasingly contended in shows where they were likely to be seen as charismatic, discrete personalities. Uno, the first beagle to win Westminster, in 2008, was imagined to strategically manipulate his fans and judges, nearly inspiring a hagiography from journalist Richard Sandomir.[31]

In the decades following Paget's writing, "Mr. Beagle," as Americans knew him, gained recognition as an ideal, dual-purpose pet: friend to the kids and hunting pal for dad, as well as, presumably, more work for mother.[32] In a fit of enthusiasm in 1946, Ohio professor L. E. Wolfe suggested beagles "contributed more thrills and healthy exercise to the lives of men" than any other sporting dog.[33] By 1950, beagles were arguably the most popular breed in America, with fans who spent weekends at "beagle derbies." California's first was held in April 1950, six months before Snoopy, history's most famous beagle, appeared in *Peanuts*.[34] Notwithstanding Snoopy's imaginary commitment to the Great War, it was hardly incidental that these changes followed closely on the culmination of World War II. A September 1944 article in the *Peoria Heights Herald* proclaimed that "the prestige of the dog in America is at a new high" thanks to outstanding contributions on the battlefield and "stimulus to morale" at home.[35] Dogs and humans comforted and conformed with one another.

But beagles also became something else during these years: experimental animals. In November 1963, Allen C. Andersen, director of the Radiobiology Laboratory of the University of California, Davis, gave a lunchtime talk on the study he managed, one jokingly referred to as "Project Hot Dog."[36] Although the research concerned atomic radiation, a topic of no small importance, Andersen opted to talk about dogs. He grouped them into three categories: there were "working dogs," such as the German shepherd or sled dogs, and "pleasure breeds," like the Pomeranian and other minimally capable laborers. But there was also a third category: "experimental dogs." These were recent entrants to the field, Andersen thought, and less technically advanced. While centuries of knowledge had accumulated around breeding capable hunters, little was known about what made a "good" laboratory dog.[37]

Andersen and others spent the middle of the twentieth century trying to find an answer. In the process, they settled on beagles, pushing them through multiple phases of modification, standardization, and expanded production. Spending years and even decades alongside these animals, scientists conformed to beagle ways of being. Although Andersen associated a dog's breed with its vocation, since chihuahuas made good companions but unsatisfactory guards, we can recognize in each of Mr. Beagle's occupations both technological tinkering and conformation. Becoming laboratory animals transformed beagles' lives, bodies, and historical trajectory.

Dog Models and Proxy Science

Science is a fundamentally narrative practice, "a rule-governed . . . craft of narrating the history of nature" in which "possible worlds are constantly reinvented in the contest for very real, present worlds."[38] Drawing on field studies of animals, for instance, scientists and popular science writers have told persuasive stories about the pacifism or violence inherent in human nature, about Man the Killer versus Man the Cooperator, with clear political implications.[39] That science is storytelling does not make it incapable of producing truth, however. Animal studies generate facts, which may be confirmed or thrown to the dustbin of the outmoded, just as they yield stories about what human beings are and what differentiates us from the rest of the natural world.[40] The scientific use of nonhuman animals is creative, even autobiographical.[41] It can be science *and* science fiction all at once.[42]

In research terms, a good nonhuman animal model is one that closely conforms to certain human systems or characteristics.[43] Pigs are said to have hearts like ours, while ferrets might share a respiratory disease burden. Although typically expressed in rigorously technical language, the similarities and differences highlighted in various cases can reveal a great deal about a researcher's or lab's assumptions and preoccupations. When scientists standardize and transform creatures to more closely approximate human beings and then use them to model behaviors or processes, they are frequently also telling stories about imagined pasts, redefining what is "human" or not in the process.[44] While nonhuman primates have long offered mirrors for differentiating "man from ape," the same is true for many other creatures, from fruit flies to hogs and dogs.[45]

Dogs, in particular, have served for centuries as vehicles for imagining, experimenting with, and constructing social orders. Breeders in Victorian England expressed their hopes and fears about social status through breed classification systems, Harriet Ritvo has shown.[46] By knowing dogs, some believed, one could know the order of things: dogs accept their place in the hierarchy of living beings, just as colonized peoples should—or so the troubling thinking went. Thomas Laqueur notes that many European painters from the eighteenth century onward used dogs as surrogates, even doppelgängers, for themselves, as canine looking became a visual metaphor for the social world and the human desire to see and be seen by others.[47] Similar interminglings of knowledge about dogs and theories of social order appear almost everywhere dogs are found, including literature and film: Mikhail Bulgakov's novella *Heart of a Dog* offers the fate of street mongrels as a metaphor for the violence of Soviet bureaucracy, while Wes Anderson's *Isle of Dogs* (2018) uses the banishment of pups in a futuristic Japan to reflect on modern xenophobia.[48]

American writers, in particular, have long taken dogs as stand-ins for humans, representing social and environmental transformation as well as "correct" behavior.[49] The same holds within the world of dog training: American manuals printed well into the twentieth century connected "intelligence" and "trainability" in breeds with their proximity to whiteness and conventional masculinity, one of many reasons that pit bulls face far more discrimination than poodles.[50] Dogs serve as working and imaginative devices for understanding our past and present, in the lab and in society writ large.[51]

But experimental dogs are also "future generating machines" that reveal new directions for scientific programs and human collectives.[52] At Andersen's radiobiology laboratory, for example, one set of beagles ate radioactive, strontium-laced chow, developed debilitating cancers, and suffered profoundly before euthanasia by laboratory technicians. If minuscule amounts of strontium proved deadly to dogs, entire aspects of American life might have been reformulated accordingly. Meanwhile another group, in close contact with their hot brethren, ate better than any household pet in the country, received exemplary medical care, and surpassed expectations for canine longevity. The two cohorts served scientific purposes, but they also offered imaginative affordances: one group was living out the worst visions of an apocalyptic future devastated by nuclear conflict, the other thriving in a utopia in which the bombs never fell. I call this practice of imagining nonhuman laboratory organisms as living exemplars of past, present, or future forms of life *species projection*. More than anthropomorphism by another name, species projection indexes the inventiveness in experimentation with nonhuman organisms and demonstrates the imaginative work involved in processes of conformation.

The last century of research with beagles, indeed, is evidence of continuous renegotiation of the degree of similarity between humans and other animals: Are our genes similar? What about our hearts, or our capacities for memory? All, or none, of the above? Recent studies, for instance, suggest that dogs and people share profound genetic similarities: "Man's Best Friend Becomes Biology's Best in Show," punned one article title on the subject.[53] That idea, however, might have surprised many earlier researchers. "Becoming human in the right ways" is central to survival as a dog, Eduardo Kohn wrote of the dogs he met during fieldwork among the Runa people of Ecuador's Upper Amazon, and beagles had to become human in the right ways to survive as laboratory dogs, a process involving extensive human intervention.[54] In doing so, they became *proxies*, tools that stand (or sit) in and "function as the necessary forms of make-believe and surrogacy" to "enable the production of knowledge."[55]

As proxies, they were not merely "model organisms," a now-ubiquitous term first used in its modern sense around 1946 to describe the flatworm *Planaria gonocephala*.[56] By calling *Planaria* "models," scientists meant that the simplicity of flatworms allowed them to easily study processes that also occur in more "complex" organisms. Nowadays, "model organism" also usually implies genetic modification, such as mice predisposed to breast cancer. Because introducing a successful new model organism is a path to funding and scholarly recognition, the last century's scientific publications are suffused with an ever-growing array of them. But until the late twentieth century, experimental organisms were models in a more diffuse sense. Dogs modeled processes, but rarely as "model organisms," a term that struggles to capture the diverse things that creatures do in laboratory settings.

Proxy, on the other hand, ties together the work of modeling and imagination.[57] Schematically, the following chapters explore the use of dogs as *breeding* proxies, *physiological* proxies, *abstract* proxies, *affective* proxies, and *cognitive* proxies. Chapter 1 shows how some early scientific dog colonies were motivated by eugenic theories of heredity that dogs were thought to uniquely exemplify. The flexibility of canine forms seemed to support racist visions of controlled human breeding. When beagles appeared in nuclear research, on the other hand, they were explicit stand-ins for humans exposed to radiation because of their similar physiology, as chapter 2 reveals. Chapter 3 explores the gradual abstraction of the proxy relationship, as beagles became standardized commodities and computer models in pharmacological studies, which enabled interspecies comparisons of prescription drugs and a burgeoning transnational traffic in beagles. Chapter 4, on the other hand, shows how the validity of scientific studies of cigarettes came to rely on ideas about the emotional life of beagles. Finally, chapter 5 tracks how beagles became proxies for something they had been largely considered incapable of doing: advanced cognition. In studies of Alzheimer's, dogs were reconceptualized as individuals with mental capacities like ours.

In each instance, beagles were living beings integrated into and imagined as complex tools. To make an experimental beagle into a Cold Warrior, a smoker, or a drug user required inventive devices that augmented a dog's normal capacities, technical prosthetics for dogs that were already scientific and imaginative prostheses as well. A smoking machine designed in the 1960s, for instance, allowed beagles to inhale smoke directly from a lit cigarette. In other circumstances, dogs were "debarked" or "devocalized," euphemisms for the surgical modification of their vocal cords, to limit the collective cacophony of their kennels.[58] These cyborg augmentations of the beagle highlight how the laboratory dog was a new kind of biotechnology from the leisure dog and reveal how species projection and proxy science rely on significant technical tinkering.

But none of this is to say that proxies always work as intended. "Have you ever seen a dog?" asks a character in Robert Musil's novel *The Man without Qualities*. "You only think you have. What you see is only something you feel more or less justified in regarding as a dog. It isn't a dog in every respect, and always has some personal quality no other dog has. . . . Even in the lab, things never behave just as they should. They diverge from the ideal course in all possible directions, while we keep up a fiction that this is to be blamed on our faulty execution of the experiment, and that somewhere midway a perfect result is obtainable."[59] Many beagle studies failed in their stated aspirations while succeeding in revealing unexpected knowledge. Dog colonies, which almost universally ran out of funding before their intended conclusions, were magnificent places for generating incidental knowledge about dogs, such as nutritional requirements or cage design.[60] Paradoxically, too, many contemporary canine pharmaceuticals emerged out of work originally promising insight into human health and wellness. The chapters thus chart a shift toward the view that dogs are worth studying in themselves, rather than as means to human ends. Thanks in part to the emergent market for companion pharmaceuticals and genetic testing, studying dogs as proxies has been partly replaced by studying dog *as* dogs.

Sacrifices for Friends

Popular discourse about "man's best friend" suggests a primordial and timeless relation of amity between humans and dogs. The Latin word *familiaris*, which distinguishes domestic dogs from other canids, effectively defines them by a relationship with people: *familiaris* implies they are of the family and household, that they are servants, familiars, and intimates.[61] But the history of beagles in scientific research reveals the complications of that friendship. Relations are not static, Marilyn Strathern reminds us, but "reinvented, or rediscovered, at moments of new knowledge-making."[62] Although dogs have a sensitivity to human communicative acts that may enable rapid conformation to laboratory settings, using a dog for an experiment is not as simple as turning on a microscope, as anyone who has dragged a stubborn friend away from the park might guess.[63] Instead, encounters between unique living beings, laden with affective and emotional ties, shape whether experiments work.[64] That beagles are pliable, cheery, and loyal was part of their value and utility. But the story here is not an entirely rosy picture of cooperative companionship. Beagles were not just good to think with, or to work with, but also to exploit.[65]

Since the turn of the twentieth century, when the term was popularized by French physiologists, the killing of laboratory animals has been described as "sacrifice" (today occasionally shortened to "sac").[66] With dogs, the word conjures long-standing associations of loyalty, including Old Yeller's death protecting his family from a rabid wolf, while critics see it as a misleading genteelism, a way to justify killing without guilt.[67] The persistence of sacrificial language within laboratories illuminates the tensions in our close relations with other animals, what Radhika Govindrajan calls "interspecies intimacy." Although "intimacy" suggests care, it frequently also involves violence: there can be care in sacrifice, Govindrajan notes, exploring religious practices in the Himalayas, where sacrificial animals represent significant commitments of time and labor and are understood to be valued members of the family.[68] The same holds in many laboratories, where carefully bred and expensive beagles, with whom researchers often develop emotional attachments, are killed in a gamble on valuable discoveries. To see "sacrifice" as only a euphemism for killing obscures how it is simultaneously a recognition of value. If knowledge production alters relations, sacrifice is one avenue for doing so.

Studying sacrifice in this way helps us to grasp how the story of dogs in science can include both a catalog of horrors and caring scientists who valued little above the well-being of their animals. "In the agony of death a dog has been known to caress his master," wrote Charles Darwin in *The Descent of Man*, "and every one has heard of the dog suffering under vivisection, who licked the hand of the operator; this man, unless the operation was fully justified by an increase of our knowledge, or unless he had a heart of stone, must have felt remorse to the last hour of his life."[69] Darwin, a lover of dogs and staunch critic of laboratory cruelty, was also a public defender of animal experimentation.[70] Scientists have gone to extravagant lengths to protect their animals, and quite a few contemporary researchers or veterinary technicians adopt beagles after they participate in experiments, a pathway that the Massachusetts medical lobby asked of the Beagle Bill.

On the other hand, whether through callousness or resignation to sacrifice's inevitability, many experiments involving beagles appear unacceptable, even barbarous. One example makes this exceptionally clear: in February 1972, researchers at the Armed Forces Radiobiology Research Institute in Bethesda, Maryland, published a report about the performance of beagles subjected to lethal doses of radiation. It is difficult to read with a neutral eye. Following similar analyses of monkeys, miniature pigs, and rats, beagles restrained in small plexiglass boxes were trained to flex their legs in response to an auditory cue.[71] The dogs performed admirably, despite being easily distracted by sudden changes in their surroundings—a residue of the ancestral chase of rabbits. Each beagle was then exposed to a powerful pulse of gamma-neutron radiation and tested "at least once a day until death." None were allowed outside during the studies, and all died like other victims of history's nuclear catastrophes: vomiting, losing control of basic digestive functions, and convulsing from central nervous system damage. Heartbreakingly, multiple dogs tried desperately to flex their forelegs despite the convulsions.

A historian's task is to understand how bewildering activities from the past made sense in their time and context, and this book shows how proposing, funding, and running such a study came to make sense to researchers and funding agencies. Explanation is not justification, however, and the book is also a meditation on our responsibility for the ways that dogs, beings that seem so similar to us, have had to live. In historicizing these conformations of life, we can perceive alternative possibilities for living together, as humans and as dogs. "The problem," wrote Donna Haraway, "is to learn to live responsibly within the multiplicitous necessity and labor of killing."[72] But doing so also requires questioning both that necessity and that labor. Close encounters between dogs and humans can offer proof of more ethical modes of companionship, but they can equally support Clive Wynne's claim that human-canine relationships are structured by "super-dominance," a "total social subordination . . . combined with humans' control over resources including food, movement, shelter, and sexual access."[73] Even the most well-treated companion dog is, Marc Bekoff and Jessica Pierce insist, fundamentally a "captive."[74] Science has made use of super-dominance, just as breeders and trainers have, and it does not require a rejection of the possibility of truth to acknowledge that many studies carried out with beagles should never have happened.

Whose Beagle Is It Anyway?

Beagles were initially a rather British animal, but the beagle as experimental subject was a distinctly American product. Even before the dogs arrived at labs, British breeders worried that Americans had begun to build a better beagle.[75] In the second half of the twentieth century, however, American beagles became an international standard, used by researchers in countries around the world. "The breed of dogs most used in the Anglo-Saxon countries since 1954, then in Europe in order to be able to compare the results of identical experiments, is the Beagle," explained a French researcher in 1972.[76] To tell the beagle's story, the following chapters move from New York and Davis to Hamburg, Paris, and Chiba, with pit stops along the way.[77]

The dogs that traveled the world were not only laboratory devices, but locally and later globally mass-produced products, connecting unexpected places through circuits of exchange.[78] In traditional accounts of lab animals, an early twentieth-century period of shared research materials gave way, very recently, to aggressive commodification: what historian Robert Kohler influentially called the "moral economy" of fruit fly researchers became today's world of trademarked, genetically modified mice.[79] Lately, however, Kohler wondered whether "political economy" better reflects the power and politics that always shape scientific labor, and this book builds on that suggestion by charting the continuous entanglement of laboratory dogs and modes of production and exchange.[80] It shows how beagles became scientific commodities, lively animal capital that supported novel, transnational movements of knowledge and wealth.[81] The history of biology, it argues, is already a history of forms of capital.

Many nineteenth- and early twentieth-century scientists were content to collect dogs off the street, as Pavlov did for his conditioning experiments. Laika, Belka, and Strelka, the famous Soviet space dogs, were found this way. Being a stray was part of their value, because that life had made them hardy.[82] The first half of the twentieth century, however, saw heightened interest in so-called normal or standard dogs.[83] Every dog was a little different, and as laboratories sought to reproduce findings in different settings, the consistency of experimental dogs became a key concern.[84] The obvious answer, for many, was to control access to dogs at the local pound, saving the best for science. In American cities such as Chicago and Birmingham, pounds operated like proto–supply companies, converting surplus dogs into lab and teaching tools. Some were bought, others freely taken, as dogs hovered between natural entity and experimental commodity. We can name this system *salvage accumulation* of experimental animals.[85]

The growing interest in standard scientific dogs drew additional focus to their conditions of production: if fruit flies or rats could be bred at centralized hubs, scientists reasoned, something similar had to be possible for dogs. Yet the slow pace of canine reproduction constrained a Model T approach, and few researchers had budgets for expensive dogs.[86] Instead, a medium-sized, *internal colony* model thrived—for example, a group of Irish terriers at Columbia University—with populations sustained by occasional purchases from local breeders. Although some researchers aspired to larger production levels and sold dogs for a marginal return on costs—a first step in explicit commodification—the market for "standard dogs" was limited in the 1930s and 1940s. Profit was secondary, for the largest beagle colonies, to use.[87] At the same time, however, dogs underwent a dramatic commodification in midcentury American society. Costs grew as pet ownership and the market for pet products exploded, turning the attention and finances of wealthy people to the science of dogs.

The internal colony model gradually became unsustainable. The absence of institutional funders and the emergence of stricter animal care regulations following the 1966 Animal Welfare Act meant that private breeders hesitated about selling to labs. In walked large-scale, for-profit breeders. Over time, these companies produced and advertised their dogs as explicit commodities: the Beagle® rather than beagles.[88] Increased demand, particularly for pharmaceutical and toxicological studies, meant that established breeding companies cornered a significant market. This third, *monopoly production* model, which emerged alongside the broader growth of monopoly capital in the American economy, continues into the present, as a small handful of consolidated breeders have largely replaced internal colonies.[89] These breeders, with international outposts sequestered in isolated locales, also offer an additional service: discretion and invisibility, shielding dog researchers from organized protest, which began to seriously threaten their work in the 1970s and 1980s.[90] Transformations in the practice and politics of animal experimentation responded to and influenced the global market for laboratory animal commodities.

At no point were laboratory dogs free from the influence of money. But at each phase, beagles represented a peculiar commodity: one produced through a mix of practices from traditional pet breeding and agriculture, yet subject to the conditions of scientific "biocapitalism." What most differentiated beagles from cell cultures or radionuclides, which all moved transnationally and accrued value globally, was that dogs interacted emotionally with those who kept them. Beagles had "encounter value": it was worth something that they were enjoyable to work with.[91] In turn, their use value came to depend on the consistency of such encounters, because only sociable, psychically nourished beagles could be counted on to produce "good" science.[92] This trait of the beagle commodity was partly a result of centuries of intentional breeding by hunters, but it also had to be instilled by producers. Dogs were carefully bred, and even more carefully housed, in specialized kennels; their health, so crucial to the success of any experiment, was a matter of assiduous attention. One company, Marshall Farms, promised that all their beagles were held daily as puppies to ensure quality. Encounter value was thus a labor matter and extended beyond laboratory practice.[93]

In turn, research that supported the importance of caring for dogs became a driver of efforts by pharmaceutical corporations to market drugs to dog owners. The cognitively and psychically complex dog that lived in your home deserved the latest treatments—even if, and perhaps especially when, those products were failed human drugs repackaged for the pet market. As the use of specialized laboratory beagles became more morally and politically problematic, researchers have returned to the salvage of pet dogs, but this time volunteered by owners in a model that I call *companion science*. By highlighting these intersections of intimacy, production, and markets, the book demonstrates how the emphasis on *care* and *entanglement* as a response to instrumentalizing approaches to the nonhuman world is inadequate.[94] Care can enable the work of controlling and killing lab animals just as it supports their entry into new markets, even if the beagle's encounter value has now begun to erode its exchange value.

A Historical Beagling

In 1929, Danish scientist August Krogh, who won the Nobel Prize for his work on the regulating mechanism of the capillary network, laid out a dictum concerning the choice of experimental organisms. For any scientific problem, there exists "some animal of choice, or a few such animals, on which it can be most conveniently studied."[95] "Krogh's principle," as it was popularized by biochemist Hans Krebs in the 1970s, was a more descriptive formulation of what Claude Bernard, a founder of experimental physiology and infamous vivisector of dogs, once called "the happy choice of an animal."[96]

The history of modern science is littered with examples. Using fruit flies, Thomas Hunt Morgan and others were able to elucidate the basic mechanisms of Mendelian inheritance, in part because flies reproduced so quickly that they frequently generated mutants.[97] Midcentury molecular biology benefited tremendously from *C. elegans*, nematodes whose transparency allowed development to be followed at the cellular level and whose small size made complete anatomical description manageable.[98] Cats, rats, zebrafish, pigs, rabbits, nematodes, macaques, and mice all participated in research projects of profound significance.[99] A historical litany of support for Krogh's vision, the list goes on and on. In 1923, Paget applied a parallel principle to the breeding of hounds, citing the British poet Somervile: "A different hound for every different chace / Select with judgement."[100] Yet what is striking about the history of the laboratory dog, and of experimental beagles specifically, is the wild diversity of their application. While beagles were popularized in postwar radiobiology projects and pharmacology, they were also used to study behavioral genetics, electrocardiography, rabies vaccines, periodontitis, the inhalation of tobacco smoke, and more. We could reverse Krogh's principle: for those who can afford to use them, there exists some problem that can be studied conveniently with beagles.

Why, then, were they valuable experimental animals? No single explanation suffices. Beagles were useful because they were larger than rats or mice; because their happy temperament made them more cooperative; because they would tolerate inhalation masks; because they responded to contraceptives; and because they lived long enough to display cognitive deterioration. Beagles were useful, often, because they had been useful before—but in another place and often for different reasons. Prior utilization pushed the breed from one scientific discipline to another. Rather than being ideal for one or more discrete reasons, as scientific explanations might have it, momentum from earlier experiments frequently became its own justification, creating a long-lasting species path dependence that continues to structure research in the present.[101] Wherever twentieth-century scientists tried to understand human health, one tends to find a beagle study.

To tell the story of how dogs came, sat, and stayed in the lab, it is inadequate to follow one group of researchers or a single a branch of scientific inquiry. The story of experimental beagles is one of branching research programs and cross connections.[102] In some ways this is fitting: when beagles hunt, they rarely do so in a straightforward way. They do not race directly for their prey but scent them out. They also tend to meander, which is why early beaglers warned of the scourge of backtracking dogs. Beagling, historically, was a leisurely sport: huntsmen would halloo and coax their dogs as they scampered back and forth across marshy fields, through woods and over hills, in pursuit of a hare's scent. The point was to follow the dogs as they went, enjoying a nice view and brisk wind along the way.

But a good beagler knows the dogs do not always get it right by themselves. One must recognize when to push the dogs and when to stand back, leaving them to their work. After a number of years of research, I came to feel that beagling *historically* should function similarly.[103] Meandering through the history of science over the last two centuries, never losing sight of the dogs, one finds a patchwork of strange and fascinating stories that reveal a great deal not just about conformation and species projection, but about the infrastructure and practice of science and medicine.[104] Despite their familiarity, beagles have a way of defamiliarizing well-known scientific stories. Not every interesting anecdote about dogs could fit into the following pages, but when the howls and yelps that lie between the lines of old reports and archived correspondence point the way, quite a few surprises emerge.

At times, this presents a challenge. Dogs leave few letters, live for short periods, and make less-than-willing interview participants. It is easier to imagine stealing the Declaration of Independence than seeing firsthand the beagle facilities of some contemporary companies. "Writing about animals," notes Etienne Benson, "depends on tracks, trails, or traces . . . those often-unintentional indexes of a now-absent presence."[105] Animal history is beagling, in other words, and this book draws on interspersed material from traditional archival collections, much of which is used for the first time, as well as ephemera and interviews with surviving scientists. "If thought searches," wrote Gilles Deleuze, "it is less in the manner of someone who possesses a method than that of a dog that seems to be making uncoordinated leaps."[106] Uncoordinated leaps, however, often possess a logic of their own.

What follows is history at the mesoscale: multiple sites and moments knit together by beagles and their handlers.[107] Through unexpected means, the bodies of beagle dogs became a medium from which scientific truth could be generated, tested, and confirmed despite geographical and disciplinary differences. Results were "true" because experiments were carried out on the same kind of animal, a convenient one for the problems—even if one of the consistent challenges was knowing whether one beagle was really the *same* as another. Beagling through history reveals unifying and disunifying movements in the unwieldy constellation we call "science" and also in the expansive realm of knowledge beyond it, because so much of what we know about dogs floats in the purgatorial space before the gates of science.

A Path through the Woods

Each of the following chapters covers a relatively distinct scientific discipline and may be approached individually as interests dictate. The book, however, was written to be read from its front to its back. The chapters proceed chronologically, with occasional overlaps, and build on each other. Individuals and institutions introduced early on tend to reappear later. The rabbit is sighted and then vanishes; the dogs howl into the underbrush and out again.

Chapter 1 charts how dogs became essential research organisms for American scientists in the first half of the twentieth century. It starts with the story of the beagle's Americanization: how a defiantly British breed was resignified as an American hunting companion. The chapter then explores the breed's patriotic reemergence around World War II, when the renewed popularity of hunting made beagles into midcentury America's most popular dogs and a canine symbol of American identity. As growing dog ownership coincided with attention among scientists to the question of "standard" dogs, the second part of the chapter explores early efforts to build large-scale dog colonies, from Charles Stockard's eugenic Dog Farm to Cornell University's Research Laboratory for Diseases of Dogs. These projects drew on analogies between dog breeds and human races and set the foundation for subsequent experimental dog colonies, establishing the beagle as a key laboratory organism.

Chapter 2 follows Atomic Energy Commission (AEC) projects at the University of California, Davis, and the University of Utah, known as "Project Hot Dog" and "Beagleville," respectively. Reeling after the first use of atomic weapons in World War II, America's nuclear planners were desperate to understand more about the effects of radiation on longevity. Because mice and rats failed to live long enough, they turned to dogs, particularly beagles, to study the risks of continuously ingested, injected, and inhaled radiation. These laboratories literally wrote the book on using beagles in experimental settings, producing both epistemic and institutional momentum that carried the dogs into later projects. But these projects were also experiments in keeping canines, and the chapter explores practices of transparency, exchange, and humor within the labs.

Chapter 3 turns to the place of beagles as pharmacological testing animals. Alongside the AEC projects, pharmaceutical and toxicological research represents their most numerically significant scientific usage. Where the AEC projects were national undertakings, pharmaceuticals made beagles into "global" scientific organisms, and the chapter explores the expanding production of beagles as laboratory commodities. Charting transformations in international testing regulations at midcentury, when beagles became drug users, the second part of the chapter explores the history of pharmacokinetics, the study of how drugs move through the body. It shows how physicists and computer scientists made a home for themselves in biology, as beagle data set foundations for mathematical and computational models as well as laboratory protocols, making the dogs tools for *interspecies scaling* amid a new vision of *biological* relativity.

Chapter 4 explores the role of beagles in the long twentieth-century fight against tobacco, telling the story of how pathologist Oscar Auerbach looked to beagle dogs in an attempt to prove that experimental animals in a controlled setting get cancer from smoking cigarettes. Auerbach and others noted that beagles became addicted to smoking, just like people, but tobacco industry critics charged that the experiments were not reproducing real human smoking at all. The battle, in other words, concerned whether beagles *enjoyed* their smoking, and the chapter focuses on the tendentious role of emotion and pleasure in laboratory animal experiments. By analyzing a competitor beagle-smoking study that emerged from earlier AEC radiobiology work, the chapter also explores how animal experiments can both close and sustain controversies in public discourse.

Chapter 5 traces efforts in the late twentieth century to make beagle dogs into models for Alzheimer's disease and reveals the unexpected connection of this research to the market for pet pharmaceuticals. Because Alzheimer's is a human-specific disease by conventional definitions, the notion of a canine model required more capacious understandings of the condition's social and environmental influences. The chapter focuses on research into Anipryl, a dog-specific drug modified from a human Parkinson's treatment, which became one of the first blockbuster products in the market for companion-animal pharmaceuticals. In the background of changing ideas about the cognitive capacities of dogs, the chapter also shows how expensive, specialty-bred beagles have lost value over the last few decades as researchers turn to volunteered dogs amid the rise of companion science.

The book concludes by looking at the lingering scientific question of how similar human and beagle bodies really are, as well as at the role of beagles in recent political debates over laboratory experimentation and even the handling of Covid-19. Returning to notions of species projection and conformation, the book ends by reflecting on what nearly a century of research with beagle dogs has meant for science, scientists, and everyone else. There will likely be beagles in the future of science, it concludes, but perhaps finally shed of their association as *the* laboratory dog.

With that, the dogs line up at the brace, heads to the ground, blades of grass massaging their noses. A scent weaves forward, into the trees and out again.

J. Otho Paget. Portrait by Cuthbert Bradley. 1902. Chromolithograph. © National Portrait Gallery, London.

CHAPTER 1

A New Breed of Science: American Beagles, Eugenics, and National Fantasies

In February 1946, distinguished scientists and New York City notables were invited to attend the preview of a new exhibit at the American Museum of Natural History, "*Really* Man's Best Friend." Along with an introduction to the exhibit, the preview included a presentation ceremony for that year's Whipple Award, named for Nobel Prize winner George Hoyt Whipple and intended to recognize "outstanding services to mankind"—not by humans, however, but by dogs: the first recipients of what newspapers called the "highest prize of dogdom" were dalmatian mixes Josie and Trixie, a mother and daughter aged four and one, respectively. Norman Kirk, surgeon general of the US Army during World War II, presented their prize collars.[1]

Josie and Trixie were the only winners of the Whipple Award, which was never presented after 1946, but their accolade is revealing about the centrality of dogs in midcentury American science.[2] As descendants of dogs that Whipple himself used to study pernicious anemia, work that found that diets rich in liver and apricots could reverse the condition's course, Josie and Trixie were evidence of the successful scientific use of dogs. Yet they also exemplified animal experimentation's future promise as blood plasma donors in ongoing research carried out by Whipple's colleagues at the University of Rochester. They had, one reporter claimed, "saved literally millions of lives."[3] The blood flowing through Josie's and Trixie's bodies symbolized the canine continuity of the medical breakthroughs of the past and the much-hoped-for discoveries yet to come.

In addition, Kirk's conferral of the prize tied the dogs to medical interventions that had saved the lives of American GIs overseas. Soldiers who received life-saving transfusions after battles at Normandy or Guadalcanal had dogs to thank, according to a barrage of postwar newspaper articles. "The American soldier received the benefits of these studies; they were applied to him, not tested on him," Kirk noted.[4] His stoic appearance in some photographs even suggested the dogs were receiving battlefield commendations—one newspaper, having missed the joke, described them as "military awards."[5] On the contrary, however, the Whipple Award and the exhibit it accompanied were part of an explicit campaign of public advocacy and education by American scientists. Funding for the exhibit came from the Friends of Medical Research, a national organization of lay supporters of animal experimentation, and each panel in the show stressed that research with dogs was essential to medical and scientific breakthroughs that benefited both species. Insulin for diabetes, the Blalock-Thomas-Taussig shunt for "blue babies," and vaccines against rabies: all of this came from cooperative scientific work of men and dogs.

In particular, "*Really* Man's Best Friend" was designed to encourage visitors to be critical of antivivisectionist protests against canine experimentation, the precursors to today's animal rights movement. Resisting animal experimentation, the exhibit argued, meant placing soldiers and loved ones in direct danger. The italicized emphasis on the truth of human-canine friendship was also important: sacrifices for science were meant to prove a profound species companionship. Dogs were inextricable from the very idea of American biomedical progress. Science was improving the lot of both man and his best friend, the story went, and great things were still to come from the partnership.

Eight years later, the exhibit seemed like an understatement. "Working in the interest of dogs today are research forces that did not even exist a decade ago," proclaimed Harry Miller, director of New York's Gaines Dog Research Center, in January 1954.[6] The center itself was only a few years old, an effort of the Gaines Dog Food Company to educate owners about canine health and to expand the market for dog chow. As proof of his claims, Miller pointed to three recently opened facilities. The Jackson Memorial Laboratory in Bar Harbor, Maine, was carrying out an extended study on the inheritance of behavior in multiple dog breeds. The Cornell Research Laboratory for Diseases of Dogs, in Ithaca, New York, was developing vaccines for notoriously destructive conditions such as canine hepatitis and distemper. And the dog nutrition program at the University of Wisconsin, Madison, was providing new insights into canine dietary requirements, including the function of the vitamin B complex. Together, the three projects reflected a shift in the relationship between dogs and science. Throughout the first decades of the twentieth century, stray dogs had appeared in varied experiments as tools to explore human physiology or demonstrate medical techniques for trainee physicians, but the organized application of science to dogs was something differently entirely.

Around the same time, another important change occurred in American dogdom. After seventeen years of uninterrupted rule, the cocker spaniel lost its spot as the American Kennel Club's most popular dog to the spirited beagle in 1953. The AKC numbers offered only a partial picture of American dog ownership, since many owners bypassed the fees needed to register their dogs, but beagles had been on a steady climb across the country for a number of years. According to Leon Whitney, a breeder and veterinarian in Orange, Connecticut, whom we will meet again later on, this popularity could be attributed both to increased interest in field trials, as more and more men spent weekends traipsing through the woods with their dogs, and to a burgeoning suburbanite desire for "a small, easily cared for dog" that lacked the woolly coat of cockers.[7]

The evolution of canine science and the popularization of beagles were intertwined and somewhat uniquely American developments. Ultimately, they would rearticulate the relationship between humans and their supposed best friends all across the world. For that to happen, however, beagles first had to cross the Atlantic and find prominence in the United States. A community of dog lovers also had to develop the activities eventually known as beagling. "This thing called beagling," noted North Carolina reporter H. W. "Frog" Hayes in 1952, "is certainly not a sport which originated during this so-called Atomic Age."[8] Beagling's history was indeed far longer and coextensive, in many ways, with the story of American dog ownership itself.

This chapter shows how elite dog breeding in the Gilded Age, democratized pet ownership amid the Cold War, and persistent attention to the quality of experimental dogs led to the first sustained canine research colonies. Many of these efforts were shaped by a view of dogs as proxies for breeding and heredity, often motivated by racist and eugenic concerns. Twentieth-century Americans knew that dogs came in an endless variety of sizes, shapes, colors, and abilities. They also knew that careful breeding could produce especially useful or capable kinds, what everyone eventually called "breeds." It was not such a far leap to wonder whether the same might also be true for human beings: experimental dogs offered a powerful vehicle for reimagining American society, sustaining fantasies of a perfected human race. Talk about dogs is frequently also talk about us.

While dog owners are typically imagined, today, as natural opponents of animal research, the chapter also shows that much of the early science of dogs was supported by pet owners, whether business magnates, scientists, or regular Americans. Of particular importance were members of the Rockefeller family, whose various philanthropic organizations, especially the Rockefeller Foundation, played a key role in funding early experimental inquiry into the dog, starting in the 1920s. But the chapter also reveals the importance of a broader network of dog supporters in sustaining canine studies, an early example of citizen science that united participatory research and industrial interest.

Initially, scientists chose their dogs relatively arbitrarily. By the 1930s, however, most became convinced that someone, somewhere, needed to select a standard breed. It had been done with rats; why not with dogs? Crossbreeding experiments by embryologist Charles Rupert Stockard at first appeared promising, but their premature conclusion left the issue unresolved. Then, starting around 1939, work at Columbia University suggested the Irish terrier as a possible choice for centralized production instead. Yet terrier boosters faltered, interrupted by the arrival of World War II, and no institution could commit to hosting the much-yearned-for stock center for dogs. Instead, smaller projects such as the dog studies at Jackson Labs and at Cornell emerged, quietly redirecting scientific attention, by 1950, to the beagle—at precisely the moment the breed was in wider cultural ascent. While scientists agreed on certain traits for the ideal research dog, the chapter shows how quotidian familiarity, whether from hunting or pet ownership, was equally important.

The chapter begins in the nineteenth century, tracing the slow popularization of beagles through field trials and bench shows in the northeastern United States. It analyzes changing notions of dog ownership between 1870 and 1950 and emphasizes the significant role of urbanization and industrial capital in making beagles "America's dog" after World War II. Next, the chapter explores the largest early effort to breed scientific dogs: Stockard's eugenic Dog Farm, which opened in 1925. It follows the emergence and premature termination of that project and tracks its lasting implications for the Rockefeller Foundation's focus on experimental dogs and the possibility of a standard breed.

The chapter ends with the Cornell Research Laboratory for Diseases of Dogs, which opened in 1951 and adopted beagles as its laboratory breed. I show how the Gaines Dog Research Center, an early sponsor of the lab, helped to popularize the idea of a science of dogs through popular newspaper and radio programs. The early prominence of the beagle as a scientific dog and the notion of dog science were inseparable, the chapter shows, from the uses that beagles were bred to fulfill and their changing associations in the American cultural zeitgeist.

The Live Beagle Club of the World

The earliest recorded beagles reached America by the early 1870s. Isolated individuals likely preceded them, but only in that decade did newspapers and journals regularly feature reference to beagle hounds—often lost, occasionally murdered. A representative joke from the period asked about the difference between a bald head and a starving beagle: "One is found without hair; and the other is hound without fare."[9] In 1862, P. T. Barnum, the famous showman, hosted America's first National Dog Show in New York City, the "centre of the canine trade for this continent," promising foxhounds and beagles.[10] But at the time, "beagle" designated a variety of hound-type dogs rather than a definite breed standard: what we now know as the dachshund was occasionally referred to, for instance, as the "German" beagle. "Sufficient it is to say that more than one-half the dogs called English beagles are veritable mongrels," bemoaned a Philadelphia devotee writing under the pen name Rusticus in 1883. The man was pleased, nevertheless, that the dogs were increasingly seen as fitting companions for sportsmen, not just "pot-hunters, boys and negroes."[11] Like many writers of the era, he saw the value of a breed in its ownership by the right (white) kinds of people.[12]

Toward the end of the nineteenth century, sporting journals such as *Forest and Stream* were frequent hosts to debates on the beagle question: whether a woeful lack of standardization and inconsistent dog show judging could be resolved by an official organization. One supporter, who adopted the pen name Razor, urged "lovers of the beagle" to keep faith in "the idea of being united in one vast body," chasing their melodious hounds through the scrub brush of a more perfect union.[13] Not long after the violent conclusion of the Civil War, that vision of unity had obvious political resonance. In 1883, this dark night of the soul ended: in November, a circular in *Forest and Stream* announced the American English Beagle Club (AEBC) and invited applications from those interested in "the English beagle."[14] The club's first president and secretary were Pennsylvanians, but prominent fanciers around the country signed on, including celebrated Connecticut breeder Norman Elmore, whose dog Ringwood sired many of the era's champions, and General Richard Rowett of Illinois, traditionally considered the first American beagle importer, although others may have beaten him to the punch.[15]

The club's name exposed tensions over what beagles were understood to be. Geographically, they were American, but culturally they were English, a problem for patriots and Anglophiles alike. As a compromise, two days after Christmas in 1883, O. W. Rogers of Billerica, Massachusetts, suggested simply naming the organization "The Beagle Club."[16] But "English" was necessary, countered another, because it excluded merely "American" beagles, creatures he hesitated to even describe, from the organization's purview. Many agreed: owning an *American* beagle meant keeping an inferior hound, a mere rabbit dog. True beagle breeders, instead, were aiming at something they labeled "perfection," and "English" was as much a reference to quality as it was to origin. Eliminating "American" from the group's title, to tie the dogs more directly to their Englishness, however, was also rejected. The Revolution had not been for nothing.

The AEBC published its first official breed standard in April 1884, and inaugural president W. H. Ashburner exhorted members to make "a grand display" at bench shows in New York and Philadelphia. Public display was crucial to acquaint the wider world with the best of beagles, because there was neither a central depository of information nor a book on the breed.[17] Linguistic debates raged and grew tiresome: "We beg to suggest that further discussion of the proposed name will be purposeless," pleaded *Forest and Stream*'s editors that January.[18] Nonetheless, "English" was finally dropped six years later, in 1890, and a report on the unanimous vote explained simply that the "American descendant" now decisively outclassed "its English brother."[19]A British writer, reflecting on declining beagle entries at the Birmingham Dog Show, had prophesied the change a few years earlier: "If Englishmen are so callous to the merits of one of our most charming home breeds, I trust that the inherited sporting instincts of our American cousins may insure the merry beagle an appreciated future."[20] Cousins, brothers: the newly rechristened American Beagle Club designated Americans as the true caretakers of the breed, at least in their own imaginations.[21]

For years after, club members focused ad nauseam on defining the breed standard and hosting bench competitions. Centered in the Northeast, the organization struggled to accommodate a geographically disparate membership, and its early focus on shows seemed to miss a growing locus of enthusiasm: hunting and field trials.[22] "As the upland game birds of our Eastern and Middle States are becoming more and more scarce and lengthy journeys are required to insure even passable sport," explained Philadelphia's *Times* in 1885, "many gentlemen of Philadelphia who are fond of shooting and whose business will not permit of their being long from the city are devoting their leisure hours to rabbit hunting near at home with beagle hounds, not a few of which dogs of the highest breeding have been imported from England."[23] Beagling was, in part, a compromise leisure activity in the face of settler ecological destruction. Later Beagle Club president Herman Schellhass made the point explicit: "As certain parts of the country become thickly settled and the feathered game exterminated," he explained, "lovers of field sports, who have heretofore devoted their time in the field to bird-shooting . . . , finding the game so nearly exterminated as to destroy the pleasure of seeking it, discard their bird dogs in favor of the Beagle."[24] Unlike its vanishing avifauna, America's rabbits and hares were omnipresent. Schellhass, who lived in Brooklyn, celebrated the ability to find droves of them within only a half hour's trot.

The same year that the American Beagle Club adopted its slimmer name, nearly a dozen Massachusetts men founded a new organization dedicated to a competitive version of these outdoor hunts. Arguing that the other club had "done practically nothing" in advancing the beagle, they called their rival group the National Beagle Club (NBC).[25] An informal meeting of "a number of Beagle men" was held in Boston's Mechanics Hall on April 3, 1890, and they announced their existence in the *Boston Herald* that May.[26] By October, the club had thirty members with two hundred dogs and promised the first field trial of beagles "ever given in this country."[27] The *Herald*'s coverage of the competition, however, acknowledged limited public familiarity with the breed. "About everyone knows a mastiff, a Newfoundland, a collie or a terrier when he sees one," but few would recognize a beagle on the street. To aid the reader, a picture captioned "A Typical Beagle Dog" was included.[28] Comparison with "Little Duke," a "representative of the best type of the beagle hound," who appeared the previous year in the *Philadelphia Times*, reveals significant flexibility in the breed's representation.[29] The *Herald* added that beagles should have "a soft, pleading expression," the affective pull that would later be key to their scientific appeal.[30]

Boston was perfectly situated as a national epicenter of beagling. Emerging from the Civil War primed for urban and industrial expansion, the city held sprawling new suburbs and wealthy men with an interest in leisure. It was also home to the first metropolitan park system, including landscape architect Frederick Law Olmsted's acclaimed "Emerald Necklace."[31] The parks, begun in 1878 and expanded in the 1900s, were meant to preserve lands deemed "cathedrals of the modern world," but they were also explicitly framed as spaces for leisure and recreation, used "to foster a healthy democracy and inculcate the masses with a middle-class appreciation for nature and beauty."[32] Metropolitan parks and large wooded areas offered ideal beagling opportunities. The National Beagle Club's first field trials in nearby Hyannis were a success, and membership swelled.[33] In December 1890, leadership considered a 364-acre game preserve where they hoped to erect a two-and-a-half-story clubhouse with sleeping quarters and model kennels capable of holding forty to fifty dogs.[34] For all their confidence, however, NBC members had a chip on their shoulders. Against predictions by "the leading sporting papers of this country" that the club would be a spectacular failure, president O. W. Brooking averred, "We have proved them wrong."[35]

Formed in opposition, the two clubs ultimately confirmed the narcissism of small differences. Granting divergent emphases on dog shows and field trials, the true tension appeared to be geographical. Schellhass once counted about nine hundred individuals owning beagles, with most names in Pennsylvania, followed by New York, Massachusetts, and Ohio.[36] And while some founders of the Boston club were also members of the original, quite a few were not. By January 1891, momentum was decidedly with the men of New England: "We are the live Beagle Club of the world," asserted Brooking.[37] Western beagle owners, who were fewer in number but spanned more than half the country, felt left out of both organizations, however. Writing in San Francisco's *Breeder and Sportsman* in 1889, one hoped that a regional club might spring up while acknowledging that "many people on this coast do not know what a true beagle is."[38] Two years later, a beagler from Pontiac, Michigan, suggested forming a group "in the West" on the lines of the NBC.[39] Such a subdivision was considered, but leaders decided instead to simply open membership to all who might be interested.

With the American Beagle Club struggling, executives of the two groups arranged to meet at the Westminster Kennel Club show in New York on February 24, 1891, to form a "harmonious amalgamation." This expanded National Beagle Club, precursor to today's National Beagle Club of America, would continue to host field trial competitions, but it would also provide judges for bench competitions and maintain the breed standard, the traditional tasks of the older organization.[40] Although the term "beagling" was still rare in the United States, almost exclusively a reference to activities in England, hunting with the hounds was gaining social cachet.[41] In February 1892, W. G. Gates of Chagrin Falls, Ohio, laid down the whopping sum of $1,000 (close to $35,000 today) on champion Massachusetts beagle Frank Forest.[42] In 1901, a large spread about the NBC in the Knoxville, Tennessee, *Journal and Tribune* noted growing interest in beagle field trials—"the American Squire's pastime"—which seemed poised "to become as popular among our wealthy citizens as [they were] in English rural life."[43] Beagles had found an institutional infrastructure, and Americans had made beagling their own.[44]

One Nation, Under Dog

While some saw beagles as a workingman's dog, early American beagling was dominated by the upper classes.[45] Even with the requisite leisure time to devote to field trials, just acquiring the dogs was expensive.

While domestic stock existed, bred continually from the earliest arrivals, many of the best sporting beagles were imported from British breeders into the 1930s. An 1881 blip in Pennsylvania's *Evening Dispatch*, for instance, noted the arrival of "a very fine pair . . . of the best strain of blood in England."[46] *Forest and Stream* also featured routine "Importation of Beagle" announcements, steamship arrivals of dogs hand-picked by British experts.

Just as it originally suited a settler culture, beagling also matched transformations in work and wealth during America's Gilded Age. In 1901, Ernest Gill of Baltimore, the president of Martin Gillet, one of the oldest and largest tea importers in the United States, explained that he tried not to turn a profit on his lauded kennel because beagling should remain "a safety valve and a relief from the business cares of everyday life."[47] An article on the first field trials at the Burlingame Club outside of San Francisco called them, in 1916, "more of a tonic than a sport."[48] Field competitions were entertainment that enabled proximity to nature and established distance, however imaginary, from the tightening bonds of work. This fell in line with the shifting place of dogs in turn-of-the-century America: no longer accepted as laborers in major cities, most breeds were conceived primarily as objects of leisure.[49]

As beagles grew in popularity, their connections to the wealthy classes deepened. A fad of beagle raising swept New York and New Jersey in 1900, exemplified by Harry Payne Whitney's plans to erect a major kennel on his estate.[50] Whitney, the multimillionaire husband of Gertrude Vanderbilt best known for his family's success raising thoroughbred horses, kept a large pack of beagles at "The Manse," his Long Island home, and frequently invited enthusiasts over to talk shop.[51] The Rockefellers, too, were avid dog fanciers, including scion John D. Rockefeller, but his nephew, William G. Rockefeller, became most notable in the area: William's Rock Ridge kennels bred a number of prized beagles based on esteemed English stock.[52] He and Whitney were known to talk shop in what a newspaper account called "beagleology."[53]

Beagles had "the widest circle of upholders in good society," noted a newspaper in 1906, including younger "millionaires of New York, Boston, Philadelphia, Baltimore and Washington."[54] In 1913, beagle trials were extolled as the "latest society fad" in Philadelphia, just a few years before a "fashionable concourse of society people" began popular weekend hunts at Burlingame.[55] An overly enthusiastic report on the first trial there declared it "one of the great events in club history," adding that "golf links and polo might as well not exist as long as the beagle fad lasts."[56] Maximilian Foster's 1917 novel *Shoestrings* satirized this San Francisco elite obsession through the character of J. Lester Tams, who pursues his aspiration to recognition as a gentleman by, among other things, owning a pack of beagles. At one point, Tams muses about ordering quality dogs express from good kennels on Long Island.[57] The beagle fad reached from sea to shining sea.

Acquisition was not the only barrier to mass participation, however. Serious competitors relied on staff to manage their growing kennels. Alexander Henry Higginson, a grandson of biologist Louis Agassiz whose inheritance allowed a lifelong devotion to hunting, noted in 1903 that a good beagler should expect to own up to ten dogs and consider hiring his own "kennel man" to care for them.[58] Railroad executive and financier George Jay Gould was said to have a pack of beagles in North Carolina that matched Harry Whitney's in quality but relied on assistants to care for them because he could visit his estate there only once a year.[59] Despite aspirations to leisure, keeping beagles was often a business of its own.

Like many trends, beagling lost its allure, but the dogs themselves resisted the fickleness that made and sank the fortunes of other breeds. The German shepherd, for example, took America by storm following its introduction in the 1910s, helped on by charismatic cinema star Rin Tin Tin, yet fell off dramatically by the time of Rin Tin Tin's death in 1932.[60] A parade of exotic-seeming breeds, such as chow chows, followed, but beagles held steady: "Half a century ago, in 1885, the beagle was the fifth most popular of the 24 breeds then recognized by the American Kennel Club, and today, this grand little hunting dog is in exactly the same place among the 103 varieties listed by the governing body," explained West Virginia's *Bluefield Daily Telegraph* in 1936.[61]

After the German shepherd lost favor, aided by a souring of public opinion about many things German from war overseas, the Boston terrier held top honors at the AKC until 1936. It was then that cocker spaniels, hunting dogs with "melting eyes," became America's most popular breed, reaching record registrations of 79,507 in 1947.[62] Cockers spent a groundbreaking seventeen years atop the AKC list, spurred by frequent advertising portrayals and the popularity of woodcock hunting, but they were finally beaten, in 1954, by beagles. That year, 52,262 beagles were registered, roughly one for every seventy-seven children born, and 45,398 the year before. Cocker registrations, on the other hand, steadily declined.[63] Importantly, both breeds were celebrated for their emotive appearance.

Trends in dog registrations were manifestly capricious, but a few factors explained beagles' sustained success. For one, heavy importation from England ceased by the 1930s, encouraged by protectionist 1935 regulations of the US Department of Agriculture, which required three generations of breed records to certify imported dogs.[64] Breeders soon deemed American beagles equal or superior to their British rivals. A new International Beagle Federation (IBF) also formed in February 1931, popularizing trials in the spring as well as fall and making beagling a year-round, nationwide activity (the group largely represented regional associations in America and Canada, despite its grandiose name).[65] In addition, pet hair became a matter of concern as more dogs spent more time indoors, more moms spent more time vacuuming, and more kids fell prey to allergies.[66] Many Americans preferred dogs that shed less, and boxers, another short-haired breed, made significant gains in 1953 as well. Shifts in the perception of hunting, however, were especially important for beagles.

While hunting in America has a long, varied history, including colonial charters that guaranteed the right to pursue it on land imagined as a sportsman's paradise, the aggressive, masculine image of frontier hunters popularized by Teddy Roosevelt had reached an apex early in the twentieth century and waned dramatically by 1941.[67] The war years, however, rehabilitated hunting's place in the American cultural mythos, with sport and gaming magazines connecting the fight for freedom abroad with dogs and the backwoods hunt at home.[68] "Should your subscription copy of *Field & Stream* be a few days late . . . that's just one more peeve you can chalk up against the axis," noted editors in October 1942.[69] Forming a beagle club was touted by the National Recreation Association as an effective "For Men Only" program for war and industrial workers, and a 1945 advertisement for the Life Insurance Companies in America features a man who saves and invests in war bonds to eventually open the beagle kennel of his dreams.[70] "In spite of the number of sportsmen now in the Service, the volume of mail concerning dogs received by *Field and Stream*'s Kennel Department is showing a decided increase over even prewar days," wrote the magazine's editors in 1943. "It seems a safe bet that actually more dogs are being bought now than in the days before certain egotists in Germany, Italy and Japan decided to see just how much hell they could raise before getting their ears slapped down."[71] To sportsmen, war was dog training by other means.

As soldiers returned home, dogs were highlighted as therapeutic options for transitioning back into civilian life. Seeing-eye dogs aided veterans who lost their eyesight, while programs such as "Beagles in Reconditioning" at Camp Ellis in Illinois offered field trials as physical exercise and recreation.[72] One alumnus eventually won a local competition, evidence that rehabilitation could morph fluidly into a peacetime hobby.[73] While World War I had increased the general popularity of dogs and stories of heroic canines, the beagle's popularity, in particular, "began to soar after World War II when millions of returnees took to hunting," noted columnist John Lyons.[74] Beagles had failed to make the final cut in the US "Dogs for Defense" training program, losing out to German shepherds, but their peaceable nature may have formed part of their postwar appeal.[75] Upon returning from the European front, "many a man wants a business that will permit him an outdoor life," explained *Field and Stream* editors in February 1945.[76]

The ultimate effect was significant. "Beagling is no longer a struggling, secondary sport," Glenn Black, former secretary of the IBF, wrote in 1949. Field trial entries had grown by an average of 30 percent annually since 1944.[77] That expansion was aided by ever-larger percentages of vehicle owners, and beagling manuals frequently described how drivers could store their dogs effectively in car trunks for trips. The increased availability of trains was equally helpful: "Let's take a ride in the comfortable club car on the Erie Railroad train leaving Cleveland at six P.M. for Pittsburgh," Black offered as the hypothetical start for a beagling journey.[78] Trials were now a sport that less wealthy individuals participated in with regularity, especially in the Pennsylvania–Ohio–West Virginia nexus, where the IBF's competitions largely took place.

For some, beaglemania was more fundamental. The dogs were "reliable, durable, intelligent enough to work out a twisted trail." A beagle was a hard worker, "and even when the going gets rough he whips his tail about merrily, though it be tinged with blood from beating through the brambles," wrote John Lyons. "What could be more American?"[79] At a moment of ascendant patriotic nationalism, beagles assumed the mantle of the properly American dog, despite their origins elsewhere. In this way, they joined a lineage of imported organisms, from cherry trees to mustangs, that were imaginatively integrated into the broader nationalist vision.[80] This role had been foreshadowed in newspaper coverage of a popular Philadelphia breeder in 1917, when one writer mused, "What the terrier is to an Englishman the beagle may almost be said to be to an American."[81]

By the early 1950s, the "almost" had dropped away. A magazine anecdote about an American sergeant and his wife, who take in a small stray in Austria, has the husband explain his decision to add another dog to the mix by affirming, "Beagle is an American dog and he has priority. . . . No Kraut dog is going to make my American dog sleep out of doors."[82] Beagles symbolized the men who had defeated their enemies abroad and planned to build on those successes at home. Typically accorded masculine pronouns, beagles were also associated with a normative, presumptively white masculinity. According to reporter

Jack Melder, a beagle was "one dog that is truly American, if there is such a thing as a truly American dog."[83] This straightforward projection of nationalistic self-adulation onto the breed was unsurprising to some: "As with his gods, man creates his dogs in his own image," remarked Nelson Foote, director of the Family Study Center at the University of Chicago, in 1955.[84]

Estimates from the period suggest that Americans owned twenty-two million dogs by 1954, a gain of 200 percent over thirty years and roughly one dog for every seven people.[85] The early baby boom was thus coextensive with an equally notable canine boom. What we now think of as the pet industry ballooned in response to this burgeoning market, as dogs became items of quite conspicuous consumption.[86] Dog food companies sold nearly a billion pounds of canned food in 1951, doubling that total in 1954. Expensive advertising and an array of novel small businesses, such as canine beauty parlors and training schools, all confirmed that the animals were worth spending hard-earned money on, even as newspaper columnists occasionally treated them with a degree of suspicion.[87] As historian Lizabeth Cohen notes, mass consumption during the era was not simply a personal indulgence, but "a civic responsibility designed to improve the living standards of all Americans."[88] Beagle ownership sat at a unique nexus of interests in this consumers' republic, improving the living standards of dogs as well.

The breed's image thus shifted from 1930 to 1950: no longer a faddish hobby for the well-to-do, beagles "belong[ed] to the country and the small town," and their popularity surged in the South and West accordingly.[89] "Beagling," which entered the American cultural lexicon just before the Roaring Twenties, became commonplace by the late 1940s and took on broader associations, such as references to "the beagling bibliophile" or "word beagling" by syndicated New York columnist O. O. McIntyre.[90] Most importantly, "humble" and "hardworking" beagles were reenvisioned as ideal, multipurpose animals, equally at home in the fields and the city, where increasing percentages of Americans dwelled. A beagle perfectly completed the nuclear family—father, mother, children, and dog—and their charm was invariably highlighted in discussions of the breed. "It's this [gentle] disposition of the beagle that represents a real 'extra' to the owner," explained one writer.[91] America's dog, proxies for ideological efforts to articulate American life, specialized in encounter value, that benefit that emerges from the relationship between two living beings. A few decades earlier, laboratories had begun to make the same discovery, and it would change the breed's lot dramatically, pushing them into a new relation as proxies for scientific theories of breeding.

A Nightmare Dog Show

On December 16, 1922, Leon Fradley Whitney, president of the Northampton, Massachusetts, Fruit Growers Supply Company ("Just what the name implies"), opened his heart to a scientific hero. "I wish to discuss a matter which is of great importance to me and my future," Whitney wrote to Charles Davenport, director of the Cold Spring Harbor Laboratory in Long Island, New York. "Frankly I believe that the work that you are doing, if I understand it thoroughly, is the greatest cause today that a man can undertake to work at."[92] Whitney would be leaving for New York on Christmas Eve, and he hoped Davenport could spare him an hour to talk things over. He confessed that he would happily stay a week.

The cause Whitney referenced was eugenics, the program of improving the "quality" of the population, or race, via controlled reproduction and restricted immigration. A course on the subject taught by the vibrantly mustachioed economist Robert Sprague had entranced him while Whitney studied at the Massachusetts Agricultural College in Amherst. And despite a comfortable postgraduation job, Whitney could not shake the pull of race improvement; he was a true believer and hoped to dedicate his life to the cause. Davenport, on the other hand, was eugenics' leading scientific representative, having founded the Eugenics Record Office, which collected and analyzed genetic records of American families. He did not have a week, so the two met at the Harvard Club in New York City on December 26, a mild winter afternoon.[93]

What drew Whitney to eugenics was a fascination with breeding in the widest sense. After growing up in New York, he had studied reproduction in college and, while finishing his degree, began raising dogs for fun, selling puppies to his neighbors in Northampton.[94] As he explained to Davenport, he was driven by "one big question": if humans had raised horses and cattle to higher levels of usefulness through careful breeding, why could the same not be done for humans themselves? Whitney was hardly alone in seeing animal breeding, especially that of livestock and dogs, as both metaphor and practical evidence for eugenics. Francis Galton, Darwin's half cousin and an early leader in eugenic thought, had felt similarly.[95] As far back as Plato, ideas about the breeding of hunting dogs had supported arguments for selective mating.[96] Nor was it uncommon for twentieth-century American dog manuals to associate beauty and trainability in dogs with breeds imaginatively closer to white Europeans.[97] The temptation to analogize dog "breeds," which were still commonly called "races," to human groups was a powerful one.[98]

Davenport, who never fully concealed a distaste for Whitney in their decades-long acquaintance, cautioned that the human problem was "not so simple as that of breeding a horse." Human breeding was complicated, more "than that of all breeds of dogs, cattle, poultry, and fruit of all species added together."[99] Unfazed by the jest at his business experience, Whitney committed to exploring dog breeding scientifically, and when his mentor could not refer him to a treatise on the subject, Whitney decided to write his own. A year after their first meeting, he appraised Davenport of his crossbreeding experiments so far: "German police and Pomeranian, two Beagle hound and airedale, Norwegian bear dog and foxhound, water spaniel and fox hound, bloodhound and fox hound, Gordon setter and shepherd, bull terrier and fox hound, and pointer and foxhound, St. Bernard and German police and several others of less striking contrast."[100] Whitney's inconsistent capitalization of breed names reveals how the concept of breed remained a work in progress, but what was most important to him was "striking contrast." His attempted matings highlight how aesthetic differences in dogs served as proxies for human difference.

Davenport was less than enthused by the work but nevertheless connected Whitney with Yale economist Irving Fisher, the first president of the newly founded American Eugenics Society (AES). On the strength of Whitney's earnestness and enthusiasm, attested by friends and former teachers, he was hired as the AES's first "Field Secretary," dutifully submitting records of his family's traits to Davenport and relocating to New Haven, closer to Fisher. The work of the AES, including fundraising and voluminous correspondence, kept Whitney busy, but dogs remained constantly on his mind. He bred multiple generations of his beagle-Airedale mixes to explore the inheritance of size, even though he lacked space to keep them and annoyed his wife in doing so.[101] In May 1925, Whitney drafted a funding proposal and asked Davenport whether the Carnegie Institute, which supported the Eugenics Record Office, might also support his dog-breeding experiments. Although a response appears lost, Davenport penned an emphatic "No!" on his copy of the letter.[102]

Whitney turned instead to Yale primatologist Robert Mearns Yerkes, a member of the AES Advisory Council and one of the most well-connected scientists of the era. "The races of dogs are just as numerous as the races of men," Whitney explained in October. "I believe that so much of eugenical and genetical importance can be found out from dogs."[103] Yerkes looked favorably on Whitney's proposal to start a small colony at Yale, as did Wesley Coe, curator of the Yale Peabody Museum and another member of the AES Advisory Council.[104] But unknown to

Whitney, another scientific dog study had already taken shape earlier that year.

In March 1925, Charles Rupert Stockard, an anatomist at Cornell Medical School specializing in embryology but with wide-ranging scientific curiosities, proposed a new research project on endocrine modifications and the "inherited constitution" to Wickliffe Rose, president of the General Education Board (GEB). The GEB, an educationally focused nonprofit funded by John D. Rockefeller Sr. and overseen partly by his son, John D. Rockefeller Jr., was eager to support the expansion of American teaching and research in the basic sciences, and Stockard's project fit the bill.[105] Building on a "vast amount of breeding of dogs by fanciers" not yet "utilized" by scientists, Stockard envisioned a specialty laboratory outside of New York City where he could carefully raise dogs to explore how the endocrine system related to genetic malformation.[106]

From the start, Whitney had suffered in the eyes of his AES colleagues, bereft of either an advanced degree or an Ivy League pedigree. And what Stockard lacked in dog-breeding experience (he had virtually none) he made up for in academic credentials, including recent elections to the National Academy of Sciences and the American Philosophical Society.[107] Yerkes and Coe, who knew of Stockard's plans, urged Whitney to consult him before going further, all but ensuring Stockard would maintain his first-mover advantage and become the first to construct a large-scale experimental canine lab in the United States.[108] To a general eye, however, he was far from the obvious choice to do so.

Born in 1879, close to Mississippi's border with Arkansas, Stockard attended the Mississippi Agricultural and Mechanical College (MAMC; now Mississippi State). There, under the guidance of entomologist Glenn Herrick, he fell for ornithology. Following the outbreak of the Spanish-American War and a predecessor's departure for active duty, Stockard, the ranking officer of his class, was appointed commandant of the college.[109] The position distracted him from studying natural history, but Herrick, who considered Stockard "a naturalist of splendid promise," encouraged him to persevere and apply for a fellowship, in 1903, to complete his graduate studies at Columbia University.[110] "The South needs strong men in every line of teaching," wrote MAMC president John Crumpton Hardy in a recommendation letter. "I believe that, if you will honor Mr. Stockard with this appointment, he will in turn honor your institution as well as his Alma Mater."[111] At Columbia, Stockard studied with the influential biologist Edmund Beecher Wilson and, later, with Thomas Hunt Morgan, a fellow southerner on the cusp of major breakthroughs in the study of fruit fly genetics. After graduating, Stockard took a position teaching anatomy at the Cornell Medical College, where he would stay for the rest of his career, a Mississippian successfully ensconced in one of the North's great educational institutions.

With a southern accent he never entirely camouflaged and a trim military bearing, Stockard could be imposing; even friends considered him imperious and demanding. His academic prominence, however, was well established, especially for work on critical periods in embryological development.[112] Like many biologists of the era, Stockard studied normal development in part to grapple with the perceived dangers of abnormality and degeneration, and he understood his research as support for the cultural program of eugenics. He served on the AES Advisory Council and its Committee on Research but avoided further entanglement, including an offer from AEC president Clarence Cook Little to become New York chairman in 1928.[113] Stockard nonetheless corresponded regularly with Davenport, who sold him chicken eggs, and first proposed siting the dog lab on land near Cold Spring Harbor.[114] For Stockard, as for Whitney, studying development in dogs was a way of understanding, and eventually correcting, abnormal development in humans.[115]

The GEB was slow to take up his request, but after prodding from Cornell president Livingston Farrand, who had helmed the Rockefeller fight against tuberculosis in France and stressed his university's full support for Stockard, the board agreed in June 1925 to set aside $20,000.[116] The money would go to purchasing a farm, hiring permanent staff, and building facilities for research and habitation, and the sum was significant enough to dissuade Stockard from accepting a job that year from his alma mater, Columbia. "I am still 'licking my chops' over the fine opportunity," he told the GEB's Abraham Flexner in June.[117] His new laboratory would be perched on a plot of land up the Hudson River near Peekskill, New York. Davenport had in fact secured space at Cold Spring by November 1925, but Stockard thought Westchester County preferrable to the "damp" conditions of Long Island. Although he named his new facility the "Experimental Morphology Station," most knew it as the "Cornell Anatomy Farm" or, simply, the "Cornell Dog Farm."[118]

Once operational, the Dog Farm was meant to offer confirmation of Stockard's working theories about the "endocrinic basis of constitution," which both he and his funders understood to have fundamental importance for eugenics. The topic was of interest to more than just scientists: John D. Rockefeller Jr., for example, agreed to give the AES $5,000 the same year Stockard's project started. Endocrinology, which grew significantly during the first half of the twentieth century, opened a new avenue of attack for eugenic projects: both fields stressed the hereditary or "constitutional" aspects of human nature, but endocrinology also offered a way of conceptualizing novel forms of intervention.[119] "On the basis of this analysis," Stockard explained, "we hope to be able to devise treatments by injections of proper extracts, gland transplantation, etc., that will enable us to control and direct the development of constitution."[120] The "races" of purebred dogs, Stockard thought, offered direct parallels for human beings by approximating "corresponding human types, e.g., giants, dwarfs, acromegalics, etc."[121] Dogs were endocrinic proxies and paralleled "better than any other mammals . . . the various modified and distorted growth conditions which are exhibited by human beings."[122] Careful breeding of dogs could, in this sense, lead to the improvement of human races.

Stockard struggled to overcome a number of early obstacles, however. The original grant was insufficient to construct suitable laboratory space, so he requested an additional $8,000 in October 1926. More challenging still, the farm was meant to carefully hold dog populations representing three different "types," along with numerous crossbreeds, something few had ever really tried to do.[123] Stockard's own experience was with smaller animals, such as guinea pigs, which he imagined to be mostly similar.[124] But dogs could not be kept inside, because of exercise needs and limited temperature control. Living outdoors, they fell victim to parasites. In February 1928, Stockard wrote to Flexner about the devastating effects of hookworm: "As quickly as the animals are cleaned of hookworm they become reinfected by larvae eggs from the ground."[125] Equally dangerous were fleas and lice, so he appealed for further funding to support a trip to Europe, where he could learn from existing experimental facilities.[126] Constantly thinking of the dogs as human analogues, in 1931 Stockard compared the laboratory's disease challenges to those of urbanization: "It is much the same sanitary proposition which a human community faces in growing from a sparsely settled village condition into a thickly populated town."[127]

Stockard, who took the train from Grand Central to visit the farm on Saturdays but was often absent otherwise, found that the majority of available commercial canine diets were nutritionally insufficient, forcing the farm to develop its own mixed feed and to supplement diets with "wheat germ, yeast, bone meal, carrots, spinach and cod liver oil."[128] The various breeds also seemed to have different feeding requirements, complicating issues further. "A good deal has been learned about desirable diet for dogs," wrote Alan Gregg, director of the Rockefeller Foundation's (RF) Medical Sciences Division, which took over funding the project in 1930. Visiting in 1932, Gregg noted that one potential benefit of Stockard's project might be the general "care of experimental animals," since the first five years at the laboratory primarily yielded findings in canine husbandry, rather than endocrinology.[129] Two years later, Gregg suggested to Stockard that the lab's results might contribute to a commercial enterprise breeding "a standard dog for experimental purposes," but Stockard did not take up the suggestion.[130]

For over a decade, the workers at Stockard's farm interbred dogs to try to prove that inherited endocrine disorders shaped development. In his lectures, published as *The Physical Basis of Personality* (1931), Stockard summarized some of their first results, which he anticipated would take another ten years to bear fruit. "It is of great significance that certain human freaks practically parallel in their growth and form these diversified canine types," he noted.[131] Hounds, pointers, shepherds, and huskies were "normal" types, in Stockard's view, while French bulldogs or great Danes were "abnormal," displaying glandular defects that emerged at "certain critical stages." In his Beaumont Foundation lectures from 1927, Stockard was even more explicit, and brutal, in the comparison: "With no attempt to make a serious subject frivolous," he began, "it may be added that a most striking and convincing comparison is obtained on holding a fine pointed French bulldog up on its hind feet by the side of a human achondroplastic dwarf." The point could be made equally "by placing both the man and dog down on all fours."[132] Dog breeds were not exactly equivalent to human racial groups, which Stockard saw along seminational lines—a "German" type compared with an "English" type, for instance—but their peculiarities revealed how "normal" humans in those groups could develop "abnormally." An "African pygmy," he argued, could be seen as "a not fully metamorphosed large negro."[133] Here, dogs emerged as practical and metaphorical materials for racist visions of human development. As Stockard argued in 1931, "Strangely enough, dog fanciers have been unconsciously selecting and preparing a splendid array of material, most ideally suited for a scientific investigation."[134] With newspapers frequently reporting on Stockard's findings, these ideas reached a sizable portion of the reading public and shaped popular conceptions about dogs.[135]

In 1938, Stockard began work at his home in Woods Hole, Massachusetts, on a final monograph about the dogs, one his Rockefeller funders eagerly awaited. A year later, however, he died from lung cancer, ending the hugely ambitious project before even provisional conclusions.[136] Fearing the loss of years of work and their own investment, RF officers agreed to fund Stockard's assistant, Ellen Paaske, to assemble his notes into a book, in consultation with Cornell colleagues.[137] It was released, following years of discussion and substantial revisions, through the American Anatomical Memoirs series of the Wistar Institute in Philadelphia in 1941.[138] The result was a "mine of information," wrote University College London biometrist Hans Grüneberg—not exactly as a compliment.[139] Frequently lacking background data to make experiments legible to readers, the book overflowed with analyses of canine "hybridizations," attempts to locate genetic bases for shortened legs or contracted snouts. Stockard had been unable to resist seeing his results as support for the possible elimination of human abnormality, but through a glass, darkly. "Is it surprising," wondered Grüneberg, a German Jew who had escaped Nazism to England, "that the startled author of this nightmare dog show should utter warning after warning against 'mongrelization' of human races?"[140] Despite early predictions that Stockard's book would be "a scientific publication of high quality," its intellectual influence was marginal.[141] By the end of 1942, Stockard's name had all but vanished from the public eye. The work, Grüneberg concluded, "remained a torso," offering a sad plea to change our lives.[142]

Like his findings, the fate of the Dog Farm was put in the air by Stockard's death. There were still nearly three hundred dogs as well as multiple junior researchers, including experimental physiologists Oscar Anderson and William T. James, who had not yet completed their studies. The year of Stockard's death, James even turned down an offer from Cornell psychobiologist Howard Liddell to return to Ithaca in favor of a pay increase from Stockard, who "does not seem to feel that this project will be discontinued any time in the near future," James explained.[143] It would prove a career-threatening miscalculation. As discussions expanded about what to do with the farm, it became clear that the project had served primarily as a vehicle for Stockard's aspirations. Gregg had already fretted years earlier that the Dog Farm "worked out very unsatisfactorily in point of training afforded to younger men or collaboration with Stockard's colleagues."[144] Few at Cornell knew much about it.

In a detailed, postmortem proposal, James argued that continuing the studies might still reveal much about constitutional differences and personality types, including the genetic foundations of human neurosis.[145] He recommended either converting the farm into a long-term research station for those working with dogs or using the facilities to breed "pure types of dogs for experimental purposes" on a commercial scale.[146] Horsley Gantt, one of America's foremost Pavlovian dog researchers, encouraged Gregg to continue support of the site.[147] The problems with doing so, however, were twofold: Cornell, which owned the land and facilities, had no full-time faculty who could manage the operations; the RF, on the other hand, had cut its medical sciences budget to roughly one-third of the 1930 levels.[148] There was neither capacity for, nor interest in, providing additional funding. Liddell, who also received support from Rockefeller, once more offered to bring James to Ithaca in 1939 to complete his work on some of the dogs, and around thirty animals were transferred there for future use at Liddell's own Behavior Farm.[149]

In 1941, some of James's findings about the dogs received unexpected public exposure after an Associated Press report, gracing the front page of many American newspapers, explained that he had discovered the "nerve" origins of civilian resistance to bombardment during World War II. "Why Bombed Public Can Take It," ran one headline, describing a study in which James had put dogs through "two kinds of 'life,' one resembling peacetime, the other war."[150] A group of stolid basset hounds and excitable German shepherds were conditioned to expect minor shock after the ringing of a bell, the air raid siren in Pavlovian miniature, and James found that it was actually shepherds, imaginatively associated with nervous humans, who could tolerate the suspense. Here was species projection in its most literal form, and many members of the public were distressed by it. In one representative letter, Nellie Mathews Meyer, president of the Kentucky Antivivisection Society, castigated James for a pointless experiment that demonstrated his lack of "compassion or common decency towards helpless creatures."[151] New York poet Harold Faller, on the other hand, offered a more ambivalent response in verse:

> The basset hounds and the shepards, they went
> To war that day
> Up Ithaca way. . . . .
> To Dr. James they lent
> Their nerves for an experiment. . . .
> In short, says Dr. James, it proves that
> dogs and men are twain
> In suffering pain.[152]

News of these studies was, in a sense, the last gasp of Stockard's once great dog project. James would later take around twenty-five ancestors, a mix of terriers and beagles, to the University of Georgia to develop a new colony where he could study the structure of social hierarchy, but Stockard's facilities and many of the other dogs were lost.[153]

The New Little Rockefeller Project

The nightmare dog show had born strange fruit, and a final RF appraisal in 1939 was dismal: "Dr. Stockard's illness and death . . . resembles in its effect the death of a lone farmer at harvest time. This is a project which must be considered an acute disappointment."[154] For most, it was a warning of the laboriousness of running a large colony of scientific dogs. In 1949, Cornell animal nutrition expert Clive McCay would castigate Stockard for his overconfidence: "Some years ago an eminent anatomist was granted half a million dollars for studies with dogs. He established his kennels but met with a long series of failures because he knew too little about the feeding, immunization and care of dogs. This researcher was defeated in his investigations in endocrinology because he understood too little about the handling of dogs and failed to acquire staff members who were adequately trained in parasitology, nutrition and veterinary medicine."[155] Stockard's follies had revealed, in other words, the necessity for laboratories to employ a broad array of experts in canine treatment and care, prophesying the need for a yet-unnamed laboratory figure that we now know as the veterinary technician.[156] Any other way of operating such an institution would, as scientists in this area liked to say, "go to the dogs." Motivated by suggestions from Gregg and others, Stockard had even begun drafting thoughts for a book, just before his death, "on the care and maintenance of an experimental animal farm, discussing the actual set up for the care of groups of experimental animals, particularly dogs," but he never got far.[157]

Stockard's failures would largely dissuade Rockefeller from funding individual scientists working with colonies of large animals in the future. But despite this disappointment, an interest in scientific dogs survived Stockard, not least because dogs remained a vital concern for many other researchers. After decades of battles with antivivisectionist critics of canine experimentation, scientists had in fact gained a decisive public opinion advantage by the late 1930s.[158] A coalition of individuals and institutions, united under the umbrella of the National Society for Medical Research (NSMR), organized to sustain that momentum by arguing publicly that the use of dogs was essential to basic biomedical progress, especially the production of new vaccines and pharmaceuticals. Resisting the old "Ostrich Code for Medical Scientists," which consisted of silence about animal research, shady acquisition of animals, and avoidance of legal struggle, researchers instead emphasized their "care" for laboratory dogs, thousands of which were used annually in the United States and Canada, according to 1939 estimates.[159] The exhibition "*Really* Man's Best Friend" was one manifestation of this approach in practice. As Frida Robbins stressed about Whipple winners Josie and Trixie, "We couldn't work with the dogs if they were unhappy. We need their co-operation."[160] A pamphlet from the Surgery Study Section of the National Institutes of Health added in 1949, "It is a shortsighted and wasteful policy to attempt research work with dogs if proper standards of animal care are not maintained. Dogs that are not in good health will not tolerate investigative procedures as well as animals in excellent condition, and, for this reason, false scientific conclusions might be attained. An institution should give research dogs even better care than is received by pet dogs."[161]

One challenge in doing so was that there were few commercial retailers of "research" dogs prior to the 1960s. Most practitioners, instead, salvaged what they could from the unclaimed populations of local pounds. Although some pounds received funding from local universities to ensure access to the healthiest dogs for experimental and teaching purposes, the result was typically an inconsistent supply and criticism from humane societies.[162] An obvious alternative, with a clear public relations benefit, was the establishment of a well-maintained source of dogs with known genetic ancestry that could be sold directly to labs. National stock centers had formed for fruit flies and maize at Cold Spring Harbor and Cornell, respectively, and standardized rats were being produced in large numbers by the Wistar Institute in Philadelphia.[163]

One of the most aggressive early proponents of such an institution for dogs was Columbia University biochemist Erwin Brand, a German-born researcher who had developed a small experimental colony of Irish terriers in Port Chester, New York, for Rockefeller-sponsored research on cystinuria, a condition causing recurrent kidney stones.[164] The colony began when a show dog from New Jersey taken to Columbia in 1935 was found to have cystinuria, and inspired by his successes with the terriers, Brand spent much of 1938 conferring with colleagues about the need for standard dogs in laboratory research.[165] "Discussion with a number of research workers," he summarized to the National Cancer Institute's Ludvig Hektoen, "seemed to indicate a wide-spread interest in a supply of standard genetically uniform dogs for certain investigations in such fields as physiology, biochemistry, nutrition, pharmacology, toxicology, virus diseases, and cancer research."[166]

In Brand's view, a centralized dog colony could breed a few useful varieties and sell them at reasonable prices to researchers. Cornell Medical College biochemist Vincent du Vigneaud, for instance, suggested he consider producing a line of toy dogs for isotope studies.[167] But such an institution could be self-supporting only in part, as Brand and virtually everyone he spoke with insisted, so external support was necessary. He had first approached Rockefeller's Warren Weaver, hoping the foundation might fund him alongside Stockard, but Weaver was noncommittal and encouraged Brand to try the National Research Council (NRC) instead. Brand then urged Hektoen, a recognized scientific leader, to support his NRC working group proposal, having already spread the gospel of standard dogs to various luminaries: Leslie Webster, a rabies researcher at the Rockefeller Institute for Medical Research; George Whipple, the pathologist at Rochester and later prize namesake; Anton Carlson, a prominent biologist at the University of Chicago and leader of the NSMR; George Harrop, research director at pharmaceutical company Squibb Laboratories; and many others. Brand was also already selling many of his surplus terriers for $5 each to Herbert Calvery, chief of the Food and Drug Administration's Division of Pharmacology, for toxicological testing.[168]

A few years earlier, an NRC working group had studied biological stocks generally, but its wide-ranging species ambit (everything from protozoans to fruit flies and birds) generated few specific recommendations. "If a new committee is appointed it could profit by our mistakes," mused Columbia geneticist Leslie Dunn, one of Brand's strongest supporters. "A more hopeful prospect, as pertains to dogs especially is that there is now available what was lacking in our case, that is one person who is willing to take the initiative and to perform the service function of maintaining stocks for other investigators."[169] That person was Brand, and he was eager to do it. Yet even as the NRC agreed to support his committee, many within the institution, such as Edmond Long, chairman of the Division of Medical Sciences, nursed doubts. Long, who served on the board of the Wistar Institute, knew of widespread criticism of the price of Wistar's rats ($4–6 each) and that most were sold below the cost of production (around $10).[170] Producing standard dogs more cheaply than rats was unimaginable, and few researchers could afford anything pricier.

In 1939, the NRC's dog subcommittee visited a handful of laboratories, including Calvery's FDA facility and Stockard's Dog Farm, to consider the logistical dilemma, arriving at the latter not long before Stockard's death.[171] They did so quietly, because Brand insisted that the membership and work of the committee "be given no publicity" to avoid unwanted attention from critics of animal experimentation.[172] Their subsequent report emphasized the need for "a standard strain of dogs of known genetic constitution" and included a proposal by Brand and Dunn to establish a small, investigational center to breed Irish terriers at Columbia.[173] Johns Hopkins physiologist Philip Bard, the chairman of the committee, later cautioned, however, that the proposal was included in the final report largely because it was "the only thing of the kind which the Committee was able to secure."[174]

As a backup, Brand and Dunn also suggested beagles, echoing tentative efforts to develop the breed scientifically. Columbia physiologist Magnus Gregersen, for instance, had tried in 1937 to develop his own line of English foxhounds, which had "the greatest number of desirable traits," before turning to the idea of a "standard beagle."[175] After Gregersen lost most of his breeding beagles to a distemper epidemic, he opted to purchase Brand's terriers instead.[176] A third group voiced a preference for Airedales, larger terriers popular among American hunters. While the committee generally agreed that a medium-sized dog with short hair was ideal, partiality to breeds differed. Stockard had shown the value of well-kept dogs, but *which* dog was still very much in the air.

As the report moved toward publication, Brand contacted Weaver again, hopeful that Rockefeller might take up the NRC recommendations. His tenure at Columbia initially hinging on external funding, Brand hoped to carve out a permanent position by establishing a "Laboratory for Animal Research" on the grounds of the Nevis Estate in Irvington, New York, a scenic plot of land given to Columbia originally for an arboretum.[177] With Dunn's blessing, Brand proposed expanding his terrier colony to around 150 animals that could be sold to "qualified research workers in other institutions."[178] Yerkes supported Brand's plan and encouraged him to ascertain interest from commercial laboratories and inquire about the availability of Stockard's facilities.[179] Yet Weaver and fellow RF officer Frank Hanson insisted that Rockefeller would not support a dog center.[180] Confidentially, Bard also noted hesitation from committee members about Brand and his plan. Originally a firm supporter, Bard's views had been modified by the outbreak of World War II. "In short, the project has been brought into sharper relief against a background of competitive projects and I cannot help but feel that there are a good many things which deserve prior consideration," he explained to the NRC's Ross Harrison.[181]

Ultimately, Columbia administrators agreed to put up $20,000 for the facility in 1941. But much as Bard had predicted, work was permanently interrupted by America's entry into World War II. Brand's original Rockefeller grant ran out in 1941, and he shifted to wartime research with the US Office of Scientific Research and Development, distributing most of the surviving terriers to various researchers and keeping a few for his own future work. Nevis was eventually selected as the site for a new physics laboratory and Columbia's cyclotron instead.

In lieu of Brand, it was C. C. (Clarence Cook) Little—the director of the Roscoe B. Jackson Memorial Laboratory in Bar Harbor, Maine, and a friend of Gregg, Dunn, Yerkes, and nearly every other prominent American in biology—who emerged as the most likely individual to bring a dog colony to fruition. With support from Rockefeller, Little had built Jackson into the central hub for producing experimental mice in America.[182] He was also a dog lover and had begun breeding springer spaniels on a small scale to study breast cancer. Although World War II delayed major Rockefeller action, Gregg inquired in early 1944 whether Little might be interested in using dogs to study the inheritance of intelligence, a lingering obsession for Gregg, who believed educators overemphasized environmental factors and underplayed heredity.[183] Although intelligence studies had been carried out on rats, Gregg was convinced that people paid more attention to dog behavior and could better discriminate among canine intelligence.[184] Well-managed dog experiments would be more persuasive to the general public and accomplish, at a delay, what Stockard had failed to.[185] As breeding proxies, dogs were positioned as the most persuasive tools for generating scientific truths.

In 1944, the plan received a positive endorsement from Yerkes, who believed it "reasonably possible to plan with intelligence and economy such a long-continuing genetic inquiry" with dogs.[186] It also obtained, somewhat surprisingly, the blessings of the Animal Rescue League of Boston, whose representative told Gregg that an experimental dog farm "would have the approval of humane societies, animal rescue leagues, etc. since it would obviate the likelihood of medical schools maintaining a business in private pets."[187] Yerkes further recommended reaching out to Whitney, who, having given up his grandest experimental aspirations, had continued experimenting with dogs and argued in a vehemently eugenic book, *The Builders of America* (1927), that they demonstrated the heritability of mental traits.[188] Although it is unclear whether Gregg did so, Whitney and Little were already acquainted from their participation in the eugenics movement and America's elite dog circles; the former's involvement was inevitable.[189] Yerkes and Gregg agreed that coordinated efforts to breed a highly intelligent dog might incidentally result in an ideal pet that could be sold by the lab for additional revenue, an idea apparently planted by biologist Alexis Carrel during a dinner conversation with Gregg in 1935.[190]

Of all this Little was easily convinced. Noting a conviction that "Stockard's work was of very great value," despite its premature conclusion, he reaffirmed a lifelong affinity for dogs. His father had been a breeder, keeping as many as one hundred at a time, and Little appeared as a judge at the Boston Dog Show in 1943. "If we are not convinced of the importance of individual variation and if we do not understand how it arises and how to utilize it," Little blustered in a letter to Gregg, "we shall never be able to create a democracy that will have in its own make-up the characteristics necessary to criticize it and to shape its destiny as it evolves." Against previous recommendations, however, Little proposed the dachshund as "the most promising single breed with which to work."[191]

Although hosting a full dog farm at Jackson would be unprofitable, Little imagined developing a "decentralized" breeding operation that used space at nearby farms and sold surplus dogs to medical school faculty and researchers (this had been Whitney's plan at Yale).[192] Little received assurances from faculty at Harvard and Yale, including Sidney Farber, that universities would happily buy dogs for the right price.[193] After a minor delay, the Rockefeller Foundation announced a major grant in 1945, totaling $282,000 over five years, for the promised work on heredity and intelligence. In his proposal to investigate "the genetics of intelligence and emotional types in mammals," Little stressed the project's "comparative" nature and its likely value for "education and medicine." The research would build on the "work of Stockard and his associates" intellectually, but depart radically in how animals were mated and raised.[194]

Where Stockard's crossbreedings were largely aimed at producing "monsters," Jackson would carefully attend to minute differences among "normal" individuals. In an early press release, he explained that the project would "analyze the roles of genetics and of environment in determining both breed and individual differences in dogs."[195] Principal research would take place at brand new experimental dog facilities at Hamilton Station, a fifty-five-acre farm across from Jackson's main facilities, donated in 1940. The site had three barns, which could be retrofitted to house animals, and other facilities including a small histopathology lab. Even with Little's decentralized model, the size makes clear, once again, the infrastructural challenges of working with canines.

To manage the facility, Little hoped to bring in Grüneberg, whose stinging review of Stockard's work had undoubtedly caught his eye. Promising an unmatched opportunity to lead one of the grand new genetic research programs, Little wrote to Grüneberg in London and offered to bring him to Jackson. A response was slow to arrive, however, as Grüneberg had been called to serve in the Royal Army Medical Corps and took time to consider the proposal.[196] In December 1944, he returned with a laundry list of concerns, first of which was the animals themselves: as an expert on mice and rats, he worried, "If I now switch over to an altogether different organism I shall loose [*sic*] this advantage and shall have to start from scratch again."[197] Grüneberg, who considered himself a morphologist, also felt "lost when confronted with behaviour." The job security of a short-term Rockefeller project, even one that Little promised would last decades, was even more concerning, and Grüneberg feared being left high and dry in the rural environs of Bar Harbor without an institutional affiliation if things went south. Having committed to aiding postwar reconstruction of his department at University College London, Grüneberg cautioned that he would not be available to visit until 1946 under even the best circumstances.

In the meantime, Little tapped a young animal geneticist and psychologist named John Paul Scott, who had previously worked at Wabash College, as Grüneberg's first assistant. At thirty-six or thirty-seven (Little was not sure), Scott would be able to stay at Jackson for an extended duration and could be, Little felt, "used to advantage under a certain amount of broadening, philosophical influence."[198] That "broadening influence" was unstated, but it seems clear that Little hoped to steer Scott into the kind of "straight genetic" work he favored, supporting the project's original focus on the genetics of intelligence. This was part of why Grüneberg had been a desirable leader: focused on producing specific genetic mutations in mice, he was less interested in behavior and would be unlikely to push the dog work in unexpected directions. Scott, on the other hand, was acutely interested in socialization and, although not indifferent to eugenics, had critiqued it as a program for maintaining elite dominance.[199]

Little also hired Emilia Vicari, Stockard's former histologist, maintaining a direct continuity with the earlier work. Vicari had been listed as a "scientific associate" of Stockard, which probably understated her contributions, and had published findings in 1937 on the cellular structure of the thyroid glands in the farm's dogs.[200] Although much of her subsequent research at Jackson concerned mice, Vicari brought knowledge of Stockard's studies and breeding system with her when she moved to Bar Harbor in 1941, following a year on the Elizabeth Clay Howald Scholarship from Ohio State University.[201]

Stockard's work had thus set the stage for an even grander experimental program, one that would occupy researchers at Jackson for more than a decade. Grüneberg never made it, however. After his wife, Elsbeth, died during the war, he was left to care for their two children and stayed at University College until his retirement in 1974. Grüneberg finally arrived for a summer at Jackson only in 1952.[202] Instead, Scott took over leadership of the project, assisted by psychologist John Fuller. Although neither proof of the inheritance of intelligence nor a perfect pet dog emerged from the project—Little, who kept working with dogs separately, thought he was close in 1956 with a dachshund-pug combo—tremendous knowledge about the social behavior of dogs did. The two would summarize their findings in an influential book, *Genetics and the Social Behavior of the Dog*, in 1965.[203] Scott and Fuller also experimented with minute differences in canine care and maintenance, publishing their approaches in a widely circulated report, *Manual of Dog Testing Techniques*, which informed subsequent laboratory designs.[204]

During the years of its operation, the lab became a mecca for dog behavior researchers: William T. James visited for a summer of research, for instance.[205] Yet Little, who resented the independent dog behavior "empire" that Scott and Fuller had created, remained fiercely critical of their perceived lack of interest in the genetics of dogs.[206]

Most importantly for beagles was that the breed was one of five chosen for the Jackson studies, each fitting roughly within Stockard's "normal" type range: basenji, beagle, cocker spaniel, Shetland sheepdog, and wire-haired fox terrier. Beagles, cockers, and fox terriers were all highly popular dogs, while the basenji and Shetland were chosen to represent behavioral contrasts. Brand's Irish terriers did not even make the cut, and their role in science largely ended there. The line of research from Stockard to Scott established logistical and scientific foundations for future dog work, situating beagles as an obvious choice of experimental dog. In this research tradition, ideas about "normal" and "standard" dogs were connected to a eugenic view of dogs as proxies for racial difference and breeding. So-called normal dogs were also normative dogs, associated both explicitly and implicitly with presumptively superior, conventionally Anglo-European people. Some years later, once Stockard's follies were forgotten, a new Cornell dog program could finally emerge: the Research Laboratory for Diseases of Dogs. Where Stockard's scientific odyssey had crashed on the rocks of an aggressively individualistic approach, the new lab would emphasize, from its very beginnings, a broad view of scientific collaboration.

Distempered Expectations

In the autumn of 1949, Cornell alumnus John Merrill Olin traveled to Ichauway Plantation, the hunting retreat of his friend Robert Woodruff. Both men were wealthy: Olin had amassed a small fortune overseeing the expansion of his father's arms manufacturing firm, the Western Cartridge Company, into the multisector Olin Corporation, while Woodruff became one of the richest men in Georgia as president of Coca-Cola. The two were also avid hunters, and Woodruff kept a pack of prized German short-haired pointers at Ichauway, a sprawling, lush estate located in the southwestern corner of Georgia.[207] The year of Olin's visit, however, Woodruff's dogs began to succumb in large numbers to a strange new illness, one that also appeared in pointers owned by Walter Teagle, the former head of Standard Oil, who owned the nearby Norias Plantation in Thomasville. Nearly as committed to dogs as he was to Cornell, Olin recommended that Woodruff and Teagle bring down James A. Baker, an accomplished veterinarian from the university, to see what was going on.[208]

In the cause of death, Baker discerned "a new and more virulent" disease with similarities to distemper. To thank Baker for his quick work and out of a desire to prevent future outbreaks, Olin encouraged Woodruff and Teagle, another alum and major donor to Cornell, to join him in funding a new research institute that would go to work on diseases like the one that just attacked their dogs.[209] Teagle, who was well connected in the New York philanthropical world, contacted Alan Gregg for a second opinion on Baker, whose postgraduate years were spent at the Rockefeller Institute for Medical Research. Gregg confirmed a positive view of his abilities, one of the most important endorsements one could receive at the time.[210]

Stockard's Dog Farm and Jackson had set precedents for the new undertaking. But even the support of Olin and Teagle was insufficient on its own for the would-be veterinary research center. New York State funding sustained Cornell's livestock research program, but the health of dogs was not obviously a problem meriting government support. Without the deep pockets of the Rockefeller Foundation, the proposed lab needed alternative capital sources. In turn, Olin leveraged his personal and professional networks to assemble initial contributions from a stunning list of the era's preeminent figures: Woodruff and Teagle; Nicholas Noyes, another Cornell benefactor and the director of Eli Lilly; Richard Deupree, president of Procter and Gamble; George Humphrey, president of steel company M. A. Hanna and later secretary of the treasury; Gerald Livingston, a former governor of the New York Stock Exchange and president of the Westminster Kennel Club; Walter Edge, two-time governor of New Jersey; James Ford Bell, founder of General Mills; Jacob France, chairman of the Mid-Atlantic Oil Company; Elisabeth Ireland Poe, a prominent breeder; John Hay Whitney, the nephew of Harry Payne; and more. Names such as Whitney's show how Olin's pet project was enabled by widespread interest in dog breeding and dog shows among America's wealthy elite, which had grown from the beagle fascination of earlier decades.

Two donors stood out from the rest and helped usher Olin's vision into existence. The first was Geraldine (Gerrie) Rockefeller Dodge, a successful dog breeder and the youngest daughter of William Rockefeller Jr. After marrying Marcellus Hartley Dodge, chairman and owner of the Remington Arms Company, Gerrie spent much of her time pursuing passions different from those of her husband. The couple's relationship was eventually so strained that, in 1962, when Cornell needed funding, Olin warned Baker that Marcellus "stands in mortal fear of the lady" and appealing to her through him "might have just the reverse effect."[211] Although a lifelong dog obsessive, Gerrie began to invest her fortune and time into dogs only as her father's death neared, in part because he had insisted her palatial property, Giralda Farms, was better suited to horses.[212] In the 1920s, she gained national renown for her German shepherds, often hand selected by assistants in Europe, and later played a key role in bringing the cocker spaniel to prominence.[213] She wrote popular books about both breeds, led the establishment of New Jersey's exclusive Morris and Essex Kennel Club Show, became the first woman to judge Best in Show at Westminster in 1933, and bred a number of other purebred dogs, including beagles.[214] With a gift of $25,000 (today worth nearly $325,000), one-half of the new facility, perched atop Snyder Hill and overlooking Lake Cayuga, would be called the Giralda Division in her honor.[215]

A second, matching gift came from Connecticut couple Lee Garnett Day and his wife, Nancy Sayles.[216] Day, a wealthy Yale graduate, initially worked in imports in New York but traveled widely on international expeditions, including a South American trip in 1915 for the American Museum of Natural History, and later served on the staff of John J. Pershing during World War I.[217] He continued to lead collecting expeditions for American museums and commanded the Norfolk US Army base during World War II, earning himself the rank of colonel, before climbing to the vice presidency of General Foods.[218] In September 1925, he married Nancy Sayles, heiress to a vast textile fortune that briefly earned her the nickname of "Rhode Island's wealthiest girl."[219] Together, the couple raised great Danes at their Daynemouth Kennels in Longridge, Connecticut, and owned an estate in West Cornwall, known as Cobble Mountain Farm. The second half of the new project was named the Daynemouth Division.

In January 1951, after a year of construction and aggressive fundraising, the Cornell Research Laboratory for Diseases of Dogs (RLDD) opened for operations.[220] A long list of publications in the new institution's first annual report demonstrates that the laboratory, which was envisioned as one part of the university's broader Center for Animal Disease Research, represented an expansion of ongoing campus research activities rather than a truly new undertaking. But the experimental kennels in Ithaca were nevertheless significant for what they symbolized: the first institution dedicated entirely to basic scientific research into what made America's beloved dogs healthy or sick. It was, so the RLDD's first newsletter claimed, "the only laboratory of its type in existence."[221]

Dogs had, after all, appeared in a bevy of scientific projects stretching back into the seventeenth and eighteenth centuries. But few research projects had considered it worthwhile to study them as ends in themselves, rather than as disposable tools for the study of human biology.

Stockard, for example, had learned a great deal about dogs, but his interest was primarily in what the animals revealed about human breeding. Veterinary science, such as it existed, was also a largely practical field with minimal basic research at the time. "Although Veterinarians have demonstrated great competence in the fields of anatomy, surgery and treatment of parasitic infections, they as yet have little information about infectious diseases of animals," an early RLDD planning document noted, because of a "lack of research facilities and funds to conduct an adequate research program."[222] The RLDD was envisioned as a dramatic new initiative to remedy that long-standing challenge.

Gaines in Dog Research

Important as such facilities were, the world's foremost laboratory on the diseases of dogs also required something else: dogs. The Gaines Dog Research Center, a publicly oriented research institution located in New York City, swooped in with a gift of $15,000 in order to support the construction of a "disease-free isolation colony" for the RLDD's programs.[223] Rather than a number of different breeds, as Stockard and Scott had chosen, it was decided early on that the kennels would house "purebred beagles."[224] Beagles were growing in popularity at the precise moment that the RLDD was opening its doors, and their wide dispersal, particularly in upstate areas close to Ithaca, made the dogs an obvious choice on the grounds of availability. A number of the lab's prominent donors were also active beaglers.

But RLDD researchers also emphasized that there was something special about beagles. Unlike other breeds, beagles exemplified a high level of specialization without excessive inbreeding. The practice of early beaglers, who placed classified ads in issues of *Forest and Stream* and transported their dogs across the country to sire litters, meant that beagles were outbred in highly managed contexts. Beaglers kept careful heritage records, a vital index of "this country's present day almost limitless network of Beagle bloodlines," in the words of Glenn Black.[225] The knowledge base underlying that network was particularly valuable at a moment when experimental genetics had not yet explicitly reached dogs and "bloodline" remained a predominant way of conceptualizing how dogs inherited traits and behavioral proclivities.

Support from the Gaines Dog Research Center exemplified how dog research both supported and relied on industrial interests, since the center was ultimately a public relations arm of the Gaines Dog Food Company.[226] Founded in 1928 by Clarence Gaines, who developed a proprietary dog food in his Sherburne, New York, kitchen while milling feed for cattle and poultry, Gaines Dog Food helped establish the market for premade dog kibble in the United States. By the mid-1930s, Gaines advertisements littered national newspapers, promising free samples to new customers in the form of meal, pellets, or cans. In 1939, after the company won a major contract to supply dog food for the US Antarctic Service Expedition, advertisements stressed that Gaines was tried and tested under the hardest conditions.

Gaines's largest competitors early on were not rival businesses, but those who made dog food at home, just as Gaines himself had. There is "danger in many home-prepared dog foods," noted a 1942 veterinary endorsement of Gaines products, which had been proven effective by "critical biological tests" on "four successive generations" (more or less a recitation of the company's ad copy).[227] Such advertising was vital in constructing a need that the company's product could respond to, and Gaines Dog Food expanded rapidly, bringing attention and jobs to Sherburne as its chow was shipped across the world.[228] A glowing bio in the New York Veterinary Society's *Veterinary News* highlighted that Gaines created jobs and prosperity "in the typical American way."[229] To the dismay of residents, however, manufacturing moved to Chicago just a few years after Gaines was acquired by General Foods in 1943.[230] In 1950, the company began diverting most of its dry dog food bags from traditional feed stores to grocery stores, which had previously emphasized canned dog food on the assumption that consumers were uninterested in purchasing large quantities at a time.[231] Gaines and others saw a market in one-stop shopping, and the modern supermarket dog food aisle was born. By 1952, sales of prepared dog food matched those of packaged breakfast cereals.[232]

Gaines's personal interest in dogs—he bred pointers—shaped how the company related to dog owners. A report on the sixth annual McKean County Sportsmans club's dog trials in 1933 announced that a pointer named Nepken Carolina Bill had emerged victorious; the dog's owner was the Gaines Food Company.[233] Gaines sponsored trials near and far for years, entering dogs and occasionally purchasing winners, and it was common for field trials and dog shows throughout the late 1930s to award their winning entries five-pound bags of Gaines Dog Food. This close relationship with the pet world emerged out of Clarence's hobby, but it also formed a strategy to build name recognition through competitive success while collecting information on what made a dog healthy in the first place. Gaines advertisements emphasized how the original formula, a "first, great step" for dogkind, was improved over years of experimentation. It was not just dog food, but *scientific* dog food.[234]

The science angle took center stage after Walter Armstrong became president of Gaines. Armstrong had worked in the dog world since his Army discharge in 1919 and led efforts to develop the dry "Ken-L-Biskit" for Chappel Brothers, who originated the canned dog food Ken-L-Ration. Focused on the question of canine dietary requirements, in 1928 Armstrong founded what he believed to be the first experimental nutrition kennel in Rockford, Illinois, using his own fox terriers.[235] Armstrong was recruited to Gaines in 1939 and there began to implement his vision for both a research center dedicated to dogs and a full experimental kennel. "Brighter World for US Dogs, Too, Planned for End of War," announced one headline on the initiatives.[236] To oversee the new Gaines Dog Research Center (GDRC), Armstrong selected Harry Miller, a journalist and two-time secretary of National Dog Week, the celebration of canines founded in 1928 by dog writer Will Judy.[237] As research center director, Miller would "disseminate scientific and popular data helpful to both dog and dog-owner."[238] The center's offices opened at 250 Park Avenue in New York City, where the English Springer Spaniel Field Trial Association had been headquartered since at least 1930, just a few blocks from Grand Central Terminal.

Anticipating major postwar expansion, the GDRC also planned to construct research kennels and a "dog zoological garden" in Ridgefield, Connecticut. First managed by celebrated Connecticut dog breeder Harry Hartnett, the resulting Gaines Research Kennels were led, after Hartnett's early death, by Elias Vail, an experienced trainer who had worked with K-9 Corps war dogs at Nebraska's Fort Robinson during World War II.[239] Extremely optimistically, the GDRC hoped its kennels would "make available to public view prize specimens of all 167 known dog breeds in the world" and present "standards of perfection" for every breed of dog.[240] Although the site never grew quite that large, visitors flocked to Ridgefield for years after its opening. In 1948, for instance, nearly 150,000 people visited a small "dog zoo" presented by Gaines, featuring around thirty dogs, at the Danbury Fair. The Gaines Research Kennels were also meant to develop a "medium-size, all weather, super-intelligent farm dog who would be herd-dog, guard-dog and companion-dog in one"—a vision notably similar to the ideal experimental dog that Gregg and Little hoped their research might generate—but Gaines got no further in that endeavor than others had. In 1950, the facility was relocated to Kankakee, closer to the site of dog food production.

Broadly, the GDRC embraced Armstrong's mandate, pumping out content about dogs and dog ownership for the general public. Its principal public relations output took the form of multistory newspaper spreads: one representative example in the Wellsboro, Pennsylvania,

Agitator from July 1947 included sections on handling new puppies, the role of dogs in the life of Abraham Lincoln, and the growing market for dog food, along with a story honoring "America's 'Dog Methusaleh.'"[241] The threat of canine distemper and its treatment was a constant topic in the spreads, revealing both Gaines's and the broader public's interest in the work of the RLDD.[242] The GDRC also produced short films and pamphlets on a variety of issues, frequently stressing the value of purebred dogs, and sent its experts across the country for public-speaking engagements. In the Gaines view, Americans were curious but bumbling pet owners who needed a little instruction. Yet the GDRC also worked to understand what Americans wanted from their dogs, surveying owners, for instance, on what the "most useful" type was.[243]

In one sense, then, the GDRC was an earnest educational advocacy organization designed to appeal to the country's expanding demographic of dog owners. Highlighting the patriotic linkage of dogs and war, it sponsored a competition to design a monument to "dogs that fought and died in World War II" in June 1945.[244] But the GDRC also represented an intentional effort by Gaines to educate the public about the value of the company's products. One of its most popular publications, *Basic Guide to Canine Nutrition*, which was reissued numerous times, recommended a healthy diet of Gaines Dog Food for any growing pet.[245] As Miller argued in 1947, prepared dog food might even be the *cause* of America's growing obsession with dogs, since Gaines kibbles served as "work and time savers for the woman in the home."[246] Gaines was enabling dog ownership by the masses, which was enabling windfall profits for Gaines in turn.

In the 1950s, the GDRC tackled other doggy issues, releasing its popular *Touring with Towser*, a guide to hotels and motor courts that would accept guests bearing dogs. "Happy traveling to you and Towser!" exclaimed Miller in the manual's introduction.[247] Importantly, the GDRC also focused on supporting veterinary and scientific research. It hosted the Gaines Veterinary Symposium over several decades and published the resulting papers in the periodical *Newer Knowledge about Dogs*, as well as its own periodical, *Gaines Dog Research Progress*, from 1946 until well into the 1970s. Support for the Cornell kennels, which partly supplanted Gaines's own, was thus coextensive with an expansive assemblage of research into dogs taking place across midcentury America. The National Dog Record Bureau, a product of deliberations during 1947's National Dog Week, opened in Santa Monica in 1948, using the Army's K-9 tattoo identification system to register and track American dogs and return lost pets to their owners. What writer Kathleen Szasz termed America's escalating "petishism" had supported the growing market for dog food and other services, changes that made the idea of applying the newest advances in science to canines not only thinkable, but necessary.[248]

Gaines set the tone, but a host of other corporations signed on to support work at the RLDD. Food manufacturer Swift and Company, for instance, provided an early grant of $15,000 to support the distemper research led by Baker. Other feed companies, such as Kasco Mills, General Mills, Nutrena Mills, and Perk Dog Food, all gave additional funding. The RLDD was also supported by interested parties in human and livestock pharmaceuticals, including Fort Dodge Laboratories, Merck and Company, Pitman-Moore, Syracuse Pharmacal, and Squibb and Company. Without these external donors, the institution would have struggled to sustain its activities. Angela N. H. Creager has shown how the National Institutes of Health and National Science Foundation took on increasingly prominent roles as funders in biomedicine during the postwar period, particularly as philanthropic institutions such as the Rockefeller Foundation shifted their priorities, and while diseases affecting agricultural animals had clear national security implications, dog maladies were less certain government priorities.[249] Instead, the institution's corporate sponsorship exemplified persistent tensions between research into dogs themselves and research into dogs as experimental materials for human studies. For many of the pharmaceutical actors, better understandings of distemper and other canine illnesses were essential for ensuring good results in tests of human drugs. The public-focused framing of donations to the RLDD, where dogs would be studied for altruistic reasons as man gave science back to his best friend, petwashed the more instrumentalizing justifications for supporting the institution.

Testing with Towser

With endorsements from prominent dog owners and corporate interests, the RLDD's 1951 opening was celebrated in periodicals and newspapers around the country, kicking off a second wave of fundraising, this time from fans and kennel clubs. What most separated the RLDD from other dog research projects of the era was its close connection to the interests and concerns of everyday pet owners. One hundred and twenty different kennel clubs signed on to support the lab in its first year, whether through individual or group donations, a number that grew to over two hundred clubs by 1951.[250] Beagle owners were particularly enthusiastic about the laboratory—predictably, perhaps, expecting benefits from the experimental use of their preferred breed—with fifty-five clubs holding field trials to benefit continuing research that year.[251]

To fete the new institution, Cornell convened a symposium on viruses in canines in 1951. The National Institutes of Health's Norman Topping spared no drama in extolling the project's significance during his dedicatory address. Scientific teaching and research were critical to staying ahead of "the communists," Topping proclaimed, and protecting America against biological attacks. "It does not matter whether the laboratory includes the words 'dog,' 'cattle,' or 'man' in its name," he continued, heading off objections at the outset. What mattered was that "there continue to be enough laboratories, enough research, and enough competent, well-trained, young scientists coming out of our universities."[252] Amid widespread concerns about America's production of scientific "manpower," if dogs got men diplomas, that was good enough for Topping.[253] Yet despite lofty rhetoric positioning it in a grand Cold War struggle to protect capitalist America against communism, the RLDD planned to focus on three rather mundane "viral" challenges: leptospirosis, encephalitis, and infectious hepatitis. Although we now know the first is a bacterial infection and the second a symptom of other infections, the three conditions were united at the time in what Cornell researchers and others called the "distemper complex."

Canine distemper, caused by an RNA virus from the same family as measles and mumps, first rose to prominence during the early nineteenth century, when it threatened British foxhounds and, thereby, British fox hunting.[254] With the rising popularity of domestic pets, distemper received significant attention in hunting journals and the periodicals of pedigree dog breeders. As Michael Bresalier and Michael Worboys have shown, the battle against distemper was waged primarily by veterinarians, who took a new interest in small animals instead of traditional agricultural animals, and by scientists interested in the new Pasteurian studies of bacteriology.[255] In 1901, French microbiologists Joseph Léon Lignières and Charles Phisalix produced a seemingly effective vaccine for *Pasteurella canis*, their candidate as distemper's principal cause. Four years later, pathologist Henri Carré isolated another agent that he declared responsible. Many American and British researchers objected to this work, and vaccines produced based on the French findings failed to fundamentally alter the disease's course.[256]

World War I delayed further research, but in the conflict's aftermath, readers of the British sporting journal *Field* established a fund to support continued work. Drawing on that funding, Patrick Playfair Laidlaw, a Cambridge-trained biochemist and pathologist, and George William Dunkin, veterinary superintendent of the UK National Institute of Medical Research's animal facilities, began to study distemper with ferrets, who are highly susceptible.[257] Their treatment, the "Laidlaw-Dunkin process," was announced in 1928 (Stockard was one of the earliest American researchers to use the vaccine, popularizing it in 1932).[258] Yet even decades later, distemper vaccines were considered only partially effective, not least because the word "distemper," which etymologically implies an imbalance of the humors, referred to a number of canine ailments. Indeed, rather than "more virulent" distemper, it was likely infectious canine hepatitis that Baker had found in Woodruff's kennels. "While we had just one virus disease in dogs to contend with, we were quite happy," noted Buffalo veterinarian R. B. McClelland at the RLDD's opening symposium. "Our bubble burst when virologists showed us that we were grouping several diseases under one heading. . . . Now all is confusion!"[259]

Fear of the distemper complex had launched the RLDD, and Baker, who received both his doctoral and veterinary degrees from Cornell, was considered the ideal individual to lead the new institute and find an effective treatment. He had served for several years after graduation at the Rockefeller Institute and then moved to the War Disease Control Station at Grosse Ile near Montreal to work on the lab's rinderpest project at the start of World War II.[260] There he primarily assisted in developing the avianized vaccine for rinderpest, a highly infectious viral livestock disease that Americans feared would be weaponized by the Axis, but Baker may also have participated in early experimental work on weaponized anthrax.[261] The linkages between Cold War national security and the diseases of dogs that Topping drew were not, at least in Baker's case, entirely an exaggeration.

Olin, one of Baker's most vocal and ardent supporters, was not disappointed by his man: in June 1952, Baker and the RLDD announced success in producing a new "two-way" vaccine that effectively immunized dogs against both conventional distemper and canine hepatitis.[262] The new vaccine was presented at the annual American Veterinary Medical Association meeting the next month and made available to dog owners across the country soon after. Just a few years later, the RLDD, part of Cornell's growing Veterinary Virus Research Institute, would announce promising advances in vaccination against leptospirosis. In a 1957 article based on one of Baker's public talks, columnist Myles Newcomb extolled the beginning of "an era of good vaccines for dogs," nearly all of which had been tested at Cornell's kennel of purebred beagles.[263] For their early support, Olin and Geraldine Rockefeller Dodge were feted by dog enthusiasts in New York City in 1961.[264]

"A dog is said to be man's best friend," began a *New York Times* article in 1963. "The Cornell Veterinary Virus Research Institute is a dog's best friend."[265] The RLDD's success against distemper and hepatitis removed another central obstacle to dog ownership, popularizing the concept of specific diseases in pets.[266] Where "distemper" had long been a mystery to the average dog owner, with treatments that mixed nutritional supplements and home remedies for dubious curative value, disparate ailments could now be identified, named, and treated. After leaving the world of eugenics to get a veterinary degree, Leon Whitney and his son George, also a veterinarian, published *The Distemper Complex* in 1953, criticizing the perspective of "Mr. Average Man," who expects his dog to be fixed up at the vet like a car with a broken clutch at the mechanic.[267] Dog owners, they wrote, needed to become educated in disease etiology so that they could care properly for their beloved animals. The success of the vaccines also put canine medicine on a preventive footing: inoculation early in life could protect dogs from the major infectious diseases. The "therapeutic revolution" had come for man's best friend.[268]

Where early research projects used dogs as breeding proxies—experimental subjects whose multifarious forms might reveal essential (eugenic) truths about human health—dogs began to emerge as worthy subjects of research in their own right by the middle of the twentieth century. The Cornell Research Laboratory for Diseases of Dogs conducted much of the pioneering research (although many were quick to join), and the Gaines Dog Research Center helped popularize the concept of a "science of dogs." Dog shows, field trials, the pet industry, and pharmaceutical researchers united to spur basic research into a species that had previously served science primarily as a cheap and accessible alternative to human beings. Dog owners responded with surprising enthusiasm to this change, collecting money to fund research at the RLDD in an early form of scientific crowdfunding and citizen participation. Doggy people were eager to become active players in experimental science if it promised a benefit to their cherished companions.

Importantly, too, at Cornell and elsewhere, beagles took on an increasingly central role as the ideal research dog. Prior laboratory programs, such as Stockard's, drew on a variety of generic, popular dogs, each understood to exemplify a representative "type." But the RLDD signaled a new trend toward centering research on purebred dogs of the beagle breed. The midcentury was a time when the search for a "standard" dog became the obsession of many biological researchers, just as buying "purebred" dogs was an emphasis for the GDRC and others. Alternative possibilities, such as Brand's Irish terriers, were tried but eventually displaced. The Wistar Institute's Edmond Farris opened one 1945 conference—"Animal Colony Maintenance," one of the first American events dedicated to the topic—by declaring that "a nearly standardized animal or group of animals for research purposes should be the aim of most laboratories."[269] Five years later, the Animal Care Panel, an international assembly of research animal experts, met for the first time in Chicago to share best practices.[270] From about 1950 onward, the beagle would begin to reign as the laboratory dog.

The intersecting stories in this chapter illustrate how research on canines grew so quickly and with support from so many sectors of American society at the precise moment that Americans increasingly viewed members of *Canis familiaris* as emotional beings worthy of expensive care and protection. Basic scientific research with dogs, which often led to their premature death in ways that would be condemned by many present dog owners, was celebrated as a crucial project to improve the health and well-being of pets across the country. John Olin even built his own Nilo Kennels, in Illinois, to field-test findings and recommendations of the Cornell RLDD. Research with beagles was not only *not* unethical; it seemed to be another way of loving them. Glenn Black, the avid beagler who cofounded the International Beagle Federation, served for seven years as manager of the Kasco Mills dog food division in Toledo, Ohio, where beagles were used as test animals.[271] For those who knew even less about dogs or how to keep them healthy, research was a kind of moral imperative: Towser needed to be understood. Economic interests put money and energy behind that view, but dog owners were eager to play their part.[272] When he died, Walter Teagle willed a significant portion of his estate to Cornell and the RLDD.[273]

Yet as altruistic as the motivations were for many of those who supported institutions such as the RLDD, the projects still relied fundamentally on the dog's existence as both a commodity and a testing subject for other industries. Olin and Teagle were unquestionably concerned with treatments for the diseases that struck down their dogs, and so were other industrial magnates for whom hunting was a prime leisure activity. But for many, understanding dogs better was likely to increase the market for dog foods, which represented a quickly growing and profitable undertaking. For still others, the RLDD offered the possibility of assuring the basic health and wellness of a species that was increasingly prominent in basic research. Distemper had nearly destroyed Stockard's colony, and eliminating both it and canine hepatitis were key first steps toward a healthy and standard experimental subject. "Disease-free" and "specific-pathogen free" became watchwords for laboratories.

In 1956, Cornell anatomist Howard Evans published a short text in *Gaines Dog Research Progress* titled "A Dog Comes into Being." Through a series of diagrams and radiographs, puppies are shown, day by day, growing slowly larger in the abdominal cavity of their mother. While kennel records could answer questions about litter sizes and ideal breeding age, Evans argued that some questions could be answered only "by maintaining a colony of dogs for specific study," just as the RLDD did.[274] The publication was an early foray for Evans, who went on to edit the standard text on canine developmental biology, and it represented a rare public view into how dogs went from embryo to pet. But the use of radiography to reveal canine development was also a pictorial allusion to other ongoing beagle studies in this moment of heightened Cold War anxieties, research that would put the breed at the forefront of efforts to understand the atomic threat to human health. Just like Evans's work, new studies at Davis and Salt Lake City would require the creation of large colonies of purebred beagles and change the breed's lot dramatically.

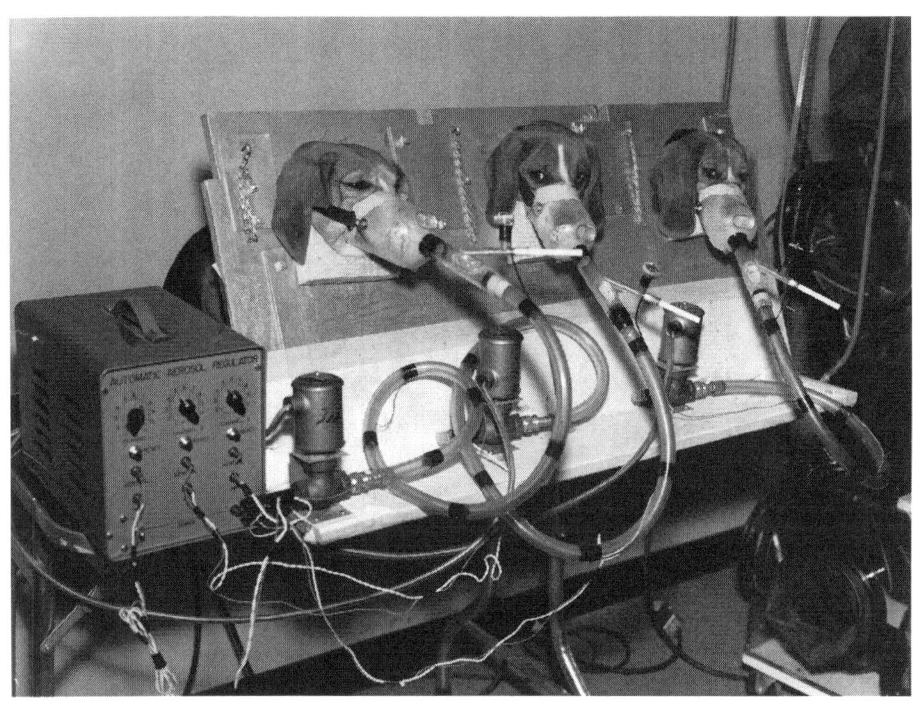

Beagles attached to a Hanford-style inhalation mask system. Courtesy of the US Department of Energy, Hanford Collection.

CHAPTER 2

Atomic Dogs: Cold War Anxieties, Normal Animals, and the Uses of Care

"There's a colony of frisky little beagles at the University of Utah College of Medicine that are destined to be among the first canine heroes of the atomic age," wrote journalist William Patrick in the *Salt Lake Tribune* in May 1952. The claim might have seemed strange to some, but Patrick assured his readers that if atomic war were ever to reach American shores, something that appeared increasingly possible after the first Soviet atomic test in August 1949, research on the dogs would save thousands of lives.[1] With any luck, the experiments would benefit the beagles, too. By their conclusion, Patrick explained, "more will be known about dogs themselves and the diseases that attack them than has ever been understood before."[2] The "hounds of Beagleville," as the *Tribune* dubbed them, were serving both their masters and themselves.

The Utah beagle lab was founded in 1950. A second colony followed it at the University of California, Davis, one year later. Each represented one prong in a multifaceted attack on the problem of radiation: research in Salt Lake City focused initially on the dangers of plutonium, the elemental foundation of the "Fat Man" bomb detonated over Nagasaki, while work at Davis centered on the hazards of full-body radiation to pilots in an eagerly anticipated fleet of nuclear-powered aircrafts and ships. "Airmen and seamen of the future may owe their safety to the hounds," proclaimed a short report in the *Sacramento Bee* in 1953.[3]

That hope proved overly optimistic, as attention at Davis eventually turned to strontium-90, a radioisotope in nuclear fallout that had been revealed in abundant, devastating clarity by American atomic testing in the Pacific. But these two large-scale radiobiology projects, planned initially by the Atomic Energy Commission (AEC) in 1949, would form the core of a growing network of dog studies, known colloquially as the "Atomic Beagle Club," which stretched across the United States. Within two decades, expensive and expansive beagle studies were also being conducted by researchers at the Hanford site in Washington state, the Argonne National Laboratory in Illinois, the Lovelace Foundation in New Mexico, the Oak Ridge National Laboratory in Tennessee, and the University of Rochester Atomic Energy Project in New York. Collectively, they aimed to contribute to fundamental understanding of the health risks of radiation, whether from future uses of nuclear weapons or civilian nuclear reactors. Dogs, who lived longer than mice or rats, were argued to be more similar to human beings when it came to long-term health effects, and beagles were chosen largely for their familiarity and availability. Because each AEC installation wanted its findings to correlate with research elsewhere, the breed became an accidental standard and was eventually affirmed as the first "experimental dog."[4] Discoveries about radiation followed, but the Atomic Beagle Club also yielded, as Patrick predicted, a surprising amount of information about the dogs themselves.

This chapter tells that story. Starting at Chicago's Metallurgical Laboratory, host to the first functioning artificial nuclear reactor in the United States, and at the University of Rochester Atomic Energy Project, it then moves to Salt Lake City and Davis. I explore why beagles were chosen by leaders in the AEC's biological research establishment and analyze the development of practices of housing and handling that allowed scientists in other fields, often with limited funding or institutional support, to later take up beagles as experimental animals. In contrast to the previous chapter's attention to the interplay between the interests of dog owners and researchers, the Atomic Beagle Club studies were sustained less by breeders and pet owners than by the demands of an active and paranoid assemblage of Cold War science.[5] One 1956 report even cautioned that news about the Davis project would be "disturbing to beaglers everywhere, and there are a lot of them."[6] Nevertheless, the atomic beagle studies gradually shed their explicit military and engineering justifications, morphing over time into something else: experimental inquiries into human and canine aging that built on research at Cornell and Jackson.

While Stockard and Whitney saw experimental dogs as proxies for the problems of human breeding, atomic beagles emerged instead as explicit physiological proxies. Beagles were miniature simulations of living human soldiers and workers, "standins for the pilots and sailors who one day will man atomic propelled planes, submarines and ships."[7] Even so, the dogs invited complicated modes of identification, which I call *species projection*. Researchers, many young and worried about looming nuclear catastrophe, saw themselves, implicitly and explicitly, biologically and symbolically, inside the bodies of their dogs. Early planning exacerbated such perceived similarities, as atomic beagles were meant to be entirely male, like the men who served in the Army or worked in nuclear reactors. Just as the fevered visions of politicians and defense planners alternately promised atomic annihilation and capitalist vitality, beagles encountered the extreme pains of radioactive exposures and decades of nearly Elysian comfort. The chapter shows how imaginative labor was just as critical as technical tinkering in making dogs function as proxies of human bodies and behavior, shaping the experiments and the knowledge they generated while being shaped by them in turn.

Reactor Breeders

Even before the first bomb lit up New Mexico's desert, radiation was a matter of acute concern in the United States. The amount of radioactive material produced prior to World War II was comparatively minuscule, with only about one kilogram of radium separated from uranium ore worldwide, but there was already serious disquiet about radium poisoning by the 1920s and 1930s.[8] A growing list of radium "martyrs," including Marie Curie, had frightened scientists, while the suffering of painters of radium watch dials brought awareness of the dangers of exposure to the general public.[9] With goals early in the Manhattan Project of producing plutonium at a scale hundreds of times greater than that of radium, many planners considered it essential to study the risks of sustained exposure. In a 1946 report with the marvelously euphemistic title *Carcinogenic Action of Some Substances Which May Be a Problem in Certain Future Industries*, atomic scientists Austin Brues, Hermann Lisco, and Miriam Finkel reflected, "The Manhattan District was determined not to encounter a delayed surprise such as that which greeted the radium painting industry when a high incidence of bone sarcoma began to appear in its workers years after exposure."[10]

At the Chicago Metallurgical Laboratory, known as the Met Lab—where the process for separating plutonium for production at an industrial scale was initially developed and the first artificial reactor, Pile-1, went critical beneath the west bleachers of the University of Chicago's football field—the Health Division had been instituted in 1942 in response to radiation concerns. Brues, Lisco, and Finkel were among the division's first members, and Brues would go on to lead biological research at Argonne National Laboratory, the Met Lab's successor. Work in the Health Division was intense, because immediate results appeared essential to the safety of major ongoing programs, and researchers often worked ten-hour days, six days a week.[11] Like most early American nuclear facilities, Met Lab biologists began by exposing mice to radiation. Jackson Labs provided two specialized strains with limited natural tumor incidence, some of the only mice that could be purchased in the thousands. The Met Lab also acquired specially bred rats after early experiments suggested they had a higher radiation tolerance.

High levels of plutonium exposure appeared to produce cancer and skeletal damage in both species. Because animal studies had previously underestimated the danger of radium, the findings provoked serious concern and demands for work with other experimental subjects. At the time, one obvious choice was human beings. Thus, between April 1945 and July 1947, AEC laboratories injected hospitalized patients, a number of whom were terminally ill, with plutonium, uranium, polonium, and americium in order to study the effect on life expectancy. Of these individuals, eighteen received injections of plutonium from Los Alamos and Rochester scientists.[12] Justified initially on the basis of needing to quickly establish radiation baselines, planners recognized limits to the studies even before they became the subject of popular controversy in the 1990s.[13] Working with terminally ill patients, many of whom gave questionable consent to participate, complicated the ability to delineate "true irradiation damage" from other medical conditions.[14]

Atomic researchers looked instead to larger nonhuman animals, such as goats, sheep, pigs, and dogs. The Met Lab, in particular, emphasized dogs in order to develop a more complete clinical picture of plutonium's effects. In 1945, physiologist Clifford Ladd Prosser and others at the Met Lab injected plutonium into a mixture of beagles and non-purebred dogs, finding declines in white blood cell count and many deaths, although a few dogs survived with apparently minimal damage.[15] The previous year, scientists in Rochester, the other major site of wartime atomic health research, had begun to expose dogs to x-rays in order to test radiation "tolerance levels."[16] Nine months into the project, they also began collecting sperm samples to determine whether radiation threatened male fertility by reducing semen counts, and Rochester radiobiologist George Casarett led a subsequent, multiyear follow-up to extend the results over a dog's full lifespan.[17] The significant investment and involved nature of the work, which involved technicians arousing and masturbating the dogs by hand, demonstrated the paradoxes of the physiological and imaginative equation of experimental dogs with male nuclear workers.

The initial Rochester studies involved nearly one hundred dogs, mostly male beagles and a number of strays added to maintain population numbers. Rochester possessed, in this sense, the first nuclear beagle "colony," although few involved seemed to consider it significant.

For the most part, reports from both Chicago and Rochester, delayed as they were by wartime classification requirements, offer little explicit comment on the choice of experimental subject.[18] At Rochester, the use of dogs was informed by the influence of George Whipple, who had brought a small canine research colony in 1921 when he became the founding dean of the University of Rochester Medical College.[19] Whipple was a lifelong hunter and kept a coterie for leisure and experimental crossbreeding. Like Josie and Trixie (the prizewinners introduced in chapter 1), the beagles used at Rochester emerged partly from his efforts.[20] There still, breed selection had as much to do with general availability and popularity as particular experimental considerations. Emerging at roughly the same time as the Jackson behavior studies, the dog work in Chicago and Rochester represented some of the first long-term experimental studies with canines and contributed to the sense that dogs were essential to biomedical progress.[21] At each site, beagles were a logical choice.

Although the Rochester studies took a number of years to bear fruit, the Met Lab research generated rapid interest after a few dogs developed delayed osteosarcomas following their plutonium exposures. This effect was largely invisible in rats, who had, one researcher noted wryly, "a distressing way of dying of pneumonia at just about the time they are ready to develop bone tumors."[22] Dogs, usually referred to in generic species terms, appeared to be better proxies for tumor development on a timeline similar to that of human beings. Clear data on this subject was critical for determining what health physicists called the human "tolerance" and "maximum permissible dose" of radiation. Both terms were euphemistic, associating radiation with medicine rather than harm, but the question of how much exposure a researcher could receive before threatening his or her long-term health was especially urgent following the deaths of physicists Harry Daghlian and Louis Slotin from acute radiation poisoning at Los Alamos in 1945 and 1946.[23]

Although multiple laboratories established their own internal standards, the AEC sought guidelines that could apply across facilities and be shared among countries. At the first conference on radiation tolerance, held at Chalk River in Ontario, Canada, in September 1949, representatives from the United States, England, and Canada agreed on tentative standards for exposure to plutonium, based in large part on work by Brues, Finkel, and others at the Met Lab. Because much dog research was still ongoing, the numbers drew primarily on data from rodent studies compared with estimates of the body weight of the "standard man," a statistical amalgam designed to represent an average male soldier weighing roughly seventy kilograms. The resulting standards were so restrictive that, had they been implemented, they would have completely halted operations at Los Alamos, the center of American nuclear weapons research.

Months of anxious communication between nuclear scientists and decision makers in Washington, DC, ensued before a tentative solution emerged from Wright Langham, a biomedical researcher at Los Alamos. On March 2, 1950, Langham wrote Thomas Shipman, head of the Health Division at Los Alamos, proposing a follow-up project to improve on the Chalk River calculations. He emphasized numerous challenges in relying on rat data: their skeletal growth differed from that of human beings, likely affecting the appearance of bone cancers; their shortened lifespans did not allow for extended, chronic exposures; and pneumonia tended to kill them before tumors fully developed. The only meaningful experimental evidence from a larger animal at the time came from rabbits.[24] "Aren't we perhaps expecting too much of the rabbit?" Shipman later joked, paraphrasing Harvard occupational medicine specialist Harriet Hardy.[25] To avoid these problems, Langham suggested comparing the toxicity of plutonium and radium in dogs, an idea he floated by MIT radiobiologist Robley Evans earlier in February.[26]

Langham chose dogs because of the work at Chicago and because the ratio of skeletal to total body weight was similar between humans and canines. The latter was especially important because the Chalk River estimates were based on statistical averages of human skeleton and organ weights.[27] Similarity with dogs made scaling those figures easier. The previous year—at a secret June 1949 meeting for the upcoming fifth American nuclear test, Operation Greenhouse, at Enewetak Atoll—ad hoc planning committee chair George LeRoy had noted a decision to focus on three "well-studied" experimental animals: mice, hairless pigs, and beagles. The mice were important for basic physiology work, while pigs were key test subjects for studies of skin exposure.[28] But as an indication of their uncertainty and unfamiliarity with the latter, the meeting's note taker wrote "Rochester beagels (sp?)," and LeRoy added a caveat: "if we could get hold of enough of them to breed."[29] By July, apparently owing to supply issues, the recommendation had changed to 120 "American fox hound dogs" for the tests. The Naval Medical Research Institute (NMRI) then constructed colonies both on Enewetak's Japtan Island and back in Bethesda.[30] Piggybacking on this momentum, Langham proposed using foxhounds and expanding the existing breeding program to support his plutonium studies.[31] Because plutonium was a matter of special concern for Los Alamos, he recommended siting the research there.

Shipman brought Langham's two-part proposal, including both a large study on injected plutonium and a companion program on inhaled plutonium, to the attention of the Biology Branch of the AEC's Division of Biology and Medicine. "To boil the thing down as much as possible," Shipman explained to the division's chief, Shields Warren, "we would like to utilize the facilities which you have already paid for at the Navy's installation in San Francisco," where foxhounds were bred before crossing the Pacific to Enewetak.[32] But Shipman disagreed about hosting the dogs at Los Alamos. Space was limited, most work was classified, and he wanted to avoid perceptions that Los Alamos staff were "campaigning for low tolerances for our own ulterior motives."[33] Paul Pearson, the head of the Biology Branch, acknowledged those challenges in May 1950 and noted the advantages of placing the dog project close to a major medical school.[34] An ideal location had, coincidentally, just appeared on the horizon.

Welcome to Beagleville

In 1946, hematologist Richard Young was recruited away from Northwestern, his alma mater, to become the dean of the University of Utah's College of Medicine. A Chicago native with strong ties to the city, Young seized on an early opportunity to return home three years later when J. Roscoe Miller vacated the deanship of the Northwestern Medical School to become the university's next president. At a time of major institutional expansion, replacing Young was an urgent priority, and one of the most intriguing candidates was John Z. Bowers, a genial young radiobiologist, working under Shields Warren, who fostered aspirations for a stable, long-term research position.[35] When the possibility of going to Utah was floated to him in the fall of 1949, Bowers expressed interest, although he worried about leaving Warren, the mentor he had followed from his doctoral studies in Boston to the AEC.

As part of Utah's pitch, president Albert Ray Olpin offered Bowers the opportunity to build a new program in radiobiology. Administrators had drawn up a list of candidates to lead such a program years earlier but never quite cleared the final hurdles. Bowers, who earned accolades as a commander in the Navy Medical Corps and even survived a torpedo attack, appeared to be the perfect man to lead an institution in transition.[36] His doing so would represent a boon for the AEC as well. The commission had finalized plans for a National Reactor Testing Station in Arco, Idaho, where experimental civilian nuclear power reactors would be piloted just three and a half hours north of Salt Lake City, which greatly intensified the attraction of getting the Utes involved in radiobiology.[37] In a January 1950 recommendation letter, Warren confirmed "it would be an asset from the standpoint of the Atomic Energy Commission to have a man with Bowers' knowledge and experience at a major medical center not too far distant from our reactor proving site at Arco."[38]

Sending Bowers to Utah was a win for everyone involved. Eager to finalize the deal, Olpin agreed to let Bowers finish a year of supplementary training in California with Joseph Hamilton, the director of biomedical research at Berkeley during the Manhattan Project, and Bowers formally accepted the position in February 1950, planning to arrive by November. Writing in April, he alerted Olpin of additional discussions concerning "possible programs at the University," including "a proposal which could result in a broad and productive program." Bowers offered to clarify those ideas whenever Olpin was available.[39]

The program was Langham's dog colony. After Los Alamos had been ruled out, Utah emerged as the next best choice.[40] Its medical school would make finding scientists easy, and Salt Lake City, a comfortable residential town that the AEC would not have to manage as it did Los Alamos, sat at the intersection of major highways, enabling the movement of materials and people south to Nevada's testing ranges, west to Berkeley, north to Arco, and east to Fort Collins in Colorado, the site of another planned reactor and several atomic agriculture projects.[41] As the dog program's director, Bowers would bring a wealth of connections and experience in the still-nascent field of radiobiology.[42] Years later, the AEC's Walter Claus framed Bowers's arrival in comically biblical terms. In the nation's capital, "upon a throne of gold sat the Great White Father, and the light from him did shine as the sun cast from a thousand polished shields—and his name was Warren. And the Father turned to his noblest son, Bowers, who sat upon his right hand, and he said: 'Go thou hence, John, and bring light unto the benighted.'"[43] On August 14, 1950, the AEC entered into a contract with Utah to "conduct research in the study of the biological effects of radiation, with emphasis on large animals." Initially, this meant comparing the result of radium and plutonium injections in dogs to data from the dial painters in order to produce a "toxicity ratio" for plutonium.[44] But Robley Evans had recently found that mesothorium, an isotope of radium, could be more dangerous than radium itself, which earned the former its own "place at the Salt Lake City table."[45] When the contract took effect, it brought dogs as well as light to Utah.[46]

For nearly two years, the attention of this grand undertaking was occupied by planning and construction. Lab buildings needed to be zoned and raised, researchers found and relocated. Equally important were the dogs, particularly ascertaining which breed made for the most useful experimental subjects. Utah was understood, very early, to be a "blueprint" for subsequent efforts at other facilities, with planners convinced that experiments of its scale were unlikely to be repeated, so the chosen breed needed to be as close to an ideal as possible.[47] Rather typically for many AEC projects of the period, plans to carry out research preceded answers on how to do so.

Langham had followed precedent in suggesting foxhounds. Usage at Operation Greenhouse produced a brief flurry of experimental interest in the breed and numerous publications from NMRI researchers.[48] Rochester researchers, too, had mixed foxhounds into their experimental colonies, because they were favorites of Whipple. "We agreed that we should use a moderately large dog and selected the American foxhound," Langham noted of early Utah planning meetings. Because of the possibility that breeds might have different susceptibility to cancer, some also argued "that a few beagles should be included."[49] If foxhounds had an unknown predisposition to bone tumors, years of results could be wasted.

Early planning documents reveal how dogs were always envisioned as proxies for particular American lives. Initially, test dogs were imagined to be stand-ins for female watch dial painters, simulating the experience of "a girl working for a year or two in a dial plant and at an age, namely 18 months in the dog, corresponding to a young girl of age 16 to 24."[50] Planners then changed their minds and decided to use only adult male dogs, because it was adult men who were "employed in the plutonium industry." They also worried that sex differences might produce "unpredictable" results, evidence of how female bodies were perceived as obstacles to pure science.[51] Procuring a sufficient number of male foxhounds within a reasonable distance of Salt Lake City was infeasible, however, and reality trumped aspiration.

The planners "agreed that there was nothing magic about the American foxhound," Langham recorded in October 1950, acknowledging that it might "be well worthwhile to reconsider the availability of other breeds, especially the beagle."[52] However limited in their magic, foxhounds remained the breed used for atomic tests such as Greenhouse, and Shipman insisted that Utah should not overlook the value of using dogs from the same stock.[53] Nonetheless, foxhounds fell away. Beagles, originally a backstop, became the main choice. A. E. Brandt, a biometrician from the AEC's New York Operations Office invited to consult on the project, promised to look into possible beagle supplies among his acquaintances in the beagle clubs of the New York area.[54]

The original Utah proposal assumed that most dogs would be bought from qualified kennels, minimizing the need for paid veterinarians or breeding facilities. But the planning committee quickly recognized what the NRC dog committee had years earlier: that "there were no large-scale, well-controlled dog breeding colonies in existence that could provide adequate numbers of any one breed."[55] As Harvard's Sidney Farber put it bluntly at a 1945 conference on experimental animal colonies, "No commercial breeder now produces dogs and cats for use in research or teaching."[56] Because breeders typically sold their animals as puppies, purchasing a large quantity of dogs at the right age would generate additional expenses. Rochester's atomic researchers, for instance, made the news in 1952 when the *New York Times* and Long Island's *Newsday* reported that "bootlegged beagles" were being acquired from local breeders at $35 a piece, compared to $6 or $10 for conventional dogs.[57]

Veterinarian Robert Bay, who joined the Utah project around the time of Bowers's arrival, instead encouraged the establishment of an internal breeding colony. A lab-run colony, he noted, would have numerous advantages, including "accurate records of age, parentage, immunization, diet, presence or absence of parasites."[58] In the spring of 1951, Bay purchased eight purebred beagles to begin the colony and, as Claus put it in his "Genesis" account, "all proclaimed that the name of the dog shall be Beagle."[59] The first dogs were ultimately not New Yorkers but West Virginians, descendants of champion field competitors.[60] Their heritage of outdoor success—their "bloodline"—was understood to imply consistency, but limitations in availability forced the lab to fill out the colony with beagles sourced closer to the university, many the progeny of Utah show dogs. Few expressed disappointment at the shift from foxhounds: nearly a century of selective breeding efforts left scientists with dogs of seemingly high uniformity. Some were even descendants of the earliest American beagles, imported by Rowett and Elmore, "pedigrees farther back than most humans can trace their ancestors," mused a Utah student journalist.[61] The beagles seemed to be purer test subjects than the men or women around them.

While construction on the laboratory facilities was completed, dogs stayed at a nearby animal hospital and local residences. Then, nearly a year after the first dogs came to the university, sixty-one beagles and their puppies moved to the new Radiobiology Lab kennels. The first was injected with plutonium a few months later, on June 24, 1952, and a newspaper headline announced, "The Hounds of Beagleville: They May Save Your Life."[62] That *The Hound of the Baskervilles* told of murders by demonic dogs did not seem to trouble the headline writer.

A Prisoners Dilemma and Project Hot Dog

Reactor workers and researchers were not, however, the only groups interested in the health effects of radiation on large organisms. So too were members of a special task force, formed close to the end of World

War II, with the goal of developing viable atomic-powered aircraft. In 1946, the Army Air Force sponsored a feasibility study, "Nuclear Energy for Propulsion of Aircraft," to be carried out cooperatively by the AEC and Fairchild Engine and the Airplane Corporation at Oak Ridge National Laboratory.[63] So NEPA, the umbrella acronym for subsequent efforts, was born. An extended follow-up analysis by MIT researchers, known as the Lexington Project for their first meeting location in May 1948, concluded that additional theoretical and experimental work could overcome the obstacles to atomic aircraft. In a classified memo, AEC chairman Robert LeBaron summarized that there was "a strong probability that some phase of nuclear powered flight could be achieved if adequate resources and competent manpower were put into the development."[64]

Even optimists recognized the difficulties ahead, however. Availability, maintenance, cost, and supply of fissionable material were all stumbling blocks, not to mention the safety of pilots, who might fly ten or more hours in close proximity to a reactor. In response, a Medical Advisory Committee was organized for NEPA, chaired by UCLA radiologist Andrew Dowdy and including Robley Evans and UC San Francisco radiologist Robert S. Stone, who had led health research efforts at the Met Lab. Brues and Warren were brought on as consultants. In January 1950, the committee recommended multiple possible animal experiments whose titles revealed the breadth of the challenges: "Physical Fitness following Total Body Exposure to Penetrating Radiations," "Long Term Effects of Acute and Intermittent Total Body Exposures to Ionizing Radiations to Ionizing Radiations," and "Psychomotor Performance of Patients following Total Body Exposure to Penetrating Radiations."[65] Studies on nonhuman animals could get only partway to the desired answers, however, so experiments with human "volunteers" over a period of twenty years were deemed essential.[66] As Stone wrote in justification: "If another war develops and the Army is called upon to occupy territory that has been previously bombed by a nuclear powered weapon, the troops will have to be exposed to greater than usually permissible amounts of radiation. . . . In general, the conclusion has been reached that, not only each species of animal differs from every other in their responses, but also various strains within a species differ."[67] Prisoners, Stone noted, would be ideal for the project, because they could be monitored over many years.[68] The expectation was to adopt similar protocols to a series of experiments undertaken during World War II at the Statesville Penitentiary in Illinois, where prisoners expected to receive lightened sentences or favorable parole decisions after "volunteering" as research subjects in tests of new antimalarial drugs.[69]

The broader program had support, but human experiments became a lightning rod. The plutonium injection studies were fresh in many minds, and so were other atrocities: Joseph Hamilton confided in Warren that the plan had "a little of the Buchenwald touch."[70] The Advisory Committee's own report noted an anonymous abstention concerning the human tests, possibly influenced by Warren's "increasingly dim view" of such work.[71] On February 1, 1950, after further hesitancy was voiced, the AEC rescinded an endorsement of the experiments. The core NEPA group remained adamant about the need for human studies, however, so the AEC Biology and Medicine Division met again in May, warily accepting the proposal only "if necessary" and carried out "under established principals."[72] The Army, the other critical voice in the matter and likeliest source of human volunteers, was unmoved, however. M. C. Leverett, of Fairchild's NEPA Division, ruefully labeled further advocacy of human experimentation "a waste of energy" in February 1951.[73] In conceding the issue, he emphasized that experiments on dogs remained a vital alternative.

Commenting on a copy of Langham's notes from an October 1950 planning meeting for the Utah dog project, biomedical researcher George Sacher wondered whether NEPA and Utah might not be able to solve each other's problems. "If both could use one breed, this would be advantageous," he wrote. "The control populations would be enlarged, to the great advantage of tumor incidence statistics, etc."[74] Brues, Sacher's boss, concurred, adding, "We feel very strongly that some effort should be made to use similar material"—namely, dogs—"in these two experiments."[75] NEPA was formally dissolved in April 1951, in part because of its connection to the "lost cause" of human experimentation, but attendees of a multi-organization meeting that March agreed that the Air Force Aero Medical Laboratory should continue some of the studies negotiated by NEPA's radiobiologists.[76] These included an analysis of radiation in homing pigeons led by Andrew Dowdy, a study of mules (who were understood to be minimally affected by radiation) at the University of Tennessee, and a possible three-million-dollar program, "Effects of Ionizing Radiation on Dogs," at the Mayo Clinic in Rochester, Minnesota.

That dogs were understood to be replacements for the imagined prisoners was clear: meeting attendees stressed that dog studies served "as a substitute for any human experimentation." Here the proxy relationship was explicit: like prisoners, dogs could also be confined for extended periods and managed minutely.[77] In both cases, captivity was experimentally valuable. The scale of the proposed funding, tripling that of other comparable programs, also demonstrated their perceived significance.[78] Following Brues's note to Langham, the AEC's Monthly Status and Progress Report for January 1951 revealed discussions of a study of "work capacity and longevity" in dogs between members of the NEPA committee, the Navy, and the University of California.[79] Mayo disappeared as a site for the desired research, and NEPA's various proposed animal studies coalesced into one program.

The result of these debates was "Project No. 4," part of a broader collaboration between the AEC and the University of California.[80] In April 1951, UC president Robert Sproul wrote to the AEC requesting funding for the new program, "Effect of Radiation and Work Capacity and Longevity of the Dog" (the first "and" was a typo, meant to read "on"), to be overseen by George Hart, dean of the School of Veterinary Medicine at Davis, the agricultural branch of the University of California system.[81] Just west of Sacramento, Davis had nearly three thousand acres of cheap land, sat within reasonable distance of Berkeley, boasted preexisting veterinary expertise, and hosted the state Agricultural Experiment Station. All of this made it an excellent location. Paul Pearson, who helped plan the Utah program, visited Davis in September 1950 during a tour of other atomic energy projects to discuss the plans.[82]

Project No. 4 was allocated $68,956 for 1951, less than the proposed Mayo funding but the largest such contract that year, nearly six times what other AEC biology projects received.[83] Today the sum would be worth over $800,000. As its title suggested, the proposal initially positioned Davis as the location for two concurrent studies, one exploring "work capacity" and another analyzing risks to "reproduction" and "longevity." The AEC office in Washington, DC, however, suggested that separate populations of dogs would be required for effective experimentation, so Project No. 4 was bifurcated: the Navy research installation at Hunters Point, near San Francisco, where Greenhouse foxhounds once lived, would host work capacity studies featuring male animals, while Davis would use female dogs for longevity and reproduction work.[84] Although the original impetus for these projects was separate from Utah's, beagles were selected as the experimental animals to ensure continuity of results between the two facilities. To run the project, Hart chose Allen C. Andersen, a veterinarian and former research assistant at the Experiment Station whom nearly everyone called "Bud."

The thematic division of Project No. 4 once again revealed beagles' paradoxical position in relation to human beings. Although ostensibly about work capacity and longevity of dogs generally, male animals were selected for work studies because it was (human) male bodies whose work capacity was in question. The "work" that a dog could do—here, running on a treadmill—was assumed to translate to the work capacity of male pilots or reactor workers, just as at Utah. On the other hand, female dogs were selected for the reproductive studies because, at least implicitly, reproduction was what women did, and female reproductive cycles were considered more "consistent." This was a notable shift from the early Rochester emphasis on male fertility and seemed to acknowledge the wider perils to human reproduction in the Atomic Era. Implicit gender ideologies nevertheless structured the experimental system and its output, subtly reifying the division of labor in the American nuclear family: men worked, women reproduced (which was not work).[85] Gender also framed the projects in other ways: the Beagle Club's originators—Wright Langham, Robley Evans, and Austin Brues—were colloquially referred to as the "Founding Fathers" and the projects as their "offspring."[86]

Those divisions partly fell apart, however, just like the original plan to use only male dogs in Utah; logistics once again overwhelmed assumptions about sex and gender. As Andersen explained bluntly in 1952, a sufficient number of beagles "could not be obtained on the Pacific Coast."[87] "As soon as the required number of dogs of one breed was noised around," he explained, "the price went up and we were offered dogs at 50 to 75 dollars per animal"—nearly double the price paid at Rochester.[88] Local antivivisectionists also intervened early in the project's planning and threatened to make it "impossible" for the lab to obtain dogs by notifying area breeders what they would be used for. Purchasing animals from farther away did not solve the problem either, because it opened up multiple pathways for disease to enter the kennels and raised costs even further. Davis soon found itself in the same position that the team in Salt Lake City had, but a humane representative assured Andersen that "all objections would be withdrawn" if he operated an internal breeding program. Alan Gregg had been told the same thing a decade earlier.[89] As Andersen later reasoned, doing so might even save money.

Project No. 4 was approved in June and began formal operations on July 1, 1951. The early years were once again focused on constructing facilities and acquiring and breeding dogs. Even more so than in Utah, the Davis program faced significant obstacles. The lab was forced to initially house its beagles in a Quonset hut, after their barking in close proximity to campus produced fears that they "might be poisoned" by irritated residents.[90] Rochester's smaller atomic beagle colony faced similar threats.[91] Yet the Quonset hut was inadequate, and more than one hundred dogs were lost because of inadequate facilities and the spread of disease. In 1953, James Baker flew across the country from Cornell to inspect the facilities and offer advice on managing disease, revealing how the RLDD provided a knowledge infrastructure for other beagle programs.[92] A steel strike further slowed construction of dog enclosures, and a much-delayed Maxitron X-Ray device, needed for the radioactive exposures, was finally installed in October 1952—only to be further delayed by the absence of a lead lining required for its monitors to function properly. The sibling project at Hunters Point never fully materialized, so work capacity studies, using treadmills acquired from Harvard's famous Fatigue Lab, were reintegrated into the Davis project as an additional factor in the studies of longevity.[93] Nevertheless, a satisfactory measure of "work capacity" was elusive, and Andersen noted that the "reproductive capability" of female beagles appeared "more promising" than exercise tests, as reproduction insisted, against certain masculine expectations, on its status as a special form of work.[94]

Challenges running the gamut from antivivisection activity to kennel epidemics thus slowed the start of both projects, but researchers remained optimistic that the delays represented learning opportunities. The 1952 annual report from Davis concludes that careful maintenance would "furnish valuable information on breeding and management of dogs in experimental colonies."[95] The next year's report was even more optimistic, claiming the project would "be the first on this coast to demonstrate the possibilities and costs of supplying experimental animals of this species in healthy condition to science groups for research."[96] Practitioners elsewhere were, in turn, highly attentive, and the lab received visitors from businesses and other universities interested in both the experimental and economic aspects of the operation.[97] Part of the reason was that the projects at Utah and Davis represented, in a very real sense, unique scientific undertakings. "This was the first occurrence in peacetime," former Rochester researcher James Newell Stannard reflected in 1988, "of a contract between a university and a government agency for a specific long-term study, at least in the biological sciences."[98] Initially vague about its purpose, the program was explained to the public three years later: the "work capacity" beagles were being tested for concerned the ability of pilots to fly airplanes fueled by small nuclear reactors.[99]

Despite early antivivisectionist criticism, dog owners were generally supportive of the projects, just as they had been of those at Cornell, believing that beagles were cooperating with humans in atomic work that would be of mutual benefit. National newspapers carried a glowing story in August 1951, for instance, about a beagle named Skipper who became the first dog to receive chemotherapeutic radioiodine, shipped from Oak Ridge to New England Deaconess Hospital.[100] The president of a beagle-breeding group from Ohio visited the Davis facilities in the summer of 1952 to offer his support and cooperation.[101] As an institution,

Davis maintained a broader commitment to publicly oriented research, and once the project lost its classified status, staff were "gratified by the comments of the many visitors including dog lovers, anticruelty groups, antivivisectionists, veterinarians, scientists and others who are constantly visiting or discussing the project."[102] In contrast to today's hidden animal research facilities, "for many people in Davis a Sunday drive includes a stop to see the beagle puppies at the western corner of the UCD campus," noted *Davis Enterprise* reporter Vicki Feack in 1968.[103] Tours were conducted during annual veterinary conferences and other major campus events. "The world-famous UCD Beagle Colony will be open to the public from 2:00 to 4:00 p.m. on Picnic Day," announced a 1962 headline in student newspaper the *California Aggie*.[104]

This attention, however, was not necessarily beneficial to experimental use of the dogs. Interest from "the public and the annoyance to the dogs after working hours" spurred the hiring of night attendants and the construction of "flood lights on the roadside and main gate and barbed wire on the fence surrounding the kennel."[105] The work at what researchers started to call "Project Hot Dog" was too valuable to be put at risk, and if the dogs had not at first struck anyone as prisoners, the increasingly carceral appearance of their home certainly suggested it.[106]

What Is a Normal Dog?

The Utah and Davis projects were meant to provide reliable information with clear lessons for protecting human beings. To do so, few expenses were spared: "We had to be absolutely sure that the best went into [the dog projects] because they would never be repeated," the AEC's David Bruner emphasized years later.[107] But in order for the dogs to work as high-quality physiological proxies, researchers recognized that more needed to be known about the animals themselves. The colonies drew on prior experience, some of it even with beagles, but much remained uncertain about the breed and how individual dogs would function in a large experimental program. If there were to be significant numbers of experimental "control" dogs, not subject to radiation and kept in otherwise ideal health, living what researchers called a "normal" life, what exactly was a normal beagle's life like?

The answer was more complicated than expected. Few of the veterinary scientists and even fewer of the physicists assembled at Davis or Utah knew much about beagle longevity, or even human longevity for that matter. "The fact is that we know very little about aging," Andersen admitted, citing aging expert Albert Lansing, who edited the most authoritative text on the subject, *Cowdry's Problems of Aging*.[108] "The causes of death in man as listed by the Metropolitan Life Insurance Company are many," Andersen continued, "but purely old-age phenomena have not been definitely determined. Little as is known about aging in man, even less comparable data are available for animal life, including the dog, and yet all species have old-age characteristics if accidents can be prevented long enough for them to become manifest."[109] The key, for experimental purposes, was to separate the "accidents" of a dog's life from "old-age characteristics," just as researchers had wanted to distinguish radiation effects from conventional terminal illness. But *why* anyone died of old age remained an open question.

Both linguistically and institutionally, the study of aging was a distinctly twentieth-century phenomenon: "gerontology," the science of aging, was coined in 1903; "geriatrics," the branch of medicine concerned with the elderly, emerged in 1909. Both were still young areas of inquiry in 1951, particularly in the United States, which generally lagged behind Europe in studying aging.[110] American gerontology had loosely defined disciplinary boundaries, and the vast majority of existing experimental work focused on short-lived species, such as rodents.[111] The first two issues of the journal *Experimental Gerontology* (1964–65), published a decade after work at Davis began, covered research on insects, rats, and mice, but nothing larger in size. Aside from exploratory research by physician William MacNider on the relation between kidney ailments and aging, Andersen and colleagues were some of the first to publish on experiments with dogs.[112] In anachronistic terms, the Davis project represented the first controlled gerontological experiment in the United States with a medium-sized translational animal model, and little was known about what to expect from a colony of perfectly kept dogs. "A difficult task," one journalist summarized in 1968, "has been to try to determine what a 'normal' dog is in order to determine an 'abnormal' one."[113]

How long, for instance, would a normal beagle live? Early estimates suggested a life expectancy between seven and ten years, although Utah's Robert Bay proposed a narrower bound between nine and ten in 1953.[114] The number nonetheless rose to 10.6 years by 1965 and stabilized just above thirteen in 1975.[115] Yet many dogs survived much longer, reaching seventeen or eighteen years of atomic trotting. While the project was meant to represent the most advanced thinking about senescence, what originally constituted "old age" in dogs was based largely on the practical veterinary knowledge that individuals like Andersen and Bay brought to their respective projects. As a later summary of animal use in gerontological research described the problem, "The gerontologist, who is by training a biologist, a biochemist, a behavioral scientist, and more, has had little experience in dealing with the problems that relate to the animals themselves."[116] And in 1951, even veterinarians were relatively new to dogs: as historian Susan Jones notes, most early twentieth-century veterinarians had entered the profession to work with livestock, developing a degree of familiarity with dogs and cats during the 1920s and 1930s but remaining relatively uninterested in ministering to canines until the midcentury mark, when dogs became big business.[117]

To better understand the typical lifespan of beagles, Davis researchers designed a field study that included giving dogs to members of the public who would report deaths or other health issues.[118] The motivation, however, was more than simple gerontological curiosity: at first, the lab found success selling its excess males to nearby facilities, but demand dried up within a few years, and a new outlet had to be discovered. Releasing the dogs to nearby residents offered a productive solution. The resulting study distinguished "natural" beagles, those held in homes, from "experimental" beagles, those confined to the laboratory, a distinction that served to naturalize domestic life. A "natural" beagle had once been an outdoor hunter, but its natural environment was now the home.

Published in 1958, the resulting paper was filled with grizzly data about ill-fated beagles. "Motor vehicle accidents" accounted for the majority of deaths, while "disease" came second. There were forty-three ailments per year among 220 "natural" dogs, but only twenty-nine for 523 laboratory dogs. In all, the majority of privately held dogs died by four years of age, while experimental beagles consistently lived quite a bit longer.[119] "Normal" laboratory dogs were not just healthier than natural dogs but existed within a radically different milieu, a kind of scientific and medical gated canine community. Later Davis director Marvin Goldman joked that the beagles received better health care than most Americans.[120] Normality was therefore circumscribed by Cold War culture, not a set of objective physiological markers.

The ultimate goal was confidence that "animals at the time of irradiation are not in a pathological state."[121] Yet even granting that a "normal" state could be distinguished from the pathological, guaranteeing that normality was no easy task. Animals had to be kept free of disease, protected from environmental threats, fed well, and offered opportunities for effective socialization. Each requirement brought its own set of challenges. When it came to housing, for example, both labs spent heavily. Yet those investments yielded very different kennels: at Davis, blessed by milder California weather, it was decided that dogs could live happily outdoors, so beagles were housed in thirty-by-thirteen-foot enclosures with elevated wine barrels, repurposed from nearby vineyards and covered in tar to protect dogs from the elements.[122] Although it was an eminently Californian solution, the Gaines Dog Research Kennels had actually recommended barrel houses back in 1945, possibly influencing Andersen.[123] He would later note that the lab had bought most of the state's used wine barrels, forcing a shift to beer barrels thereafter, which the dogs appeared equally happy with.[124] Barrel doghouses failed to ward off occasional heatwaves, however, and when summer temperatures climbed above ninety-eight degrees Fahrenheit, sprinklers had to be turned on to cool the dogs.[125]

At Utah, on the other hand, beagles lived in indoor kennels with outdoor exercise runs. Here, the opposite problem emerged: heating systems in the kennel floors struggled to "adequately protect the animals against pneumonia" during the study's first years, and quite a few dogs were lost to sickness.[126] Although each facility tried to control for extreme temperatures, dogs had been chosen as experimental animals partly for their presumed hardiness. Many pet owners of the era kept dogs outside, so facilities tolerated a reasonable degree of environmental variation: human homes were, after all, a dog's *natural* environment. Climatic differences nevertheless presented a real threat to the validity of data: seasonal shifts in weight, for instance, affected laboratory results.[127] But much of this became clear only years later. "My concern," rued one researcher in 1986, "is that all too often . . . the investigator seems to be unaware that these biases exist in his data."[128]

By emphasizing physiological well-being, Andersen and colleagues were snugly within the lineage of behaviorist study of animals. Although Ivan Pavlov had emphasized the emotional variability of his dogs, many American followers had ignored animal emotions in their experimental setups. Eliminating emotion as a variable was a way of controlling the laboratory and experimental system.[129] But Davis researchers realized that proper housing was not only a matter of comfort. Dogs also needed to be kept in numbers that enabled socialization while avoiding fighting. Originally, four dogs in each kennel seemed to offer an ideal compromise between economy and interaction, but beagles in such conditions would fight viciously even under close supervision, particularly during breeding periods.[130] Isolating dogs in their own pens appeared equally dangerous, however, because lone beagles tended "to overexercise on a given path along one side of the pen" and ate nearly double the ideal amount.[131] To ward off depressive beagles, two dogs appeared an "optimum concentration" for a pen, offering "contentment and maximum physiological well-being." The old adage, "Two is company, three is a crowd, four is too many, and five's not allowed," noted Andersen, was "certainly true of taking care of dogs!"[132]

In recognizing the social and psychic life of beagles, content or lonely in their little barrel homes, Davis researchers joined a number of physiologists who, from the late nineteenth century onward, had reconceptualized the laboratory as "an ecology of experiences and emotions."[133] The effort to create psychically stable environments for the dogs was a continuation of these efforts, but it also symbolized a burgeoning awareness of the emotional and cognitive complexity of dogs, which was as important to scientists as it was to America's dog owners. A beagle could not age normally if it did not live in proper social circumstances, because "pure" aging was inextricably contaminated by social circumstances.[134] The ultimate takeaway was a paradoxical suggestion that *laboratory* beagles were most likely to age normally, since many pets lived individually with their human families.

Conscientious efforts to construct normality for beagles generated multiple innovations in the care and housing of experimental dogs. The Davis lab published widely on their contributions: Andersen reported in 1955 on basic design approaches for large-scale dog housing and shared a "debarking" technique, or ventriculocordectomy, that surgically limited kennel noise. Two years later, he wrote about puppy production strategies and group kenneling practices.[135] Each of these techniques was communicated to veterinarians and influenced dog owners far afield from Davis. As important as the external environment was, both labs were even more acutely focused on what went on inside of the dogs. Each beagle was treated as much as possible, Utah researchers noted, "as a separate patient."[136] This approach built on the procedures designed and implemented at Jackson and elsewhere. Individual records were maintained for every dog, including information about their births, vaccinations, medical treatments, radiological injections or feed, and pedigrees.

Just as Stockard had discovered, sanitation was especially challenging. "Normal" beagles needed to be free of internal parasites, particularly ascaris and coccidia, which led to strict sanitation regimes at both kennels. Because much of the canine waste was also radioactive, researchers at Davis were propelled to design a new device, later nicknamed the "Radioactive Poop Machine" (RPM), which processed dog poop for storage.[137] External parasites, such as fleas, lice, and ticks, were considered more trivial, although occasional outbreaks at both facilities prompted additional cleaning measures. Project No. 4's fourth annual report summarized that "if the presence of internal parasites can be maintained at a low level throughout the life span of the animals, the end-point of the experiment should not be jeopardized from a parasitology standpoint."[138] A great deal of the early budget at Davis was spent on vaccinations against distemper and hepatitis from Lederle Laboratories in New York, whose researchers were also consulted on general lab issues, just as they had been at the RLDD. Yet even so, an outbreak of distemper at Davis in 1954 killed nineteen dogs, and infectious canine hepatitis, blamed on a shipment of "very poor breed quality" dogs from Indiana, later killed four.[139] Baker was, fortunately, busily perfecting vaccinations in Ithaca for the diseases.

As Georges Canguilhem has suggested, "pathological" and "normal" states are separated by more than simply the presence or absence of disease. "Anomaly," he notes, "becomes pathological only in relation to a milieu of life and a kind of life."[140] The Davis and Utah projects bore this out. Establishing the "normal" meant accepting baselines of elevated disease risk and variable social and physical environments; the normal was a *conformation* between dogs and people. Laboratory dogs, already transformed by millennia of domestication, were renormalized in relation to carefully managed experimental setups. Internal health (freedom from disease) was contrasted with external danger ("accidents"). In the implicit philosophy of life that underwrote laboratory activities, lives ran their course like rivers, interrupted only by "accidents" and "age," and dogs needed continuous protection to avoid those dangers and remain functioning scientific devices.

Distinguishing the normal from the pathological was challenging, especially when it came to aging, with few agreed-on standards or parameters to contrast the two.[141] "The advent of senility," wrote Andersen and his colleague Douglas McKelvie, "represents a progressive decline in functional capacity, associated with characteristic symptoms. Such symptoms are much more easily described than explained."[142] The beagle studies' efforts to isolate normal-but-unnatural age would mirror the other major American gerontological research project of the era, Nathan Shock's Baltimore Longitudinal Study of Aging, which began in 1958 and sought to carefully document the "normal" aging process in a number of ostensibly healthy individuals.[143] Both studies were symptomatic of what historian Kavita Sivaramakrishnan calls an eminently Western approach to aging, understanding it as a process to be "managed," isolated from "pathological" changes.[144]

To do so, the pains of normal aging had to be left alone at times. "Preventive medicine and a balance of environmental factors," Andersen noted, "must be substituted for therapeutic measures so as to avoid bias or nullification of data accumulated over many years."[145] Repeated catheterization or drug treatments, for example, could muddy experimental results, given how little AEC scientists knew about the effects of pharmaceuticals on dogs. That particular concern was paradoxical, as we will see in chapter 3, because beagles were becoming popular pharmaceutical testing subjects at exactly this time. The result was significant suffering for many dogs, and concerns about allowing such unnecessary pain would ultimately lead to Robert Bay's resignation at Utah in 1955, just two months after he received the Animal Welfare Institute's first Albert Schweitzer Medal for his careful planning of the dog colony there.[146] Philosophically, Andersen and McKelvie wondered about the beagles: "Do we truly know how to control psychological, physiological, and husbandry requirements for the animals' optimum life span?"[147]

Beagles were seen as living beings who required assistance in becoming normal. But they were also understood to be tools, and debates about normal and abnormal aging underlined the fabrication of their nature. Davis researchers even recognized that some dogs were more optimal than others for research. Years of whelping puppies had revealed large variations in adult size: male dogs weighed anywhere from eighteen to thirty-six pounds, female dogs anywhere from fourteen to thirty-four. If the dog population continued, as it had over the years, to double in weight, "time and cost for performing any experimental procedure" would increase dramatically. In response, researchers sought to ensure that the colony produced "dogs of uniform and optimum size and color for experimental purposes."[148] These were not "normal" in Stockard's eugenic sense, or even "normal" just in the lab's sense of a constructed health, but were "standard" beagles in a more explicit form than anyone had previously meant it. In 1958, Davis settled on two optimum variants: male dogs of twenty to twenty-three pounds, fourteen inches in height, and seventeen in length; and females of eighteen to twenty-three pounds, about thirteen in height and sixteen inches in length. A year later, "ideal" characteristics were even more specific: "haircoat of medium length and texture in two or more colors," an even disposition, and body weights between 9.07 and 10.77 and 8.51 and 10.21 kilograms (for males and females, respectively).[149]

A laboratory beagle was not, however, to be judged on the standards of an ideal championship beagle. While both labs had begun with championship pedigree dogs, subsequent litters grew further and further from that norm. "When the beagle was brought into the laboratory in 1950," wrote three Utah researchers, "the system of selection by which it had evolved was abandoned or disregarded. Selection of successive generations of breeding stock within the laboratory became more subjective. . . . A laboratory breeding scheme probably preserves the beagle phenotype, but preservation of stamina and intelligence is not assured because there are no means of testing for these factors in the laboratory. So it is quite possible that each generation of laboratory-bred beagles would be increasingly dissimilar with respect to the original nucleus of acquired breeding stock."[150] Where a powerful nose might once have enabled a beagle to pass on its genes, lab beagles were now more likely to survive for their docility and size. The beagle as an experimental animal had become fundamentally distinct from its predecessors, a different conformation: "experimental" versus "natural" dogs; *Canis laboratoris* rather than *Canis familiaris*. Beagles were now certifiably a living laboratory technology.

Elements of Similarity

By the late 1950s, work at Utah had begun to generate interesting findings about the operation of internal emitters of radiation, which can damage sensitive tissue in the body. The lab would host one of the first conferences on the topic in 1961. In contrast to Utah's consistency, the project at Davis underwent a significant transformation, shifting from its original focus on the safety of America's future atomic pilots to a focus on the basic health and wellness of regular, everyday Americans. The reasons were simple: atomic-powered aircraft never left the ground, but newly concerning radioisotopes were in the skies.

American nuclear planners had imagined that the early, immensely expensive atomic tests, such as Greenhouse, might never be repeated. But nuclear testing proceeded consistently and insistently from the moment bombs fell on Hiroshima and Nagasaki until aboveground tests were eliminated by the Partial Test Ban Treaty in 1963. As they did so, test detonations catapulted larger and larger quantities of newly concerning radioactive isotopes into the atmosphere, particularly after the development of catastrophically destructive hydrogen bombs. Worldwide fallout would peak in 1963, not coincidentally the year aboveground testing ceased.[151]

Strontium-90, one of nine isotopes of strontium, a nonradioactive element under normal conditions, was especially worrisome. Trace amounts were detected following the earliest atomic tests, but strontium-90's specific dangers were not immediately clear. Early Met Lab studies on dogs instead focused on strontium-89, which promised medical applications.[152] Gradually, however, the picture clarified: strontium-90 acted as what scientists call a "bone seeker," replacing calcium in the body's bones and causing cancer. How much strontium-90 could do so was uncertain, so civilian and military planners turned an urgent focus to understanding its biological effects in the early 1950s. One effort, a secretive project codenamed "Sunshine" by AEC commissioner Willard Libby, would go on to formulate the retrospectively sinister "Sunshine unit," which contrasts the amount of strontium and calcium in bones.[153]

The dangers of strontium-90, which were not initially disclosed to the general public, became infamous after a citizen-led "Baby Tooth Survey" found traces of the radioisotope in the teeth of thousands of the nation's children.[154] The sunnier a child's smile, the more ominous his or her prognosis. Medical journalist Walter Schneir's 1959 exposé on the subject in the *Nation* only furthered widespread concern: "People throughout the world will suffer death and illness from the nuclear tests conducted to date—and the effects of these tests will still be felt by mankind 10,000 years from now," he prophesied.[155]

It was Operation Castle, a series of nuclear tests at the Bikini Atoll in March 1954, that brought the issue to an institutional head. In July 1957, Wright Langham cautioned Charles Dunham, the new director of the AEC's Division of Biology and Medicine, that another test like Castle should never be held. Sixty percent of the strontium-90 produced by all tests, from Trinity to the USSR's atomic testing, had been generated by Castle alone.[156] Instead, Langham thought it essential to understand present and future risk by studying the chronic effects of strontium-90 in human bones over longer time frames. "In this case," he wrote, "one good experiment would be worth a thousand expert opinions."[157] Langham had in mind another "big dog experiment," which he proposed, after Dunham's urging, at an AEC Biology and Medicine meeting in December 1956, and asked Norris Bradbury, the director of Los Alamos, for approval in May 1957. If sited at Los Alamos, the project would finally make good on his long-held desire to host a large-scale dog experiment.[158]

But Bradbury balked at the initial price tag of $2.1 million and told Dunham as much; Langham withdrew the proposal shortly thereafter. Radiobiologists had long suffered more limited budgets than their physicist colleagues, which bridled work at Utah, Davis, and elsewhere. "I recognize that until the biomedical people find it reasonable to envision expenditures comparable with those which are taken in stride by physicists," wrote Austin Brues, "there will be no perfect experiment depicting the ultimate in effects of radioactive substances on the life span."[159] But Langham also nursed doubts about his core concept, believing that American policy makers would never be satisfied with findings from animal experiments.[160] Such concerns slowed the project rather than killing it, however, and within a few months Davis emerged as an alternative location. Andersen and colleagues were already envisioning a modest expansion of their work, so Langham encouraged them to swell their aspirations.[161] During a meeting in June 1957, Brues and Robley Evans agreed on an experimental design that involved comparing chronic administration of strontium-90 and radium to dogs.[162] The studies would support the ongoing work at Utah, where single injections of strontium-90 had already been carried out, and allow for an additional contrast between individual and long-term exposures.

The result was AEC Project No. 6, joining and eventually replacing Project No. 4 as the latter wound down in the late 1950s. Although results were considered critical, dissent emerged from representatives of the major American laboratories about how the project should be structured. Rochester's Harry Blair was firm in opposition: "The only strong conviction I have about the Davis experiment is that I don't understand it," he wrote in October 1957. Brues, Evans, and others, on the other hand, were firm in their support.[163] Disagreements primarily concerned the experiment's goal, but Blair was also hesitant about Project No. 4's own inadequacies. "I should not like to saddle Davis personnel and the Commission with another experiment like the external radiation one," he explained, which was unlikely "to yield nearly as much information as it could have."[164]

He had hit on something others quietly believed. "I can certainly see why Dave Bruner is having trouble getting the project underway," Langham mused in January 1958, "when we so-called founding fathers can't agree on what the offspring ought to be."[165] The important thing, for Langham, was to study the health hazards of strontium-90 in three different scenarios: further fallout from weapons tests, hazards from waste disposal or reactor accidents, and fallout from a possible nuclear war. Those in the third situation, he remarked dryly, "will have problems that are more pressing than Sr-90."[166] To get useful data on every metric, thousands of dogs were probably necessary, but the founders agreed that a few hundred could suffice with proper precautions. Challenges of scale tormented America's atomic biologists: NEPA planners had also worried that only thousands of experimental subjects could provide adequate information, but doing something appeared better than doing nothing at all.

Blair was mollified, and Project No. 6 began operations just prior to 1960. The year before, Dunham had appeared before a congressional committee to defend the biology programs more broadly. Atomic energy was like fire, he offered, which caused cancer when used carelessly rather than burning your house down.[167] The beagles, however, would offer "a very good idea" of the permissible exposure to elements such as radium, plutonium, and strontium-90.[168] And they would do so as proxies for two different populations: the dogs at Utah "were comparable to an adult human working in a radium plant," while the dogs at Davis were the children of Sunshine, fed strontium from their infancy. The first group would represent the effect of radiation on nuclear industry employees, while the new cohort of animals at Davis were positioned as future generations of the nuclear era. "If they can beat the law of averages and the laws of nature," wrote journalist William Arbogast after Dunham's testimony, "the Atomic Energy Commission's pack of hound dogs may become howling scientific successes."[169]

Optimistic wordplay masked larger concerns about the future of planetary life. The period of the second Davis project's emergence was, Joseph Masco reminds us, the heyday of the Cold War and the "Age of Fallout," a time of fear and anxiety as American society was recalibrated by civil defense programs and a proliferating discourse of nuclear risk. The 1960s exemplified a second, more pessimistic phase in the American public's response to atomic peril, as citizens were urged to prepare "for life under the constant shadow of nuclear war."[170] The Federal Civil Defense Administration had been founded in 1950, quickly spreading nuclear narratives and imagery through classroom drills and public service announcements.[171] "Normal" citizens lived in contrast to, and in fear of, their potentially irradiated selves. "I wonder now only when it will happen," wrote poet Sharon Olds.[172] William Dickey's poetry, in turn, conjured up an image of the era's characteristic young man: "For him it is already there / like the underwear that he puts on in the morning. / It is with him all the time, as his shadow is."[173]

Some of those men were scientists and worried about atomic annihilation. Large-scale efforts to stockpile blood and other biomaterials emerged directly out of such trepidation.[174] In turn, voluminous correspondence among members of the AEC reveals concern about fallout as well as a persistent, morbid fascination with a future devastated by nuclear war. Stone's predictions about atomic-powered jets and Langham's jest at the irrelevance of strontium-90 to the victims of atomic conflict are as revealing about the ambiguities of the scenario—who is fighting whom, why, and where?—as they are indicative of real fears. Two possible trajectories for the coming years emerge in writing from the decade: in the first, American military and scientific supremacy subverts a (presumptively Soviet) threat, allowing for continued normality; in the second, following an unpredictable exchange of atomic weapons, even those who avoid the direct, death-dealing powers of the bomb have to cope for decades, perhaps centuries, with its radioactive aftermath.

Beagles at Utah and Davis simulated these alternate trajectories within the accelerated pace of canine and laboratory life. The dogs were always, as early planning documents reveal, explicitly positioned to represent health risks to male reactor workers, pilots, or soldiers, and they were not the only animals to do so. Experimental nuclear detonations involving pigs, sheep, and mice reveal how nonhuman organisms emerged as vital ciphers for the limits and frailty of human existence.[175] "The protected body of the Cold Warrior," Masco writes, was "prefigured by the vaporized, mutilated, and traumatized animal body."[176] In skin studies at the Hanford nuclear site, specially bred pigs were also explicitly positioned as representatives of (white) reactor workers, bred to carry a similar color of skin and to weigh the same amount as the scientists who studied them.[177]

The lifespan beagle studies, then, did more than imagine a Cold War physiology. They were tests and interventions into the future. The Utah and Davis laboratories were spaces of biological simulation that complemented the more literal nuclear weapons simulations at Los Alamos and beyond, collectively generating facts and possibilities. Some beagles lived longer than ever before, developing cataracts, osteoporosis, and cancer—the maladies of old age—while receiving exquisite care and treatment. Many of their siblings suffered unimaginable pains from irradiation and ailments that few dogs had ever experienced. Control dogs lived side-by-side with their irradiated kin, utopia and dystopia lodged within the same chronology, until illness or sacrifice swept them all away. Beagles produced and lived in various possible worlds simultaneously, demonstrating with each breath and biopsy what scientists themselves could expect from the two potential futures stretched out before them.

Those alternate timelines were ultimately realized in unexpected ways by the fate of two nuclear communities: Richland, Washington, near Hanford, with a highly educated, racially homogeneous population earning stable salaries and receiving effectively state-sponsored health care; and Ozyorsk, a Russian nuclear town close to the Mayak plutonium plant.[178] In September 1957, thousands of tons of nuclear waste were released accidentally at Mayak during the Kyshtym disaster, forcing the evacuation of thousands of residents and causing nearly fifty deaths. As Roger McClellan, a prominent radiobiologist who grew up in Richland and later worked with beagles, noted, resulting disease burdens at Ozyorsk aligned with many of the findings from the strontium-fed beagles at Davis—once the additional health burden of human smoking was accounted for. In Richland, on the other hand, where there were no major accidents, children from many of Hanford's nuclear families would live long, comfortable lives. "We validated those studies at Davis," McClellan remarked.[179]

As the beagle experiments pushed into the 1970s and 1980s, the lifeways at Utah and Davis also mirrored the unequal conditions within American society: extended longevity for a select few (roughly 15 percent of the dogs), while a majority were passed over for necessary interventions. A randomized group stratification distinguished bodies that would or should suffer, and all dogs lived within a system of palliative care that kept them alive only so that they could be studied and made useful at the moment of their inevitable deaths. Goldman's joke about the beagles' exceptional care, voiced amid decades of neoliberal restriction of the American welfare system and debates over the desirability of universal health care, reveals what such treatment and attention could make possible: dramatically extended lifespans and improved health for the few. Studies of aging and radioactivity were studies of the kinds of aging that might occur within Cold War capitalism. The atomic beagles thus exemplify the work of species projection, how experimental organisms are rarely models of human physiology in the abstract without also being proxies for particular human forms of life.[180]

The very structure of the experiments demonstrated that "normal" aging, and therefore also "normal" and "abnormal" deaths, were culturally and politically determined. A normal dog was kept safe "from the increasing danger of the automobile," which represented the principal threat to canine lives, leaving the system of automobility, massively expanded since the beginning of the twentieth century, a constraint on the normality of canine aging.[181] A normal human, in contrast, would be kept safe from the increasing danger of the nuclear bomb and the runaway potentials of atomic power. This normality was framed by race, gender, and class: in gerontology, the idea of normal aging was long tied to a wealthy, white, and masculine standard, and so too was normality in AEC programs. Early planners and managers at both Utah and Davis were overwhelmingly white and majority male, yet many technicians and assistants were female, black, lower class, or a combination of these. The Davis lab consciously employed a balance of white and Asian American ("Oriental") caretakers for the dogs, which was thought to "favor the work efficiency of the men."[182] Members of marginalized populations were not, however, presumptive members of Cold War normality. Nor were residents of the Marshall Islands or American Indigenous populations living downwind of nuclear testing sites with little to no warning of the danger.

Beagles were proxies, but they were also future-generating beings, potent engines for not only producing knowledge about radiation, but also influencing military strategy, energy policy, and health and safety standards.[183] They were the soldiers that might die in atomic conflict, the pilots of futuristic aircraft, and the irradiated children of tomorrow. As direct physiological proxies for threatened human bodies, beagles helped produce the infrastructure of postwar American life, serving both as elements within experimental regimes and as symbolic embodiments of cultural transformation. After Lyndon Johnson assumed the presidency in 1963 and brought his beagles Him and Her to Washington, they became the official dogs of the White House, too. Spreading outward to answer novel scientific questions across the world, atomic beagles were important for making Cold War America, just as they were made by it.

This work directly impacted the researchers who carried it out, as dog and human lives conformed together in a looping process. "The people of the laboratory adapted to the beagle; the rhythm of planning and work was soon adjusted to time spans of canine reproduction, maturation, physiological processes, and aging."[184] Laboratory life grew dependent on the rhythms of kennel life, confirming in very literal terms what Pickering suggested about the dog kennel as both a tool and example of scientific culture.[185] As experimenters and directors came and went, the beagles remained. They were, a number of Utah researchers summarized, "a sort of governor and continuum in the work at this laboratory for 20 years."[186]

Much of the research with beagles thus had an explicitly autobiographical tenor: the projects were, in large part, studies of the ways that the researchers themselves might age, ail, and pass on. Unsurprisingly, symbolic identification with the dogs was a consistent part of life at each lab, particularly in the form of jokes. Leo Bustad, who left a job managing Hanford's Experimental Animal Farm to direct the program at Davis starting in 1965, referred to colleagues as "beaglers."[187] He also used letterhead depicting an older scientist seated at a desk with a cat to his left and a beagle rooting through a trashcan on the floor: a small sign above the scientist's head reads, "Pigs forever," an allusion to Bustad's work breeding miniature swine at Hanford. Bustad's science, the cartoons seemed to suggest, was inextricable from the presence of animals.[188] In the 1970s, the Davis newsletter header featured three beagle puppies peeking over fence posts.[189]

After Charles Schulz's *Peanuts* gained nationwide popularity in the mid-1960s, a Davis progress report from 1970 featured Snoopy on the cover, seated atop one of the lab's oak wine barrels, reading the report like a consternated human. "NEVER BUG A BEAGLE!" his thought bubble reads, identifying him simultaneously with the lab's dogs and with Marvin Goldman, who was fond of the phrase.[190] Goldman kept two folders labeled "Just for Laughs" and "Comics," which include, on the one hand, a set of poems and satirical laboratory reports and, on the other, a collection of comics from local and national papers. One clipped *New Yorker*–style cartoon portrays a photographer explaining to a corporate board member, "But don't you see, if you have *your* dog in *your* annual report picture, the other directors will want *their* pets in *their* pictures. Now wouldn't that look silly?"[191] Pet keeping and laboratory management were discursively and pictorially conflated, even as they were experimentally distinguished.[192] It should come as little surprise that Goldman's letterhead read simply, "From the desk of The Head Beagle," the title of a mysterious, all-powerful entity referenced occasionally in *Peanuts.*[193]

Jokes offered a way of making light of the difficult work of living among and sacrificing dogs. Species-crossing gallows humor also allowed researchers to make sense of the contradictory relationships of care that their work required. The Beagle Club researchers considered their programs necessary, in order to protect human beings from radiation, but also recognized the loss of canine life that they required. Most of the comics Goldman collected thus concern dogs or radiation safety. One, a play on the "Twelve Days of Christmas," describes events of the Eighth Day of Christmas as "Eight Glowing Beagles." Another, a clipping from the comic *Little Orphan Annie,* shows researchers at an anonymous laboratory worrying about safely adding more dogs to their project, a problem Davis faced for decades. The earliest cartoon, from an edition of the *San Francisco Chronicle* in 1956, depicts a frightened man and jumping dog. A sign above informs us that this is the "University of California Nuclear Fallout Research" facility. A second sign reads, "Caveat (radioactive) canem": Beware the radioactive dog.[194]

The jokes and cartoons frequently reconceived parts of human history or significant cultural products in order to forefront radiobiological beagles, as if Snoopy was ultimately part of a radiobiology extended universe. In doing so, they normalized beagle research while also framing it as an essential, if humorous, aspect of human life.[195] Such humor was a tool for managing the social world of the laboratory.[196] Yet these comics and jokes also reveal the conformations of laboratory life and the work of species projection: imagining humans as beagles, and beagles as humans, was essential not just to making the day-to-day work bearable, but also in supporting the epistemic bedrock of the research. The earnest, apparently unforced nature of much of the humor supported a belief in the fundamental similarity of humans and dogs. Beagles were like humans, humans like beagles.

A Breath of Fission Air

Utah and Davis founded the Beagle Club, but they were hardly its last members. Langham's early interest in the effects of inhaled radioactive materials was shared by others in the AEC, and concern grew with publicity about airborne fission products spread by aboveground nuclear testing. Although preliminary inhalation studies had taken place at Rochester, where basic experimental techniques and laboratory devices were developed, more expansive efforts were deemed necessary. In response, a second major program on radioactive inhalation took shape all the way across the country at Hanford, Washington, the site of the most productive civilian nuclear power reactors in the United States, where inhaling radioactive materials was more than an academic concern.

Overseeing this inquiry was a young Rochester protégé, William Bair Jr., who received a doctorate in radiation biology with the University of Rochester's Atomic Energy Project in 1954. There, while focusing on cellular biology, he had served as a research assistant for Casarett's experiments on radiation and testicular function in beagles. Highly sought after, Bair chose Hanford over a rival postdoctoral appointment at Oak Ridge; the upstate Washington setting felt fresh after decades on the East Coast. When he arrived, it was already a bustling center of postwar radiobiology, with ongoing research into the environmental impacts of radioactive materials, occasioned by pollution from early reactors, and new projects on radiation tolerance in large mammals.[197] Bair's initial work at Hanford involved research into the genetic effects of radioactive material on living cells, but he was eventually promoted to assistant under biology chief Harry Kornberg.[198]

Kornberg, remembered fondly for his accordion playing and amateur aeronautics interest, had jump-started the lab's inhalation toxicology programs early in the 1950s. As Hanford was manufacturing tritium for the yet-to-be-tested fusion bomb, trace amounts of tritium oxide were detected in the urine of reactor workers. Kornberg hypothesized that tritium oxide might be sneaking in through defective masks and developed a model for how the element might move through the pulmonary system. The vast majority would be absorbed in the lungs, his data suggested, but Kornberg was skeptical of the high values, so he followed up with an experiment. Having attached a hot water bag to some tubing, he inhaled a measured amount of the gas and tested his blood afterward; Kornberg recalled withdrawing the blood himself because an assistant was too squeamish. His predicted values were confirmed and, after a second experiment exposing his skin to tritium gas, set new standards for tritium protection at Hanford. Because he weighed roughly as much as the "standard man" reference values, the data required no correction for cross-lab application.[199]

Kornberg's tritium work made clear that inhaled radioactive gases were dangerous, and Bair was charged with follow-up studies on the risks of inhaling plutonium. To conduct such studies, he built a small inhalation toxicology unit and hired his own veterinarian, Jim Parks—Bair's "self-effacing" right hand, per Kornberg. This allowed Bair to disentangle himself from the influence of Leo Bustad's longer-tenured Experimental Animal Farm at Hanford.[200] As Bair's inhalation work developed in the late 1950s, Hanford began using beagles, rather than rodents or Kornberg's body, purchased initially from Washington State University, from nearby kennels in Kennewick, and from the Animal Supply Company of Napa, California. Hanford also acquired extra dogs from Davis. As at the other sites, Hanford researchers struggled to manage disease and parasites from purchased dogs, so they established their own colony with around 140 dogs.[201] Bair and Parks then set to work developing new devices to enable the dogs to inhale plutonium. Important changes that would reshape the program were brewing elsewhere, however.

In 1959, the Department of Defense's Defense Atomic Support Agency took over research into the physiological effects of atomic blasts from the AEC, freeing up a large portion of the latter's budget while depriving it of a traditional area of focus.[202] Eager to find uses for the money, Dunham sent David Bruner to tour the nation's laboratories and explore the possibility of an expanded inhalation toxicology program. The imagined project could use beagles to complement existing AEC work, but it would need to occur "outside of any place which was liable to give us trouble from the standpoint of contamination or dog noise," Bruner later explained.[203] Rochester seemed like an obvious destination, with existing inhalation study capacity and experience with beagles, but the possibility of bringing additional radioactive material to the area spooked administrators. Growing animal rights activity, spurred by stories of Rochester's bootleg beagles, had also dampened enthusiasm for further dog work. Bair, eager to parlay his inhalation project into a larger contract, put together a proposal to bring hundreds of beagles to Hanford.

In June 1959, however, Bruner had found himself in Albuquerque, New Mexico, visiting Clayton Samuel White, research director for the Lovelace Foundation for Medical Education and Research.[204] Lovelace had begun as a small respiratory disease clinic, run by uncle and nephew William Randolph Lovelace I and William Randolph Lovelace II, but White was hired in 1947 to expand the clinic into a full research institution after he met the younger Lovelace, Randy, during wartime work on high-altitude oxygen technology at the Air Force Aero Medical Laboratory in Ohio, where this Lovelace was director.[205] The first significant contract during White's tenure came from the AEC, a multiyear investigation into the biological effects of large-scale blasts and detonations. White's findings would influence the design of blast shelters and many of the animal experiments during American nuclear tests. Not to be outdone, his younger brother Byron became a Supreme Court justice in 1962.[206]

The Lovelace Foundation—home to existing detonation experiments and located near both Sandia Base, a key site for American weapons testing, and Kirtland Air Force Base, a major training installation for the Air Force—was an appealing site for the new dog project. The laboratory's experience with respiratory ailments also limited the need for major personnel relocation. Critically, for White, the dog studies would maintain Lovelace's strategic affiliation with the AEC, rather than the Department of Defense. Although the AEC did not offer blank checks, its funding allowed room for creative experimentation that military grants typically did not.[207] White asked Langham for advice on his proposal but finished it before Langham replied, appearing before the AEC on January 14, 1960. Bruner "sealed the bargain" during a stop in Albuquerque, agreeing to fund research facilities at a cost of more than two million dollars, making it the largest inhalation toxicology project in existence.[208] White began recruiting former Rochester personnel, including Tom Mercer, an expert on aerosols, and Bob Thomas, a pathologist.[209]

As the Lovelace work was gearing up, researchers at Sandia reached out to Hanford for support. The Air Force, whose medium-term strategic vision included routine transportation of atomic weapons, asked the Washington group to study continuous inhalation of plutonium oxide by pilots and recommended beagles for the work. Bair himself admitted that beagles were not necessarily the best choice for inhalation research: "Believe it or not," he mused retrospectively, "a horse's respiratory tract is more like man's than most of the other species."[210] His team had looked into acquiring miniature horses for inhalation work, but prices were prohibitive. Sheep, which were numerous in the rich agricultural regions near Hanford and had been participants in earlier studies of environmental contamination, got their shot, too. But, given the lack of control over their bodily functions, "it was a mess," Bair summarized. Given house-training, on the other hand, beagles cleaned up relatively well.

In 1963, the Lovelace Foundations's twelve-building installation, abutting the land of the Pueblo of Isleta, was completed. If Utah was Beagleville, the Lovelace site was a Beagletropolis: early plans anticipated that one thousand beagles would eventually be housed on site, a number that eventually doubled, along with ten thousand rats, mice, and guinea pigs. Schematics also revealed how much had already been learned from the prior beagle studies. Lovelace facilities, reported the *Albuquerque Tribune,* would be "temperature controlled" and feature "radiant heat in the concrete floors and nose-activated automatic water feeders."[211] Housing plans and socialization would follow on recommendations from Davis, and the original breeding stock consisted primarily of dogs from the other AEC laboratories. A few "normal" (nonlaboratory) beagles would flesh out the initial colony, further revealing how the "scientific" beagle had separated from its show and pet kin, even as Davis researchers had emphasized that scientific beagles *were* the only truly normal members of the breed.

Like the strontium-90 project, the original vision at Lovelace was to host lifespan studies of an extended duration. Public justifications for using beagles therefore stressed their longevity, rather than previously highlighted advantages. Unlike with the openings of other AEC facilities, and despite hopes for a quiet location, however, animal activists publicly criticized the project. William S. Brown, vice president of the American Anti-vivisection Society, pleaded to "stop this senseless, useless cruelty."[212] Yet the dogs ultimately stayed in Albuquerque far longer than expected, making the colony, which continued well into the 1990s, the last survivor of the AEC beagle projects. It was also, as other programs wound down, the place where extraneous dogs were usually sent. When Rochester completed its own beagle research, for instance, many dogs traveled down to Lovelace, a kind of old-age home for elderly beagles. As chapter 6 shows, this extended duration had unexpected implications.

Caring for Laboratory Dogs

When Lovelace opened, nearly three thousand beagles lived in AEC-sponsored facilities littered across the country. The research would be connected to a number of breakthroughs: along with basic knowledge about the retention of radioactive elements in large mammals, nuclear chemist Glenn Seaborg credited the Utah project with confirming the physical half-life of radium-228 and the unexpectedly short biological half-life of cesium-137 (another radioisotope produced by nuclear fission). Utah researchers also supported the first use of rubidium-83 in biomedical research and developed a new scintillation crystal for measuring radiation.[213] The Davis studies, on the other hand, honed dosimetry techniques for strontium-90 and clarified the risks of ingestion and fetal exposures, eventually predicting some long-term dangers of radiation. Work at Lovelace and Hanford led to the construction of new laboratory devices for studying the deposition of inhaled particles and helped ascertain risks to reactor workers, uranium miners, and many others.

Yet the ultimate value of the beagle projects remained controversial for scientists, administrators, and their congressional funders. What the studies had undeniably produced was a sprawling ensemble of papers and reports filled with scientific, technical, and veterinary knowledge about beagles themselves. Their longevity, tumor incidence, optimal nutrition, watering, dental hygiene, sexual performance, psychological well-being, and cage preferences were analyzed and tested extensively. It was initially thought that bone cancers were effectively nonexistent in normal beagles, but multiple cases appeared in control dogs over the decades. "I think the major reason for this difference," mused a Utah researcher, "is that under natural conditions beagles get run over by cars, tortured by children, and poisoned by neighbors."[214] A computerized medical record system for tracking the beagles at Davis predated many similar systems for humans at major hospitals: "Why, we've punched 2,000,000 IBM cards on these animals," Andersen told reporters in 1959.[215] Along with hundreds of papers in scientific journals, researchers at Davis, Utah, and elsewhere routinely wrote for veterinary magazines and gave talks to the public about dog ownership and pet care. Even acknowledging the immense toll that experiments took on many of the thousands of dogs who ultimately participated, it was as if hundreds of millions of dollars had been spent in a coordinated effort to unlock the secrets of beagle health.

This scattered information was first assembled in 1970, when Andersen's edited collection *The Beagle as an Experimental Dog* was published by Iowa State University Press.[216] Most of the caretaking and husbandry information came from experiences at Davis, but the book, which was funded by the AEC, featured contributions from researchers at each of the major radiobiology labs: Bair and Park from Pacific Northwest (Hanford's new name); William Clapper from Lovelace; Webster Jee and Glenn Taylor from Utah; Solomon Michaelson from Rochester; and multiple members of the Los Alamos Biometrical Research Group.[217] Also included were a number of prominent scientists with canine expertise, such as the University of Pennsylvania's David Detweiler; Jackson alumnus Michael W. Fox, who had visited Davis for behavioral studies of the dogs; and the Purina Dog Care Center's James Corbin. The only notable absence was a representative from Cornell, although Andersen's project proposal originally included funding for anatomist Howard Evans to visit Davis.[218]

The book covered the basics of the beagle, snout to tail: from the breed's history to its skeletal and renal systems, about which much and little were known, respectively. Sections such as "Special Studies in the Beagle" and "Specific Techniques Applicable to the Beagle" emphasized the nuclear origins and function of much of this knowledge: how many general researchers, after all, required a chapter dedicated to "whole-body radiation counting"? But the book also collated a significant quantity of shelved data from control animals across multiple studies, one-third of which Andersen believed had never been published anywhere. Beagles were, the book demonstrated, useful experimental dogs. But the collection also revealed a more comprehensive and developed justification for using the breed than previously offered. Where early researchers had set forth basic size aspirations and then stumbled onto beagles, Andersen articulated an extensive set of benefits that became increasingly standard in later explanations. "Among the 100 or more breeds recognized by the American Kennel Club," he wrote, "the number having qualities desirable as an experimental dog is quite limited."[219] Dogs with long hair were out, because cutting it frequently was laborious. So were exceedingly tall dogs, who ate too much, and those requiring cosmetic surgery. Extremely heavy dogs were also to be avoided: "Since one hundred 20-pound dogs weigh a ton, the mere weighing of a colony consisting of heavy dogs could be a logistical problem," Andersen explained.[220] Beagles were small and used to living and working in packs, so they adapted well to kenneling in large numbers.

They were Goldilocks dogs, and more important than any other plus was their temperament. "The Beagle's excellent disposition and gay personality are two of its greatest assets," Andersen noted, "because special handling is seldom necessary and a minimum amount of restraint is required for most experimental procedures."[221] When his beagles occasionally visited other labs, researchers were "shocked when they see a dog actually hold his leg up for the blood sample!" exclaimed Andersen.[222] The Davis team had decided early on that frequently handling the dogs kept them happier, and a pleasant canine personality made handling all the more enjoyable. This "pleasant, amenable, tractable disposition" was an artifact of earlier breeding practices—American beaglers had long extolled the breed's "gay personality"—and revealed how the beagle's prescientific history shaped its capacity to produce "truth" in experiments.[223] Veterinarians from the beagle projects were indeed quick to admit that their work represented an evolution from earlier breeding practices, not a complete departure.

This emphasis on the beagle's positivity and on the caring relationships developed between dog and researcher reveals the complexity of "care" in practice, caution scholars Eva Giraud and Gregory Hollin. The well-being of the dogs was as important for maximizing their control and scientific utility as their abstract happiness.[224] That beagles were easy to work with was another way of saying they were submissive, accepting of even painful procedures. In turn, one large divergence between atomic and traditional beaglers had to do with the dogs' voice: where hunters had lauded a beagle's melodic howling, some AEC researchers felt compelled to silence the dogs surgically to limit their noise. Yet scientists did not invent dominance over beagles, and earlier breeders had long killed, or "culled," large numbers of dogs with undesirable characteristics. With a value defined fundamentally in relation to certain kinds of people, the close conformation with beagles was built on sacrifice. "The problem for dogs," summarizes Mariam Motamedi Fraser, is that intersubjectivity and interagency "are not necessarily polite."[225] Care and sacrifice are often coextensive.

The genesis of Andersen's text, he explained, was "the need for a reference book of baseline data on the breed," which became apparent with "increasing use of the Beagle in research."[226] But wider use did not imply universal agreement about their status as a singular standard. John Paul Scott, who had left Jackson for Bowling Green State University, reviewed Andersen's book in *Science* and cited it as an exemplary text for "anyone starting and maintaining a dog research colony." Yet Scott argued that adopting beagles as "*the* laboratory dog" would be a mistake: the "unique advantage" of dogs was their variation, as Stockard and Whitney also believed. "Far from being the 'typical dog,'" Scott wrote, "the beagle is a very peculiar animal in certain respects," including its lack of aggression.[227] Andersen partly agreed, noting in public presentations that "the beagle is not the answer to all the problems."[228] His book was also based, in numerous places, on data from other breeds: "The Beagle" was still a composite of many dogs. But Scott went further, suggesting the creation of national canine research centers located at "convenient" points across the country, just as Little once dreamed, to make canine genetic diversity available to researchers rather than encourage a beagle monopoly. Despite "its close association with man for some 10,000 years," Scott concluded, "the dog is still a poorly known animal."[229]

Scott and Andersen's disagreements revealed methodological and disciplinary differences. Jackson had been a program, primarily, in behavioral genetics: Scott and colleagues relied on multiple breeds to understand how heredity and environment shaped the emergence of behaviors. What was important was the dog as a tool for understanding individual and group behavior, a breeding proxy for variation. But for Andersen and others, the point was a single dog, standardized to the greatest extent possible: a living experimental technology producing similar results wherever it was utilized. For beagles as physiological proxies, it was less important whether they and cockers exhibited identical adolescent behavioral patterns than whether the laboratory beagle was consistent, like a scintillator or a scale. With ongoing research that positioned the dogs as key models of aging, Utah's Betsy Stover predicted that a future edition of Andersen's book would include chapters on "beagle geriatrics."[230]

There would be no second edition, however. Nor, for that matter, would Scott's research centers come to fruition. A critical mass of support was still lacking. Personal matters and internal laboratory issues also delayed *The Beagle as an Experimental Dog*. Original coeditor Frances Carter, who Andersen credited with jump-starting the project, died unexpectedly not long after the initial contract was signed in 1964. Andersen received special dispensation to focus on the project, but Marvin Goldman, who assisted on multiple chapters, had to balance the work with other laboratory responsibilities.[231] Nor were other authors always forthcoming: an outline that Andersen circulated in October 1965 included a request for writers on the beagle's musculature system, gastrointestinal tract, renal system, ear, and nose. No one apparently volunteered: he covered the first three topics himself, while discussion of the ear and nose were absent.[232] A "Catalog of Data" was intended to appear at the end of the text, combining information from many laboratories, but it was sacrificed for space. While Andersen originally expected that the book would be completed by December 1966, it slipped back four more years.[233] By then, he was already on the verge of retirement.

The challenge of assembling all that was known about beagles without generating a meandering or overloaded mess would only grow. That was already clear a decade earlier when Marcus Mason, a veterinarian at the Biologics Testing Laboratory in Worcester, Massachusetts, published the *Bibliography of the Dog* (1959). Mason, who began the project in response to Tufts University pharmacologist Zareh Hadidian's interest in working with dogs, spent ten years compiling material and acknowledged that there was simply too much for one book to cover. With at least one thousand articles concerning dogs appearing each year, his subtitle, he joked, should be "FOOLS RUSH IN WHERE ANGELS FEAR TO TREAD."[234] As much as Andersen's book focused on beagles, it was also a summary of knowledge about dogs generally. Although practitioners agreed that compendia like these were useful, the work of assembling them was more than most authors could reasonably commit to.

But 1970 also marked the beginning of a steady decline for the major radiobiology laboratories. In December 1971, the AEC's Division of Biology and Medicine met to assess the "advantages and disadvantages of using large domestic animals in radiobiological research."[235] That analysis focused primarily on sheep, pigs, and other similar animals, coming down on an ambiguously positive note, but discussions surrounding the meeting revealed that shutting down or radically altering existing studies was in the air. Goldman reported to Bustad that researchers from Stanford had considered dropping their sheep studies to "go into swine or dog work."[236]

Concerns only increased after 1974, when criticism of the AEC's regulatory record led to its replacement by two organizations: the Energy Research and Development Administration (ERDA) and the Nuclear Regulatory Commission (NRC). Separating nuclear regulation from energy research called into question the justification for many existing projects. The growth of a vocal and aggressive animal rights and liberation movement—representatives of which broke into the facilities at Davis in the mid-1970s and produced a hailstorm of letters and media coverage for Goldman—also made openly funding and operating large dog projects increasingly troublesome.[237] Where the labs had once welcomed visitors with open arms, the Beagle Club became increasingly circumspect about discussing its work with the public. Tours ended. As historian Susan Lederer has shown, this guardedness aligned with a broader trend in American biology to write allusively about animal experiments and circulate fewer photographs of laboratory creatures.[238]

The Department of Energy (DOE) was created by President Jimmy Carter in 1977 to consolidate agencies such as the ERDA and NRC. Although there was a brief reprieve from earlier worries, federal science budgets became tighter across the board during Ronald Reagan's presidency—a fate that DOE biology projects were not spared. "Hard questions were asked as to what scientific purposes were served by continuation of the life-span beagle studies," summarized radiobiologist Roy Thompson.[239] In 1982, directors of the Argonne, Utah, and Davis projects were informed that consolidating the dog studies at one or two locations was all but certain. A report authored by radiobiologist Ed Wrenn and his staff at Utah voiced an assumption shared across facilities that severe budget cuts were forcing the DOE's hand. For Wrenn, consolidation would limit possible findings. Scientists at risk of losing their jobs and their animals would be too focused on seeking new employment to work effectively, and the healthy interdisciplinary cooperation established between labs would succumb to a struggle over scant resources. "The main asset of the beagle labs are not the beagles," he noted, "but the investigators."[240] Yet this was a defensive oversimplification: the investigators and beagles were entangled. "When somebody puts an entire working career into a given study, . . . it is because he gets wrapped up in the work," reflected Leon Rosenblatt of Davis.[241]

Consolidation represented the destruction of this conformation, one potentially as hard on the aging researchers as it was on the aging dogs, because "older individuals sometimes adapt poorly to new environments," in Wrenn's words.[242] Davis lab manager Ed Branam accepted that relocating their beagles was feasible, but only because animals from the facility's new study on cobalt were in good health and the surviving strontium-90 dogs were so old that they would pass on before the move. The seventy-six remaining dogs from the strontium work were between fourteen and nineteen, far older than most researchers ever expected any of the dogs to get.[243] Branam added that dogs that had been sheltered from microorganisms within a closed colony for their entire lives would be at risk moving through the wider, germ-filled world. Significant differences between beagles from various labs complicated matters further. As Wrenn explained:

> At Davis the beagles live in outdoor pens on gravel, whereas at Argonne and Utah the beagles are on concrete and spend time inside and outside. At Davis the summers are hot, while at Utah and Argonne the winters are cold. The Utah kennels are 5000 ft above sea level, Argonne is at 700 ft and Davis is near sea level. A change in altitude may possibly influence the development of hematological disorders including radiation-induced leukemia. The Davis dogs have been "debarked" by surgically cutting the vocal chords [*sic*], while the dogs at Utah and Argonne bark loudly. Placing barking and non-barking dogs together could produce unwanted physiological alterations.[244]

A single study with dogs of such dramatic variability would compromise its scientific "credibility." Each laboratory had carefully constructed its own conditions of normality, which consolidation would eliminate in one fell swoop—even if Wrenn's note confirmed that *the* beagle was a multiplicitous laboratory dog.[245]

Representatives from the DOE's Office of Energy Research visited Argonne, Utah, and Davis in the last week of March 1982. Utah's Chuck Mays warned Goldman in advance of the group's arrival to Davis that he should avoid the "overly defensive" attitude displayed in Salt Lake City.[246] The recommendation could not save the colony: in February 1986, the Davis Laboratory for Energy-Related Health Research (LEHR), a name assumed by the Radiobiology Lab to better fit with new DOE funding priorities, was informed that it would have to "close out the dog colony" by the end of September. The time frame was so short that LEHR had trouble finding new homes for the dogs. Lovelace volunteered to take some, but many had nowhere to go. A number of beagles, bred carefully and handled with obsessive care for decades, the products of millions of dollars in AEC funding, ended their lives in veterinary surgery classes.[247]

One challenge in defending the beagle programs was to clearly articulate the value of these and other large animal studies. "What can they tell us?" asked a symposium in September 1983 at the Pacific Northwest

Laboratory. The answer was dispiriting. While early newspaper articles on the beagle laboratories had dreamed of fundamental understandings of nuclear radiation, Thompson and coeditor Judy Mahaffey noted in their introduction to the proceedings that the results had "found relatively little application in the development of radiation risk factors for various organs of man" by international agencies.[248] With limited funding and difficulties in comparing data across sites, the "scientifically disappointing" possibility was that much "data may not permit confident extrapolation to the human and may afford little in the way of basic understanding." The beagle studies, described as "separate facets of a single experiment," were valuable at the end of the day because they enabled confidence about radioactive safety standards and offered a degree of "essential reassurance" for political decision makers.[249] They were not for nothing, but this was hardly the result many had hoped for.

In 1989, Thompson published an extended, internal history of the beagle work in order to help DOE decision makers establish, as objectively as possible, whether the remaining projects should continue.[250] His book came on the heels of another retrospective, James Newell Stannard's door-stopping, two-thousand-page *Radioactivity and Health: A History*, which featured the beagle work as well. By 1989, then, the beagle studies were history. Thompson wagered that at least $200 million had been spent on the projects, a total not adjusted for inflation, and at least 5,389 dogs had participated between 1952 and 1983, a number that did not include beagles from a number of ancillary studies, such as a collaborative radiological health program between Colorado State University and the US Public Health Service or nonnuclear inhalation studies at the Pacific Northwest Laboratory. Although Thompson resisted a final judgment on the beagle projects himself, his book makes clear that many had yet to meet their promised utility.[251] With lessened fears of nuclear cataclysm, most simply seemed unnecessary. The projects, Stannard felt, "sometimes support the statement 'the proper subject for the study of man is man.'"[252]

Coronavirus Fears

In the fall of 1978, beagles at Argonne National Laboratory faced an outbreak of a strange new disease. More than half of Argonne's eight hundred dogs became infected, and ten died, with over fifty sustained by fluid and antibiotics at the infection's peak.[253] A scheduled meeting of the AEC's Beagle Club, set for Argonne that year, had to be canceled—the group would never properly meet again.[254] Described early on as a dog coronavirus or a canine version of Legionnaire's, the disease spread rapidly. Initially identified in military German shepherds in 1970, it was only in 1978 that the condition, later recognized as a "parvovirus," took a menacing turn, appearing at a collie show in Kentucky before claiming the lives of tens of thousands of dogs from Australia to North America, South America to Europe and Japan.[255] Untreated animals were reduced to the appearance of "a concentration camp victim," in the words of one New York dog owner, and perished in a matter of days.[256]

Leland Carmichael, who succeeded James Baker at Cornell in 1975 and became the John M. Olin Professor of Virology, worked with colleague Max Appel to develop a vaccine. Widely hailed as a vital reprieve from the devastating illness, it was the latest fantastic success of the RLDD, which had revolutionized the biomedical treatment of canine illness. The canine parvovirus vaccine would be Cornell's most valuable source of patent income for decades.[257] By this time, dogs were intricately entwined with pharmaceutical development, and the location of beagle expertise had shifted. The vast majority of researchers using beagles outside of AEC-sponsored facilities were no longer raising them in large internal colonies such as Argonne's. That work was expensive and, given the tactics of animal rights protestors, occasionally dangerous. Nor were researchers generally purchasing dogs from small breeders, many of whom had soured on the optics of selling to labs. Andersen's book included a list of local breeders who would supply beagles, something mostly unthinkable by the 1980s. Instead, laboratory beagles were increasingly raised and sold by independent, commercial suppliers, many of whom maintained a profile as low as they could possibly manage. The emergence and evolution of these entities followed as beagles gradually became something surprising and unexpected: a de jure and de facto standard for global pharmacological safety studies.

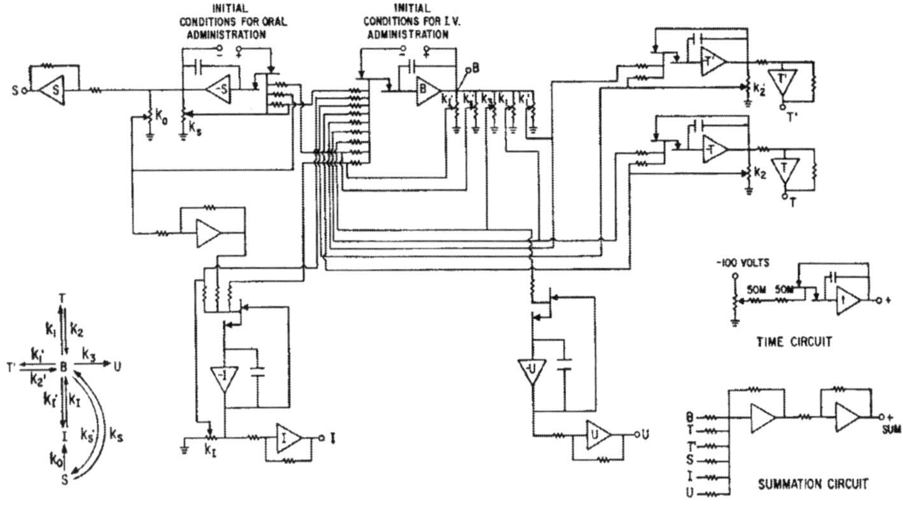

Diagram of Garrett's "electronic dog" analog computer. Reproduced from Edward R. Garrett, "The Application of Analog Computers to Problems of Pharmacokinetics and Drug Dosage," *Antibiotica et chemotherapia* 12 (1962): 228. Courtesy of S. Karger AG, Basel.

CHAPTER 3

Dog Times and Lifetimes: Beagle Commodities, Computers, and the Drug-Using Body

In February 2019, Vanda Pharmaceuticals, a self-described "global" biopharmaceutical company, filed suit against the US Food and Drug Administration over a partial clinical hold on tradipitant, its candidate for treating the stomach condition gastroparesis.[1] When news of the lawsuit broke, Vanda's stock price plunged, with skittish investors fearing failure and further rounds of costly experimental trials. While reports of inadequately tested drugs and assertions of overregulation by the FDA are typical set pieces in today's journalistic coverage of the pharmaceutical industry, Vanda's lawsuit was unique less for the involvement of tradipitant than for a secondary justification. According to a company spokesperson, "Vanda refuses to sacrifice young beagles or other animals in a study that serves no scientific purpose."[2] Vanda was against not regulatory caution, but sacrificing beagles.

These public-facing complaints were somewhat misleading. Vanda made clear that it did not "categorically object to the sacrifice of animals in the course of scientific research."[3] The company had already carried out multiple-month tests on both rodents and dogs, as mandated by FDA guidelines, all of which resulted in the death of animals. Nevertheless, Vanda called on "boards of directors, executives and employees of drug companies, animal advocacy organizations, the scientific community, and the public at large" to resist what it labeled a "one-size-fits-all approach to animal research."[4] Vanda later released data showing high levels of inbreeding in beagles, which were, they suggested, like Amish people; results were generalizable from neither population.[5] The criticisms articulated an uneasy alliance between companies, scientists, and animal-protection groups, using the rhetoric of ethical experimentation as a weapon against regulation. Editors of the *Journal of Animal Ethics* praised Vanda for "corporate bravery" after its lawsuit lost in court.[6]

Vanda's strategy was not without precedent. For multiple decades, pharmaceutical companies resisted requirements to test their drugs on beagle dogs. This chapter traces the foundations of that debate, showing how proof of experimental efficacy and safety became dependent on tests with beagles. It begins by exploring the arrival of beagles to pharmacology in the mid-twentieth century, around the same time that the atomic beagle colonies were first emerging. Fears about thalidomide and the safety of oral contraceptives would lead to the establishment of FDA guidelines concerning animal testing, including the Kefauver-Harris Amendments to the Food, Drug, and Cosmetic Act in 1962, arguably the most influential changes to federal drug regulation in the twentieth century. While the breed of dog was not explicit in early guidelines, toxicologists already possessed substantial experience working with beagles; when companies heard "dog," they knew to think "beagle." After a candidate contraceptive appeared to generate massive mammary tumors in female beagles, the FDA would make explicit that multiyear dog studies were expected for many drugs going forward. Because of the FDA's global reach, beagles became both de jure and de facto global standards, and the chapter follows them into pharmaceutical research in the United States, France, Germany, and Japan.

Around the same time that beagles were becoming integrated into the circuits of the international pharmaceutical industry, a new way of thinking about and studying drugs was emerging. "Pharmacokinetics" deployed an abstract vision of bodies as a series of interlinked compartments and enabled mathematical simulation of how drugs might traverse our organs. These mathematical models, growing steadily in their complexity, encouraged the computerization of pharmacology, a process that relied on carefully collected physiological data from living organisms such as beagles, much of it richly generated by AEC radiobiologists and veterinarians. Because drugs appeared to act "faster" in small creatures, like mice and rats, than they did in human beings, models had to be adjusted to account for drug action across species barriers, what some called "interspecies scaling."[7] This biological relativity, in turn, fostered novel ways of thinking about the time of pharmacokinetics, heralding a possibly revolutionary shift in pharmacology.

Reconstructed from circuits and signals during this period, electronic beagles emerged as abstract proxies for human bodies just as new, for-profit scientific breeders produced thousands of living beagles to meet the demands of worldwide pharmaceutical development. The chapter tracks the development of multinational company Marshall Farms to show how this process took place in practice. Beagles, once largely the local products of small business, were industrialized and explicitly commodified in the second half of the twentieth century: beagles became The Beagle™.[8] As commodity and computer, beagles were thus doubly abstract and infinitely exchangeable—at least in theory.

Some corporations sought to dethrone the beagle as a test animal, just as others were incentivized to mass-produce them for profit. Most accommodated to the new dog-testing regime, and a significant percentage of drugs taken by people all over the world were tested on beagles, whose bodies, abstracted to container flow charts and rematerialized in experimental trials, became real and imaginary foundations for the drug-using body, what anthropologist Emily Martin has called "the pharmaceutical person."[9] Imagining ourselves as defined, in part, by a long list of daily prescriptions and weekly boxes of pills required a prior effort to make beagles live in the same way. But with treatment-naive bodies possessing unique physiology, beagles displayed reactions to drugs that were not always representative of drug disposition in human beings. Understanding how beagles came to live in such a way will, in turn, reveal the ways that transnational pharma reframed health, in the words of anthropologist Joseph Dumit, as a "virtually limitless" frontier of capitalization.[10]

A Tristate Beagle Nexus

The first formal government beagle colony for drug and chemical testing was opened by the Division of Pharmacology of the US Food and Drug Administration (FDA) in July 1954 at its laboratory in Beltsville, Maryland. Paul Underwood, an FDA veterinarian, and pharmacologist Charles Durbin argued that a number of things made beagles excellent laboratory animals. Of vital importance was their "smallness in size," which meant lower overhead costs.[11] "The food requirements of large dogs can run away with much cash," J. Stuart Paterson, an English laboratory animal expert, explained poetically in 1957.[12] But beagles were also "gregarious enough to be housed in groups" and sufficiently independent to thrive when housed by themselves. Their "submissiveness," stressed Underwood and Durbin, was appreciated by those handling them on a daily basis.[13] Such arguments closely paralleled the ones offered by researchers at the AEC beagle colonies.

The FDA's Pharmacology Division was its largest user of laboratory animals in terms of both number and variety, and researchers planned to whelp two hundred beagles annually "for pharmacological and other studies," especially tests of the safety of food additives. As Herbert Calvery's participation in schemes to breed standard Irish terriers revealed, FDA scientists had been interested in a uniform population of test dogs since the 1930s.[14] Calvery helped site the agency's first dog colony, close to where the Pentagon now stands, and Beltsville also hosted a separate study beginning around 1938 about canine intelligence, led by a young animal expert named Walker Dawson from the Department of Agriculture's Bureau of Animal Industry.[15] A sort of spiritual successor to Stockard's project—but with the more utilitarian goal of breeding an ideal sheepdog, for which Dawson imported Hungarian puli and komondor dogs, known for their shepherding instincts and tightly corded coats—the study ran out of funding after World War II got underway.[16]

Beltsville's new toxicity beagles, Underwood and Durbin believed, were fundamentally different from earlier experimental canines: "Compare the dog used in research today with the unwanted foundlings or half-starved specimen gathered from the streets of large cities a few years ago." Beltsville beagles were "normal," unlike the frequently ill dogs seized from local pounds in decades prior.[17] In 1959, Tulane cardiologist George E. Burch would argue that normal dogs were the baseline "requirement for any investigation."[18] They were essential, affirmed Underwood and Durbin, to avoiding "a mass of unintelligible data" when experiments ended.[19]

The FDA could have selected another breed: Stephen Pemberton has shown, for example, how scientist Kenneth Brinkhous developed a colony of Irish setters at the University of North Carolina in 1949 to study the inheritance of hemophilia, just a few years before Beltsville opened, and with an origin story similar to that of Erwin Brand's terriers.[20] In both cases, show dogs manifested unexpected genetic diseases, a kind of ready-made experimental tool and case study. But utilization of beagles at Cornell and in the AEC radiobiology studies made a strong argument for the FDA's decision to opt for them. Significant information was emerging about how to manage the dogs, and the only citation in Underwood and Durbin's report on their colony was a paper coauthored by James Baker.

Use of beagles in drug testing partially predated the atomic colonies, however. Leslie Webster and Jordi Casals of the Rockefeller Institute for Medical Research used the breed to test new rabies vaccines in the late 1930s and early 1940s.[21] Webster had a reliable beagle source and approved of them enough that he recommended beagles to the Rockefeller International Health Division's Wilbur Sawyer for large-scale rabies studies in Alabama.[22] Beagle utilization saw a noticeable uptick after World War II, but breed choice in these early cases was primarily a matter of availability or general preference.[23] Webster, for example, offered minimal explanation for why beagles surpassed other breeds. The same was true at the American Cyanamid Company's Stamford Laboratories in Connecticut, where Diomax, an early oral antidiuretic, and a number of antimicrobial drugs were developed.[24] Toward the end of a 1947 paper on thyroid function, Stamford researcher Edmund Mayer noted, "Dogs of the same race, such as the beagles in the present study, should be given preference over mongrels in experiments on growth."[25] American Cyanamid, for its part, had originally planned to build a dog colony from Brand's Irish terriers.[26] Pharmaceutical companies were very interested in dogs, but that a dog was "normal" was more important than that it was a beagle.

In the early 1950s, as the FDA's beagle colony opened, corporate interests began to open their own beagle colonies. The New York, New Jersey, and Connecticut tristate area, where a multitude of beagle owners lived, became an epicenter for beagle toxicology as well. Scientists at New Jersey's Merck and Company were using the dogs by 1948, with a marked increase after 1951.[27] The Eaton Laboratories division of Norwich Pharmacal, located between Ithaca and Albany, began a beagle-breeding program in 1955 "when it became apparent that the quality of available dogs was not suitable, especially for toxicological purposes."[28] Lederle Laboratories, a division of American Cyanamid, bought beagles from Stamford to begin their colony in Pearl River, New York, in July 1956.[29] The Kasco Research Kennel of New York's Corn Products Company, best known for producing corn-based starches and sweeteners, bred a colony of beagles to test their products in the 1950s as well.[30] That some of these labs, such as Lederle, did business with both the RLDD and Davis highlights the entanglement of atomic, pharmaceutical, and veterinary research beagles.

The Eaton lab was not alone in feeling that high-quality dogs were needed but scarce. Researchers assembled at a 1953 meeting of the American Institute of Biological Sciences argued there was a "nationwide shortage."[31] That year, the National Research Council and National Academy of Sciences founded the Institute of Animal Resources in order to coordinate access, share information about suppliers, and train commercial producers in good breeding practices. The first edition of the institute's *Handbook of Laboratory Animals* (1954) offered a "fragmentary" listing of suppliers across the United States, designed to better facilitate "scientific logistics," but Lone Trail Kennel of Hershey, Pennsylvania, was the only explicit supplier of pedigreed beagles.[32] The handbook's list of the animal species needed by pharmaceutical companies in America, however, revealed a demand that had outgrown supply: for dogs, Ciba Pharmaceutical Products preferred "healthy beagles," while Hoffmann–La Roche appeared satisfied with beagles of any kind.[33] Breeders responded, and a 1961 directory of commercial animal suppliers from the newly rechristened Institute of Laboratory Animal Resources (ILAR) revealed a drastically altered landscape. Beagles could now be purchased from the Animal Supply Company of Napa, California; Dierolf Farms of Boyertown, Pennsylvania; the Cornell Dog Farm; Dogs for Research of Murray, Utah; Green Hills Farm in Jeffersonville, New York; John C. Landis of Hagerstown, Maryland; Lone Trail in Pennsylvania; Lu-Nor Farms in Kalamazoo, Michigan; Morris Research Laboratories of Topeka, Kansas; Professional Puppies of Cincinnati, Ohio; Stumbo Farms in Lima, New York; and Zoologicals World Wide in Arlington, Virginia.[34] Many of these businesses also supplied the atomic beagle colonies.

Because pharmaceutical production and research extended beyond New York's boundaries, so too did the breeding of beagles, now available from suppliers in each of the major regions of the United States. The RLDD and AEC beagle studies generated vital information about nutrition, kenneling, and much else besides, which made the breed approachable even for researchers with limited funding. Where Rockefeller support had once been the sine qua non for constructing an experimental colony, infusions of funding into universities after World War II allowed many academic departments to build their own small beagle colonies or convert older "mongrel" kennels into spaces for purebred beagles. Such was the case for the Iowa State University dog colony in Ames and a later colony of glaucomatous beagles at the University of Florida, although at a far more limited scale than the pharmaceutical colonies.[35]

This expansion was also replicated beyond American borders: in 1959, Pfizer's laboratories in Sandwich, England, began collecting beagles for a colony to be used in the company's pharmaceutical testing, basing their program explicitly on best practices from Davis.[36] The transition was not entirely smooth, and Pfizer researchers were still bemoaning their inexperience with the breed nine years later.[37] Douglas Appleton, an experienced beagler from the United Kingdom who published a general audience introduction to the hounds the same year that Pfizer's kennel opened, began producing pups for research purposes around the same time.[38] His strain of laboratory dogs was used prominently by scientists at Huntingdon Research Centre, a veterinary and chemical testing firm based in Cambridgeshire.[39] Many drugs in England and America would now be tested on the same *breed* of dog, if not the same *strain*.

By 1965, even before Andersen's book made it official, the beagle dog was increasingly acknowledged as the standard canine in the United States and England at private and public institutions, particularly in pharmaceutical and general toxicology. But the testing prominence of beagles was also met with growing public consternation. The Animal Welfare Institute's Christine Stevens, who previously had raised awareness about the price of Rochester's atomic beagles, helped lead protests in 1960 over the inadequate housing of the FDA chemical testing dogs. The animals, termed "Benson's Basement Beagles" by pundits in reference to their location underneath agriculture secretary Ezra Taft Benson's buildings, were said to live in "shameful, intolerable, appalling, cruel, inhumane, incomprehensible, abysmally bad" conditions.[40] But rather than demanding an end to use of dogs, critics lobbied for roomier accommodations, leading to heated debates about the construction of a larger FDA animal facility.[41] Here, the focus of animal activists on effective care rather than abolition was vividly apparent. Things might well have ended there were it not for an impending crisis concerning pharmaceutical safety that would permanently shake up the world of toxicology.

Safe Drugs and Other Dog Problems

Few medical controversies in the twentieth century loom as large the thalidomide disaster. First marketed by German pharmaceutical company Chemie Grünenthal in October 1957, thalidomide was a popular morning sickness treatment that spread rapidly around the world, becoming the third-largest-selling drug in Europe within years of its release.[42] Usage of thalidomide, however, was soon connected to stillbirths and other serious medical anomalies. In 1960, FDA reviewer Frances Oldham Kelsey refused to authorize the drug for American sale. Thalidomide's sudden reversal, from miracle treatment to poison, led many to question how the dangers could have been missed in the first place and to insist on firmer testing for pharmaceutical products.

Thalidomide had been tested on animals, but many researchers argued that the study protocols that validated it were inadequate. Leo Bustad summarized the general view in 1969: "If some of our European associates had performed toxicity studies on pregnant subhuman primates before this drug was released, they would have realized the hazards it holds[,] for they would have observed some of the same defects that tragically appeared in children whose mothers were given the drug during early pregnancy."[43] Bustad argued that a long-gestating animal, such as the miniature pig, would also have revealed the birth abnormalities. Unique in suggesting the pig, Bustad was not alone in believing that the thalidomide crisis demonstrated the need for new and different species in toxicology. Relying only on mice and rats, the consensus went, would not ensure the safety of the world's growing pharmacopeia.

American lawmakers took the first major step in 1962, when John F. Kennedy signed the Kefauver-Harris Amendments to the Food, Drug, and Cosmetic Act of 1938. Tennessee senator Estes Kefauver had begun investigations in December 1959 into the pharmaceutical industry's monopolistic pricing practices as part of the Senate Antitrust and Monopoly Subcommittee, hoping to lower the cost of prescription drugs for Americans. Over months of testimony, pharmaceutical representatives repeatedly argued that high prices were key to their basic business model: high failure rates necessitated expensive drugs, and the industry would go broke otherwise. But other testimony painted the picture of a broken system for evaluating efficacy. Albany Medical College's Solomon Garb, for instance, argued that safeguards were stricter for dog and cat food than for most human drugs.[44] Gradually, Kefauver's inquiry expanded from pricing to the question of drug efficacy, but he faced intense industry backlash over his pricing proposals. The uproar over thalidomide gave him the final push.[45] "It could have happened here," Johns Hopkins pediatrics professor Helen Taussig told Congress about thalidomide. As a *New York Times* headline announced, "Drug Curbs, Scoffed at in '60, Are Now Being Sped in Congress," in 1962.[46]

While earlier laws had targeted unsafe chemicals and treatments, what became known as the Drug Efficacy Amendments represented the first time that the American government explicitly regulated the effectiveness of pharmaceuticals, shifting the burden of proof to manufacturers before products could reach the market. This added role gave significant power to the FDA while also dramatically increasing its responsibilities. By 1970, the newly galvanized agency began to interpret "proof" to mean randomized, controlled trials (RCTs) with experimental organisms. The Kefauver-Harris Amendments were thus crucial to establishing RCTs as the pharmaceutical "gold standard" and cemented America's position as a global leader in research and safety testing.[47]

The American pharmaceutical industry, however, was unused to external oversight and initially uncertain about how to meet the newly proposed standards. The FDA had described its process for evaluating chemical toxicity in foods, drugs, and cosmetics on multiple occasions, but the agency gave few explicit requirements. Descriptions of the process were meant largely as an aid to companies planning to submit their products for evaluation. Thus, in 1949, FDA pharmacology chief Arnold Lehman and his colleagues suggested dogs for tests but specified no breed.[48] In a 1955 law review article, after the new dog colony had opened, the FDA noted matter-of-factly: "We employ beagles as our dog strain."[49] Even then, these evaluation protocols were largely aimed at determining short-term toxicity. Thalidomide's latent effects meant that the FDA needed an approach to evaluating drugs taken over extended durations.

Lehman, a well-respected toxicologist, explained before a group of drug manufacturers in 1958 that the trend for consumers to take drugs "on a maintenance basis" might soon require administration of substances "for practically the entire lifespan."[50] This raised new questions about carcinogenicity and required more expansive tests. Dogs, Lehman emphasized, would be critical for establishing risk in such cases. In 1963, after the passage of Kefauver-Harris, he expanded on the previously implicit expectations, formally separating drugs by intended clinical usage, duration of treatment, and route of administration.[51] The Lehman Guidelines, as they became known, recommended clinical trials with escalating phases and three major categories of animals for toxicological evaluation: rodents, rabbits, and dogs. Dogs represented the closest analogues to human beings, with nonhuman primates deemphasized because their husbandry remained complicated and expensive. While the breed of dog was not explicit, many American toxicologists already had experience with beagles and were familiar with FDA practice. As later FDA staff member Victor Berliner noted, industry actors offered "no objections to the beagle."[52]

Thalidomide helped to formalize the small hounds as a regulatory expectation, but it was troubles with oral contraceptives a few years later that converted their status into that of an explicit requirement. In January 1966, Merck, Sharp, and Dohme reported to the FDA on tests of an ethynerone combination, codenamed "MK-665," under investigation as a new contraceptive agent. Their findings revealed significant, troubling mammary tumor growth in female beagles taking MK-665. Because similar drugs might be used by women for nearly half of a lifetime, regulatory action was almost certain to ensue. But it also came to light that Merck had found the tumors in December and, rather than directly reporting them to the FDA, turned first to consultants to frame the findings.[53] After the FDA encouraged the Justice Department to prosecute Merck over the delay, the first post-Kefauver prosecution recommendation, other corporations felt compelled to comply.[54]

Acutely wary of a repeat of the thalidomide failures "here," the FDA and its scientific advisers recognized a need to establish clearer safety standards for long-term (or "chronic") oral contraceptives. In 1967, the agency wrote to inform manufacturers that Phase III trials could now begin only after seven-year beagle studies and ten-year macaque studies.[55] The lengthy requirements signaled a significant market opportunity for commercial beagle breeders. The formal guidelines were published in the *FDA Papers* of November 1969, but they remained largely outside of public awareness until reports, in 1970, that other birth control treatments, including Eli Lilly's C-Quens and Upjohn's Provest, were being taken off the market because of fears that they also posed a risk to users. According to reporting in the *New York Times*, chlormadinone acetate, in C-Quens, and medroxyprogesterone, in Provest, both produced mammary nodules in beagles.[56] Similar findings had caused the removal of Squibb and Sons' Verton from the market in January.[57]

Reporting cautioned that "species differences" could mean that the effects were not necessarily present in humans, leaving significant questions about just how similar a female beagle and a human woman were. "Nonetheless, it is prudent to act on information which differentiates these drugs from other oral contraceptives," explained Charles Edwards, the commissioner of food and drugs. Such prudence was undoubtedly painful for Lilly and Upjohn, companies who previously served up to 20 percent of the nearly eight million women taking contraceptives in the United States and now had no other contraceptives on the market.[58]

Further questions about how predictive beagle data really was followed in 1973, when the FDA reported that Upjohn's Depo-Provera, an injectable birth control treatment, also produced mammary tumors in the dogs.[59] Although previous drugs were removed immediately from the market based on similar findings, Depo-Provera had already been in use for thirteen years. Its continued availability was justified by the drug's initially small and "unusual" patient population, including institutionalized women whom the FDA's Marion Finkel described as "mentally deficient" or incapable of taking a daily pill.[60] Resulting uncertainties about what beagle findings meant left many dissatisfied. Citing MK-665 and other drugs with dangerous side effects in dogs, Connecticut senator Abraham Ribicoff, a former secretary of health, education, and welfare and supporter of Ralph Nader's consumer advocacy movement, charged the FDA with basic failures to regulate pharmaceuticals in September 1973 and encouraged a stronger hand in determining safety.[61]

Companies, on the other hand, pushed back against the beagle requirements. Robert Hill and Kenneth Dumas, researchers at steroid manufacturer Syntex, argued vigorously at an international conference in 1973 that "the preponderance of the evidence" did not support a continued requirement of beagle studies.[62] Upjohn took the same perspective in their multidecade defense of Depo-Provera. Interested parties repeatedly pointed to data showing that beagles experienced spontaneous mammary tumors, statistics drawn from data on beagles at the UC Davis colony, yet such claims were disputed by Berliner and others at the FDA, who had their own experience with the breed to rely on.[63]

Accepting the existence of frequent spontaneous tumors would threaten the foundations of the beagle's status as a test standard.

While formal arguments about dissimilarity between humans and dogs typically emphasized divergent study results, commercial concerns were equally paramount. By the mid-1970s, multiple large pharmaceutical firms quit development of contraceptives entirely, complaining of diminishing profits.[64] As pharmacologist Michael Briggs lamented in 1977, "Russian roulette with beagles seemed unwise!"[65] Objections reappeared continuously for years afterward, evidence both of corporate interest in destabilizing the beagle's testing position and of the tensions between knowledge produced about the dogs in government and private laboratories.[66] The still limited scientific understanding of dogs bemoaned by many had come back to bite both the drug industry and its regulators. Pharmacologist Gerhard Overbeek of Dutch pharmaceutical company N. V. Organon, makers of the oral contraceptive Lyndiol, accepted in 1973 that "the beagle is an extremely useful animal for all kinds of toxicity studies," but he hesitated with respect to carcinogens: "I do not know," Overbeek admitted.[67] By 1979, many pharmaceutical researchers consolidated around the claim that the mammary tumors demonstrated previously were "species-specific" effects in the dog that bade no ill for human women.[68] In 1992, Depo-Provera would be partly vindicated after an FDA panel supported its approval, discounting the beagle findings.[69]

Debates over contraceptives continued, but expanded use of beagles in international pharmaceutical research during the 1960s and 1970s supported the breed's increasingly central place in toxicological testing. This happened even in the face of an unexpected controversy: in 1973, Wisconsin representative Les Aspin helped unearth US Army plans to test new nerve gases on beagles, just as the Air Force was setting up studies of the effect of jet engine exhaust on the dogs. The National Anti-vivisection Society brought suit against both parties, and an outraged public demanded an end to the studies or a search for alternatives. As Ryan Shapiro notes, letters protesting the military beagle studies surpassed even those against the war in Vietnam.[70] Yet a committee of the National Academy of Sciences argued in 1974 that beagles were "indispensable" for studies of toxic chemicals, and public interest died off by 1975, following a string of political defeats for American antivivisectionists and a campaign of military obfuscation.[71] To some extent, national controversy was little match for the internationalizing beagle: in 1976, the Human Reproduction Unit of the World Health Organization concluded that "the beagle is a reasonably good breed for carcinogenesis testing," a finding that firmed up its place in long-term toxicological studies.[72] The formerly implicit was explicit: new substances were safe when they were safely taken by beagle dogs, sacrificed in substantial quantities for the promise of healthier human lives.[73]

Internationalizing the Beagle

Even before thalidomide, beagles had begun to spread among pharmaceutical researchers outside of the United States and England. In 1958, contemplating the state of research with canines, French researcher Camille Carpentier reflected that, of all the laboratory animals, it was the dog that had served science the longest.[74] Man's best friend had been his best research subject as well, and Carpentier, a retired military veterinarian who consulted on corporate lab animal breeding, expected that the number of canines appearing in a "holocaust" of physiological, toxicological, pathological, and pharmacodynamic experiments would only continue to grow.[75] Gone were the days of Claude Bernard chloroforming stray dogs on the streets of Paris, however. Between 1940 and 1960, French scientists spoke more and more, like their American counterparts, of the necessity of working with "le chien normal."[76] Healthy, high-quality dogs were desirable, but they were no more accessible than in the United States. A commission of the Centre national de la recherche scientifique had debated establishing a French canine breeding center but opted to focus on rats and mice instead for the nation's first laboratory stock center, opened in Gif-sur-Yvette in 1953.[77]

As a candidate "chien de laboratoire" that was easily bred and already available to researchers, Carpentier suggested the fox terrier, with which he was familiar.[78] An older breed, widespread in England and France, fox terriers gained international popularity during the 1930s thanks to Asta, the jovial canine assistant to Nick and Nora Charles in the Oscar-winning Thin Man series. Around the same time, Milou, the loyal canine sidekick of beloved comic reporter-adventurer Tintin, brought additional attention to the breed in the Francophone world. *Tintin* creator Hergé called Milou "roughly a wire-haired fox terrier, although a bit of a bastard," and is thought to have selected the breed for its British, cosmopolitan connotations.[79] Because of the similarity between fox terriers and Irish terriers, it is possible Carpentier had heard of Brand's experiments, but he made no explicit reference, and any particular breed preference was ultimately mild. He did not differentiate "wire" or "smooth" terriers and was quick to admit that they were aggressive and nippy in his experience, which made handling them a challenge. Whether they were "purebred" enough for true standardization was also in question.

A few other plausible alternatives, however, presented themselves. Carpentier was aware, for example, of efforts by Leon Whitney to breed and sell beagles, raised with his son, George, at their Whitney Veterinary Clinic. After plans for a full dog laboratory at Yale fell through, the father and son Whitney had developed a strain of "uniform" beagles to sell to local medical students and laboratories in the late 1940s but discontinued the undertaking after their price of $25 per puppy was rejected.[80] Carpentier acknowledged that his terriers were produced at roughly the same price, but felt that beagles were unwieldy, with long tails and dangling ears at risk in quarrels. During a roundtable discussion on scientific dogs, one colleague noted that he was himself investigating setters, Brittany spaniels, and German shepherds as possible research animals.[81] These breeds, rendering beagles positively compact by contrast, hardly satisfied Carpentier's concerns about unwieldiness. Surgeons wanted larger dogs, chemists wanted smaller ones, but what was most important overall, added veterinarian Jacques Nouvel, was "the pharmacological point of view" because that field was the "greatest user" of dogs.[82] What pharmacologists found workable would end up deciding the issue, concluded Clément Bressou, the former president of the Académie Vétérinaire de France.

They were correct, and in little more than a decade, the situation had changed dramatically. "Let us imitate the Anglo-Saxons and introduce beagles into our breeding," exhorted a representative for France's Société de l'élevage, de recherches et d'expérimentation pharmaceutique (Society of Pharmaceutical Breeding, Research, and Experimentation) in 1966, promising a full transition to beagles by the next year.[83] In 1969, researchers Pierre Escuret and Jean-Pierre Vaissaire of pharmaceutical firm Roussel Uclaf announced in a three-part article series that beagles were the laboratory dog "par excellence," drawing heavily on work by Andersen and others at Davis.[84] Understanding and working with them was so important that Vaissaire followed the articles with a 1972 book, *Le chien, animal de laboratoire,* which focused primarily on beagles. The breed had been the dogs of choice in Britain and America since 1954, he explained, citing the date of the FDA colony, and were quickly replacing mixed breeds in France out of a desire for experimental results that compared across the English Channel and the Atlantic Ocean.[85] In a review of *Le chien,* veterinary microbiologist Pierre Goret echoed Scott's criticism of Andersen's book by noting that there were reasons to regret the singular use of beagles, given differences in behavior among breeds, but he acknowledged they were undeniably the baseline for widespread research.[86]

The rise of beagles came at a moment of astounding growth in French experimentation with dogs. Nicolas Aspiotis, a veterinary scientist working in Greece, noted in a French journal in 1960 that by "an overwhelming proportion, it is dogs who serve as laboratory animals in the various hospitals, physiological institutes, pharmaceutical industries and general medical research."[87] Escuret and Vaissaire estimated that the number of dogs used yearly almost tripled between 1957 and 1966, from thirty-five hundred dogs to over ten thousand.[88] This was still only a fraction of the number of dogs used in American experiments (370,000 in 1968, according to one survey), but by the mid-1960s, the Centre national de la recherche scientifique planned to expand its original breeding center to produce the widely desired dogs.[89] The growing use of dogs for testing pharmaceuticals and the border-spanning influence of the American FDA were key.

In Germany, the situation was similar, except that unlike in France, where beagles and beagling were reasonably well known, it was pharmacology that more or less introduced the breed. Although thalidomide's impact was more dramatic in Germany than in the United States, Arthur Daemmrich and Georg Krücken have shown that the regulatory response was ultimately slower and uneven.[90] Despite this, many German companies moved to synchronize their testing methods with new American regulatory expectations. Researchers at the Marburg-based pharmaceutical company Behringwerke AG, a subsidiary of chemical and pharmaceutical company Farbwerke Hoechst AG, started breeding beagles in 1960 on the basis of the animal's popularity among American and British pharmaceutical concerns.[91] Theirs was likely the most advanced beagle-breeding operation in the country and, although it built on preexisting company experience with dogs, required substantial transformation of former animal care practices, including the construction of new kennels.

In 1964, Behringwerke's Siegfried Kamphans dedicated an article in veterinary journal *Die blauen Hefte für den Tierarzt* to an explanation of beagles. The breed was virtually unknown in Germany, he noted, but in England, France, and the United States, the dogs were loved and respected.[92] German-language references to "the beagle" at the time were almost exclusively about Darwin's ship, the HMS *Beagle*, and Kamphans explained the difference partly by the relative popularity of hunting with hounds. Although horseback hunting was legally curtailed throughout Europe, France possessed nearly eighty active hunting packs, while Germany had only four. Just one, near Lübeck, was composed of beagles.[93] Yet it was neither hunting nor sport that invited interest in the dogs, Kamphans affirmed, but scientific and medical research, the "beagle as laboratory dog."[94] Behringwerke's "Labor-Beagle" colony was based on a foundational stock of imported French hunting dogs with "good genes," which Kamphans argued was critical for their uniformity as test subjects for German pharmaceuticals.[95] The lab would later breed English and American laboratory beagles into its colony.

By 1965, Badische Anilin- und Soda-Fabrik (BASF) also had its own beagle colony.[96] While German physiologists and others had carried out experiments on dogs, typically with mixed animals, for centuries, the initiation of multiple, large-scale breeding programs demonstrated just how important beagles appeared to be for global toxicological work. That popularity spilled over to other disciplines, such as cardiology, where the dogs had long been popular in the United States.[97] In 1968, Wilhelm Schumacher and Herbert Strasser, two researchers at Hoechst, summarized that, while the breed was "to a large extent unknown as a pet dog in Germany," it had taken over in experimentation because it was inexpensive, friendly, short-haired, and compatible with group kenneling.[98] That focus on the beagle's positive attributes glossed over the obvious external influences on the breed's selection: the majority of cited publications were written in English and came from researchers at Davis and Cornell, because direct knowledge about handling beagles was largely restricted to experts in England and America. "The question is frequently asked as to why it is precisely that this breed is selected for experimental work," Schumacher and Strasser note, suggesting that German uptake was neither entirely uniform nor immediate.[99]

A parallel transformation occurred in Japan, where the pharmaceutical industry expanded following a period of postwar reconstruction. American Occupation authorities had emphasized pharmaceutical development to improve Japanese public health, which was seen as a barrier against both communism and the resurgence of militarism, but also as necessary to supply overseas American GIs with drugs, especially penicillin.[100] Support for Japan's drug producers increased with the arrival of the Korean War and the escalating strategic utility of America's military and naval installations on the archipelago. While Maki Umemura has argued that Japan's pharmaceutical product standards, introduced partly in response to thalidomide and other dangerous drugs, were slow to match those in other countries, researchers were quick to take up the beagle.[101]

Japan's Central Institute for Experimental Animals (CIEA) was founded in 1952 in order to increase the availability of lab organisms and raise awareness about the value of standardized animals. Initially, its focus was on expanding production of the laboratory mice raised by young medical researcher Tatsuji Nomura, his mother, Masuko, and elder sister, Michiko.[102] The CIEA, struggling against disinterest from local scientists, survived originally thanks to a major purchase order for experimental animals from the US Army 406th Medical General

Laboratory in Tokyo, which also arranged a five-month tour of American animal breeding facilities for Michiko in 1953.[103] Modeling their work on British laboratory animal production standards, Japanese experimental animal advocates focused for much of the 1950s on improving available rodent strains and developing monkey-breeding operations. As international collaboration grew on the question of laboratory animal production, propelled by the emergence of the International Committee on Laboratory Animals in 1957, Japanese researchers came to see Americans instead as the examples to follow.[104] In 1959, Koji Ando, the godfather of Japanese lab animal breeding, traveled to the United States as a CIEA representative and attended a meeting of the ILAR, where he learned of the growing significance of beagles in both the United States and England. Smith, Kline, and French's pharmaceutical labs, Ando heard, used only beagles.[105]

Two years later, researcher Junichi Kaneko traversed the United States as a representative for Takeda Pharmaceuticals to observe the production of research animals more closely. His observations, published in Japan's leading journal dedicated to experimental animals, emphasized "beagles," which appeared in English like other terms lacking recognizable Japanese characters. After seeing limited use of the breed at Pfizer's Groton Laboratory, contacts at the National Institutes of Health informed him of four major American labs breeding beagles: Jackson, the Utah and Davis radiobiology programs, and the Ralston Purina Dog Care Center in St. Louis.[106] At another stop on the trip he met Charles Durbin, from the FDA's beagle colony, and learned that emphasis on breed purity was still limited in the United States—even the FDA occasionally used mongrel dogs as well.[107]

The beagle should be used in Japan, Kaneko summarized, because experience in the United States had shown it to be the "best" research animal. Significant work would be required to build proper disease-free experimental kennels, however.[108] The same year, Ando bemoaned that most experimentation in the country was still based on stray dogs, rather than specialized beagles, but he stressed that many universities and corporations were beginning to use the breed in order to avoid wasting time with uncertain results from hybrid dogs of unknown ancestry.[109] Pharmaceutical companies, as elsewhere, ultimately led the country's shift to beagles: a discussion in the Pharmaceutical Society of Japan's *Farumashia* in 1966 between Tatsuji Nomura, now director of the CIEA, and University of Tokyo pharmacologist Teruhisa Noguchi revealed that Japan's first dedicated purebred beagle colony had recently opened at Chugai Pharmaceutical Company facilities in Hokkaido, in the country's far north.[110]

Chugai, which found international success in the 1950s with Guronsan, a popular health tonic addition, and Varsan, a home insecticide, acquired its dogs, unsurprisingly, from kennels in New York in the mid-1960s.[111] But Nomura noted that other groups were also interested in beagles and called dog breeding perhaps the most important question in experimental animal production for Japan. Over sixty-eight thousand dogs were used for research in the country in 1970, with "beagle" the only breed specified by researchers.[112] The CIEA's Chuhei Yamauchi visited the United States from April to June that year to learn more about the specifics of American beagle production, visiting the NIH animal facilities and Davis beagle colony, among others.[113] At the Nineteenth Symposium on Laboratory Animals in 1972, representatives from Chugai, Fujisawa Pharmaceutical Company, and Sankyo Company all shared their experiences with breeding and purchasing dogs, indicating that importation from America was still the dominant source of beagles.[114] Each, however, stressed a desire to build up domestic production. Two years later, writing in the *Japanese Journal of Pharmacology*, Kazushige Sakai and colleagues from the Department of Pharmacology at Chugai noted that beagles had become a central research animal in the country over the last few years.[115]

The concurrent transformations in France, Germany, and Japan make clear that the growing popularity of beagles occurred particularly as pharmacologists shifted consciously away from a reliance on cheaply available stray dogs toward a single standard and sought to synchronize their research animals transnationally. A survey carried out by Jackson Labs in 1961 found that twenty-six institutions and individuals were maintaining purebred dog colonies for research in the United States alone, a marked increase since World War II.[116] A separate survey of leading research by Duke physiologist Bodil Schmidt-Nielsen in 1960 found that dogs appeared in 28 percent of all physiological studies, the most of any species, with rats and nonhuman primates at 22 percent and 21 percent respectively.[117] A further statistical breakdown revealed that dogs were particularly valued for studies of blood circulation, which made them logical choices in research on the movement of drugs through the body. Beagles were, in this way, the particularization of an existing, popular research subject as well as symbols of a threshold change in quality.

Justifying the use of beagles in the Netherlands, Dutch animal expert Nicolaas Adriaan van der Velden noted in 1968 that private and public kennels producing beagles had expanded drastically across the globe following the lead of the United States.[118] The development was both akin to and distinct from the popularity of laboratory mice earlier in the twentieth century: Karen Rader has shown how C. C. Little's mass-production plant for made-to-order experimental mice transformed both science policy and biological research across the board.[119] As we saw, the cost of raising purebred dogs made them initially resistant to the enormous production regimes of experimental rodents. Transporting dogs across significant distances was also an endeavor fraught with peril. The AEC laboratories were able to raise significant numbers because they could shoulder the initial investment burden, but few other institutions could maintain large internal colonies of their own. With growing profits to be made from safely tested drugs, however, the necessity of defensible testing data had changed the profit calculations.

The beagle's increasing transnationality did not, however, imply that the dogs were absolutely the same. Even American researchers experienced trouble replicating the apparently more standard beagles in use at certain facilities: University of Kentucky periodontologist Stanley Saxe found that the dogs available to him differed from the specialized beagles used for AEC projects.[120] It was well known, further, that American, English, and French beagles diverged in meaningful ways, because of both national standards and local breeding practices. "In their current form," explained French dog expert Paul Daubigne in 1953, the nation's beagles "are certainly differentiated . . . by some particularities."[121] The German researchers at Hoechst had found blood chemistry values in their English beagles that diverged from what Andersen and colleagues had reported, and Japanese researchers were incentivized to breed their own beagles partly because of discrepancies between imported and local dogs.[122]

The problem of different breeding and management practices was commonly debated at international conferences. Robert Marsboom of Belgian pharmaceutical company Janssen Pharmaceutica, for instance, inquired in 1967 about acceptable mortality statistics in French colonies compared with American ones, something German researchers were also attentive to.[123] Dissimilarities emerged even in how the dogs were visualized across borders in scientific illustration: contrasting journal diagrams from the period, such as a French physiological map and a Japanese perfusion system diagram, revealed a pictorial and imaginative flexibility concerning what exactly a beagle was and looked like.[124] Even as they increasingly represented a kind of universal nonhuman foundation for studying drugs in the body, beagles remained partially foreign entities outside of the United States.

That unfamiliarity, however, could not stop the American selection of the beagle from generating international momentum. After French and German researchers felt compelled to mirror the standards set by American and English toxicological research, researchers from the

Netherlands, Belgium, Japan, and beyond sensed a requirement to match the animals in use elsewhere. Hans Hurni of Switzerland's Tierfarm Sisseln made the linkage clear at an international conference, "The Future of Laboratory Animals," in 1966, explaining that beagles were important because "the F.D.A. recommends [them]" for chronic toxicity studies.[125] Australian researcher Robert Hodge noted that his country's regulations "date from July 1969" and were "virtually identical with those of the FDA."[126] Representatives from Japan and Sweden confirmed that their own safety rulings relied substantially on FDA decision-making. If companies wanted to compete in the world market, they had to gain experience with and test drugs on the little hounds. The regulations, once settled, were difficult to dislodge.

Becoming properly "global" took a few more decades, but beagles were reportedly imported into China for use at Shanghai First Medical College's Experimental Animal Department by 1984. They appeared in Korean publications throughout the early 1990s and were common in toxicological research in India by the new millennium.[127] Around this time, circumspection about discussing laboratory animals and their commercial sources in scientific publications makes it harder to track the movement of beagles across borders. In 1987, World Health Organization guidelines reversed a decade of precedent by eliminating the requirement for seven-year dog studies of contraceptive steroids, giving a major victory to pharmaceutical firms and replacing the old rule with shorter rodent study expectations. The FDA, however, would retain a modified three-year beagle test requirement until 1992.[128] Dogs, especially beagles, remained a preferred nonrodent species for toxicological testing even after that. Thus, the oral toxicity testing guidelines of the US Toxic Substances Control Act continue to suggest beagles as a nonrodent animal model.[129] While other substitute breeds might have served science as well as beagles did, individually they were little match for the momentum of a progressively interconnected pharmaceutical industry and border-spanning legal regime.

The Beagle Commodity at Marshall's Farm

Scientific narratives about beagles frequently obscured the multiple causality underlying the breed's growing prominence, one that can be better understood as a complicated result of expanding biomedical production during the Cold War, interconnecting regulatory apparatuses following the expansion of American pharmaceutical capital, and a degree of chance. The dispersal of beagles among pharmaceutical companies did not, after all, imply that stray dogs vanished from research or teaching, even at well-funded institutions in the United States. Stray dogs may have been, to beagles, what bouncing alpha particles on gold foil were to a cyclotron, but they were also far cheaper. Harvard Medical School, for instance, relied on random-source pound dogs for some projects and teaching exercises until at least 1982, when physiologist Clifford Barger was informed of a new cost structure for acquiring live animals. "For several reasons beyond our control, the availability of animals from pounds has been markedly reduced," began a letter from the university's Animal Resources Center, referring to a new Massachusetts law against pound seizures. "This has necessitated the Animal Resources Center to obtain many dogs and cats from other sources, mainly dealers."[130] The letter to Barger came with an attached flyer that depicted a happy beagle. Large blue text announced the jovial dog as "The Marshall Beagle," a product of Marshall Farms.

Promotional language on the opposite side of the flyer promised "Good news for your budget!"—a transparent ploy aimed at cost-averse researchers and administrators who believed that shifting to specially bred dogs would exhaust their already limited funds. As we have seen, that aversion had a long history: the Whitney family's beagle business ran aground on the Scylla of this parsimony half a century earlier, as did Brand's expansive vision for a centralized dog-production center. But Barger's receipt of a Marshall flyer in 1982 revealed that behind the budgetary storm clouds came a structural overhaul in the political economy of laboratory animal supply and provision. Disappearing were the days of cheap dogs from pounds and large-scale laboratory colonies such as those at Davis or Utah.[131] In their stead, new institutions had reared up to fill the gap in supply: large commercial kennels that bred dogs and other organisms specifically for scientific use.

The story of Marshall Farms, one of the world's largest retailers of scientific beagles, with facilities on multiple continents, offers insight into how systems of mass production that appeared unsuited to canines nonetheless integrated beagles. As one of the secretive actors operating in the background of the scientific, corporate, and regulatory transformations covered in the preceding sections, Marshall followed a trajectory that reveals how for-profit breeders benefitted from stricter post-thalidomide regulations, new animal care policies, and growing research interest in higher-quality test subjects. Situating Marshall's journey in light of these broader changes shows how corporate actors and their production of laboratory commodities are central to the history of research organisms during the second half of the twentieth century. That account returns us, once more, to upstate New York.

W. Gilman Marshall was born in 1917 near Rose, New York, a small town nestled below Lake Ontario, roughly halfway between Rochester and Syracuse and just over an hour north of Ithaca by speeding car. Marshall attended North Rose's public schools and excelled in public speaking, winning an honorable mention at the New York State Fair Young Farmers' oratory competition in 1935.[132] From North Rose, he matriculated to Syracuse University, where he married his lifelong partner, Ina Stevens, who had taught for years in a one-room schoolhouse in Butler, New York. After graduating in 1939, Gilman, a short, charismatic man, returned to North Rose with hopes to make a living at the town's canning factory.

On the side, he raised New Hampshire chickens and founded a small ferret business with Ina. Ferrets make for useful hunting companions, both because they will tail rabbits into burrows that dogs cannot reach and because they serve as effective alternatives to cats for rodent control. Marshall purchased his first in 1938 and began to mate the ferret with others, selling the progeny to local farmers. After local rat populations were cleared out, however, many simply returned the ferrets, and he soon found himself with more than could be pawned off on nearby hunters. Instead, Marshall, a thoughtful and persistent advertiser throughout his life, placed small notices in local newspapers and magazines. In 1941, he publicized ferrets that were "Gentle, Excellent Hunters" in Rochester's *Democrat and Chronicle*; a few years later he advertised "Sturdy, Healthy, Vigorous Hunters" in *Field and Stream*.[133] The market for hunting ferrets was limited, but Marshall's saving grace, from a business standpoint, was the unexpected discovery of strong interest from nearby pharmaceutical and veterinary products companies.

The ferret was first introduced as a formal experimental animal in the 1920s for studies of distemper. Laidlaw and Dunkin had recognized that the virus was effectively reproduced in the cheaply acquired mustelids and tended to concentrate in their spleens, allowing researchers to easily extract samples.[134] Soon, American law required tests of canine distemper treatments on ferrets. Marshall's first shipment, in 1940, went to the American Cyanamid Company's Lederle Laboratories, which produced a vaccine of the Laidlaw-Dunkin method. By selling ferrets to Lederle and others, Marshall started to earn a meaningful income to supplement that of his full-time job. Gradually, the business of breeding and selling nearly sixty-five hundred ferrets a year consumed his life.[135] Lederle's Norman Pyle might have been describing Marshall when he noted in 1940 that laboratory use of the ferret had "completely revolutionized" the ferret industry.[136]

To ensure steady profits, Marshall planned a two-part business model: selling ferrets to researchers and breeding a select number of minks for clothing manufacturers. He was hardly the first to seek a fortune from the small critters, with "ferret ranches" spread across the United States during the early twentieth century, but Marshall's focus on the scientific market and his eye for promotion paid off, generating steady growth for his operations.[137] He took out small notices in publications such as *Science*, trumpeting his ferrets as the "ideal animal for distemper and influenza work." Unlike other breeders, Marshall also claimed his ferrets were "reared especially for laboratory use."[138] The next year, Marshall's farm, a family business run with help from his wife, nephew, and teenage sons, gained a large concrete block structure to house the expanding ferret colony. By 1960, the family was shipping ferrets across the continental United States and as far away as Japan.[139] Later in the decade, the Gilman Marshall Ferret Ranch began to appear in scientific resource manuals as a commercial breeder capable of supplying corporations and laboratories, and Marshall published a small pamphlet, *Ferrets in the Laboratory*.[140] Today, Marshall remains the largest ferret breeder in the world, and pet shop ferrets are frequently traceable back to North Rose.[141]

In 1962, however, Marshall added a new animal to the mix: starting with two dogs, Gilman began to breed beagles. He may have feared that ferrets would not sustain business or have heard about the need for beagles from his customers at Lederle, but Marshall was responding, first and foremost, to the same shifts in toxicology testing standards that propelled beagles into the spotlight more broadly. He purchased his first dogs the year of the Kefauver-Harris Amendments and a few years ahead of the Animal Welfare Act of 1966. Initially reticent about using dogs for research, Marshall, a civically minded Methodist, concluded it "was justified because you've got to test drugs on something and it's better to use dogs than humans."[142] Early advertisements touting both ferrets and beagles appeared at the back of the National Society for Medical Research's journal *Bio-Medical Purview*, in 1963, where Marshall claimed to be the only laboratory retailer with AKC-registered beagles and emphasized his ability to safely ship dogs across long distances. To offload extra beagles, Marshall continued to post notices in Syracuse's *Post-Standard* and Rochester's *Democrat and Chronicle* throughout the mid-1960s, promising "peppy, slick friendly pups of all ages from champion bloodlines" to those who called Gilman Marshall.[143] In November 1964, the *Post-Standard* reported that Marshall had inked a deal with the National Cancer Institute to sell 120 beagles for nine thousand dollars, a costly $75 per dog compared to his public price of $25 to $50, as stock for drug toxicity screening, transplant surgery, and radiation studies.[144] A few years later, Marshall's dogs traveled down to Oscar Auerbach's Veterans Administration lab for smoking studies, as we will see in the next chapter.[145]

Prior to 1962, the largest laboratory beagle producer in the region was the RLDD, but Marshall Farms soon supplanted them in scale. In 1966, a photograph of the company's property appeared as part of an advertisement for Marshall's booth at the fiftieth annual meeting of the Federation of American Societies for Experimental Biology in Atlantic City, New Jersey. It revealed a modest farm set on sprawling lands with Quonset-style buildings for dogs and ferrets, possibly modeled on published accounts of the atomic kennels, a little away from the road.[146] Marshall's ambiguously named business appeared in the 1966 ILAR sourcing guide as the "Gilman Marshall Ferret & Beagle Ranch," but by the 1970 directory, it had a new, decidedly more official name: Marshall Research Animals, Inc. That year, supply company Affiliated Medical Enterprises of Princeton, New Jersey, specified in the ILAR report that its own beagles originated from stock of Marshall Farms. At the start of the decade, Marshall's was a name that could be trusted for scientific beagles and ferrets.

As he had with ferrets, Marshall was quick to see the value of branding his dogs as *scientific* commodities. "These dogs are better than those you would buy for pets or hunting dogs," he explained to a newspaper interviewer, emphasizing the importance of purebred, pedigreed dogs whose entire life history was available to researchers.[147] Marshall was not simply selling beagles to laboratories, as many had before, but breeding them for laboratories. By 1967, beagles were big enough business that Marshall Farms closed its colony to outside dogs and began breeding all animals internally. Where atomic labs had recouped costs by selling extra dogs, Marshall beagles represented scientific "breed wealth," both stock for future breeding and capitalized commodities, as Sarah Franklin referred to Dolly the cloned sheep.[148] The 1966 passage of the Animal Welfare Act gave the enterprise a further boost by pushing many researchers to invest in purebred research canines like Marshall's. Always primed for expansion, the farm also began selling scientific minipigs in 1969.

While it lost a competitive bid in 1975 to supply 350 beagle puppies to the US Army for the controversial chemical weapons tests, Marshall Research Animals sold 6,595 dogs to researchers across the United States in 1976, with sales totaling around $1 million.[149] That figure made Marshall the largest retailer of scientific beagles in the United States, with four principal competitors: White Eagle Farms in Pennsylvania, Ridglan

Farms in Wisconsin, Hazleton Laboratories in Virginia, and Laboratory Research Enterprises in Michigan.[150] All but Ridglan had joined the National Association of Life Science Industries, an advocacy group, by 1977. Collectively, the five companies produced more than twenty-five thousand beagles each year for research in the United States, with nearly 75 percent of Marshall beagles going to pharmaceutical companies. The numbers were expected to grow, as the Toxic Substances Control Act of 1976 required new tests of chemicals on live animals, and prices rose to match: the Army paid $80 each for four hundred beagles in 1973 but received bids up to $127.50 per dog in 1975.[151]

Marshall's advertising, in turn, grew grander. In 1978, a full-page spread in *Laboratory Animal Science* touted their beagles and ferrets but placed the dogs first, a clear sign of their shifting market priority. "When you order a Marshall Beagle," the ad explained, "you order a healthy, friendly dog with a quiet temperament who will readily adapt to the life in your laboratory."[152] Where Utah researchers once saw beagles as their governors, the promise was now about easy adaptability to each laboratory's culture and needs. The cheery personality that Andersen highlighted as a benefit also became an explicit component of the pitch of the "Marshall Beagle," framed as a commodity by its capital letters. Beagles, like other products, had use and exchange value, but they also possessed what Haraway calls "encounter value," that additional value that emerges from the relations between living organisms.[153] The company claimed that its beagles were carefully socialized from birth: "A handled dog is more trusting," explained Gary Marshall, Gilman's son.[154] A mass-production system further meant, according to advertisements, a "continuous supply" of dogs to ensure "more meaningful data from your laboratory testing procedures."[155] While earlier experimental dog supplies had been subject to seasonality and shortages, Marshall promised constant availability.

Whether there was truth to Marshall's claims of higher quality, or they simply made for good ad copy, the company clearly believed in its product. Similar ads from the early 1980s, this time featuring only beagles, recommended that those not using Marshall's dogs ask someone who already was. "Our customers are our best salesmen," it claimed.[156] In 1985, flyers just like the one Barger received lauded the "Marshall Beagle" as a "suitable and most affordable alternative to mongrels in pharmacological and surgical studies." Positioned against the unreliability and expense of earlier dogs while refusing distinctions between breeds that were more useful for some projects than others, "Marshall Beagles" were universally applicable and available to even penny-pinching researchers.[157]

Shipping ferrets to Japan in the early days of business was a sign that Marshall would not circumscribe its customer base to the borders of the United States. From 1980 into the first decade of the 2000s, that principally meant sending dogs by air from Marshall's extensive New York breeding facilities. By 1990, Gilman's original small barns had become thirty buildings scattered across 105 acres with nearly twelve thousand beagles. Marshall's ad campaign that year claimed the company was "building a better beagle," touting a supposedly high-tech computer system that tracked each dog's genetic background.[158] "Although it's true that Marshall Farms didn't build the first beagle," the ad conceded, "they have definitely improved upon the original model."[159] The campaign made explicit what had been implicit before: lab beagles were living, useful technologies. All was not flat and simple in the world market, however: in 1992, CEO Gary Marshall was taken to court after a shipment at John F. Kennedy International Airport, destined for toxicology researchers in Switzerland, was stopped by inspectors over fears that the cages might collapse midflight.[160] Marshall avoided paying a $220,000 fine by agreeing to host a training for all of his employees in the safe international shipment of dogs.[161]

Switzerland was only one of their many destinations: Marshall beagles were traveling to nearly every continent for toxicological and other research by the early 1990s. The beagles were "a global standard—with the papers to prove it," announced Marshall in 1998.[162] The next August, just before the dawn of a new millennium, the "Marshall Beagle" became a trademarked entity, and the multinational company became Marshall BioResources.[163] Both moves seemed to cement an ontological shift in the beagle's status as biocapital. Marshall was no longer a farm raising animals, but a factory producing products defended by intellectual property claims. Where previous researchers had worried about national variations in the beagle, Marshall now promised universal consistency and compatibility from a single, trustworthy product that would look and behave the same everywhere.

Under family management since its founding, Marshall is smaller than well-known animal suppliers such as Charles River Laboratories, but in 2015 it shared an estimated 45 percent of the scientific beagle market with multinational contract research organization Covance, the former Hazleton Laboratories.[164] Despite being targeted by animal rights activists, the company has carefully avoided media attention, resulting in a quiet yet significant international expansion over the past three decades. Its first international branch opened in 1994 in Lyon, France, operating under the ironic corporate name Utopia, after an initial defeat of efforts aiming at Montbeaugny.[165] Marshall's Lyon facility was followed, around

2002, by the acquisition of a breeder in northern Italy known as Green Hill, and a joint venture in China—Beijing Marshall Biotechnology—that began to compete with established beagle breeders in Asia. In 2009, Marshall started operating a facility in Tsukuba, Japan, and purchased a British competitor, B&K Universal, to access its facilities and expertise in England. Marshall and B&K opened a controversial beagle and ferret breeding facility in Grimston, near Hull in northeast England, in 2013.

The movement to expand physically into new countries was likely a response to tightening restrictions on air transportation of mammals destined for laboratory research, which initially aided Marshall's expansion when they first hampered air shipments of primates from India. The controversy at JFK was only one instance of multiple decades of friction, including a decision by Air Canada to stop Marshall's shipments in 2007, following protests from animal rights groups that culminated in bans from most major airlines in 2012.[166] The last two decades have been a fractious time for the international supply of laboratory beagles, with Marshall often behind the headlines. Beijing Marshall Biotechnology, a relatively silent partner, came to the fore in 2012 when a shipment of beagles destined for Indian contract researcher Advinus was caught on arrival in Chennai by activists. Marshall had sidestepped Cathay Pacific's policy against transporting lab mammals by listing them on flight logs as "pets."[167] In 2016, Marshall opted to sell off the Green Hill breeding facility, one of the larger dog suppliers for drug research in Europe, after investigations led to charges against Green Hill management for violating strict Italian animal testing laws.[168] Despite setbacks, however, Marshall's American facility held an estimated twenty-two thousand beagles in June 2018, close to the entire market size for the dogs in 1976.[169] While recent analysis suggests that 1970 was the highwater point for the American laboratory animal industry generally, with a nearly 40 percent decline in animals supplied estimated between 1968 and 1978, beagles have appeared partially resistant to the larger trends.[170] Nearly fifty thousand were produced in the United States annually from 1985 to 2015.

The story of Marshall Farms, one of humble origins and major global expansion, exemplifies the transformations in the political economy of laboratory beagles. Many breeders, like Marshall, began as small-scale dealers of animals that could be bred cheaply in rural environs. But as testing requirements on pharmaceuticals and toxic chemicals increasingly demanded beagles, the dogs became vital commodities in intricate chains of transnational science and capital. Expanding regulatory apparatuses were central in producing momentum for the market of purebred laboratory dogs, as nations synchronized their drug safety expectations. "No wonder so many multinational pharmaceuticals have selected the

Marshall Beagle to harmonize their research," read Marshall's 1998 ad, highlighting the globalizing power of the beagle. As pharmaceutical biocapitalism grew, the sites of production for dogs shifted from the small farms of yore or internal laboratory colonies to massive facilities, largely indistinguishable from other industrial factory farms. Once catering to a largely northeastern market, Marshall BioResources extends across borders and supplies researchers on nearly every continent with standardized and specifically bred dogs . . . unless the airlines ask, in which case they are still potential pets converted into research subjects.

Organs without Body in Pharmacokinetics

The arrival of beagles into pharmacology was meant to solve a basic problem: ascertaining the safety and efficacy of a drug or chemical. But a parallel conceptual challenge had plagued the field for even longer, ever since its explosive development in the nineteenth century: how does a drug move through a body? The problem received one of its first explicit responses in the 1930s, when Swedish researcher Torsten Teorell, seeking a mathematical model for the distribution of substances administered to the body, offered a new approach he called "kinetics." Over time, the discipline that grew out of Teorell's initial work, "pharmacokinetics," morphed into an advanced, computer-centric approach to pharmacology, generating new dilemmas for animal testing of pharmaceuticals and, eventually, to the computerization of the beagle.

On May 9, 1933, at twenty-eight years old, Erik Torsten Augustinus Teorell defended his medical thesis in physiology before the faculty of the Karolinska Institute, the prestigious biomedical research hub near Stockholm famous for its role in selecting recipients of the Nobel Prize in Physiology or Medicine. Teorell's thesis was, by his own acknowledgment, a predictably physiological approach to analyzing the mechanism of hydrochloric acid secretion in the stomach. According to a friend, the inspiration was autobiographical: a painful gastric ulcer.[171] Readers responded favorably, granting the work a "laudable" rating, but Teorell was dissatisfied with his findings.[172]

What he lacked was engagement with physical chemistry, which Teorell considered essential for studying secretion problems in the body. Typical Swedish physiological training at the time covered the subject only glancingly, so there were few local avenues to acquire what was missing. Instead, in 1934, Teorell won a fellowship from the Rockefeller Foundation to study with two renowned physiologists, John Osterhout of the Rockefeller Institute and Harvard Medical School's Walter B. Cannon.[173] Teorell, known as "TT" to his friends, made a favorable impression, with Osterhout noting that he was "1 of the best men" to ever work in his laboratory. During Teorell's stay, he met a number of leading American researchers, including immunologist Michael Heidelberger of Columbia, who would win the Lasker Award in 1953.[174]

In October 1935, following a one-month extension to finish a study of electrolyte diffusion, Teorell's time in America came to a close. Rockefeller's support, however, did not. One month after taking a new position in the Department of Physiological Chemistry at Uppsala University, Teorell received an additional research grant of $1,400 (around $32,000 today), partly under the assumption that he would eventually take over his department at Uppsala.[175] A year later, Rockefeller provided additional funding, this time to finance Teorell's study at Cambridge University with chemist Eric Rideal, a breakthrough four-month visit in which Teorell "learned more in 3 mos." than in all his time with Osterhout.

In the summer of 1937, just before his thirty-second birthday and application for promotion, Teorell published a two-part article, "Kinetics of Distribution of Substances Administered to the Body," in the storied *Archives internationales de pharmacodynamie et de thérapie*. It was an expansion of work begun at Rockefeller, where Teorell became fascinated with the possibility of formulating a "quantitative" theory of the permeability of membranes.[176] Researchers had studied how drugs affect the human body, he noted, but they typically emphasized dosages or methods of administration rather than "*kinetics*, i.e. the time relations of the drug action."[177] If "general mathematical relations" for drug action could be attained, Teorell hypothesized that pharmacologists would soon be able "to find out which is the best way to dose and to administer a drug, or even how to *change* the drug" in order to produce a desired effect.[178]

Because Teorell's aim was to describe kinetics in the body generally, he did not emphasize specific avenues of administration: showing how long, for instance, a certain amount of alcohol might linger in the blood. Instead, he invited researchers to imagine the movement of any substance, through any means of administration, as a series of linear steps. A diagram, or "scheme," in his article depicted an injection, but it was intended to illustrate processes at a more abstract level: any drug in the body might move from the subcutaneous fat "depot" to the blood, from the blood to the tissues or kidney, and then finally be inactivated in the tissues.[179] In Teorell's quantitative vision, the body partially disappeared, replaced with a simplified blood circulation hippodrome mediated by tissue boundaries; the body as compartments. In a modest concluding sentence, Teorell noted, "These and other conclusions may have bearings upon practical pharmacology and therapeutics."[180]

The intervention was groundbreaking, although not strictly unprecedented. German pharmacologist Walther Gehlen had developed a "mathematical treatment" of intravenously administered drugs "as a function of time" a few years earlier.[181] Teorell appeared unaware of it; Gehlen's writings had made only a small splash, and at first Teorell's seemed doomed to the same fate. Fewer than ten articles in major journals cited his kinetics work between 1937 and 1950. One of the exceptions came from Italian pharmacologist Emilio Beccari, who commended Teorell for arriving at conclusions akin to his own, yet few had heard of Beccari's research either, not least because he wrote in Italian.[182]

Even worse for Teorell, the kinetics papers would count against his application for promotion at Uppsala because they seemed unrelated to his position.[183] In later years, he would call the kinetics writings his "youthful iniquity." Nevertheless, in October 1940, a year after World War II overtook Europe, Rockefeller's bet paid off, and Teorell became a full professor. Despite Sweden's neutrality in the conflict, the war limited Teorell's potential impact further, as fighting redirected scholarly focus and decimated academic communities across the Continent. Rockefeller, whose internationalizing view of scientific cooperation was shattered by the conflict, could not support Teorell again until 1943.[184] After his unfortunate foray into drug kinetics, Teorell turned to other topics in the years that followed, but the trail was not lost for good.

In 1953, Friedrich Hartmut Dost, a pediatrics professor at Humboldt University in Berlin and director of the Charité children's clinic, published his first extended monograph, *Der Blutspiegel* (Blood level).[185] The book grew out of a series of short papers that Dost, a former member of the National Socialist German Doctors' League, had written in the years after his time as a prisoner of war in World War II.[186] "Die Clearance," for instance, had analyzed how pharmaceuticals left the blood.[187] *Der Blutspiegel*, however, was more expansive than the earlier articles. In it, Dost noted that most clinical tinkering with drug dosages relied on isolated blood samples, which told practitioners "little or nothing" of use because they were not predictive of how a substance would move through the blood. Such predictions were what clinicians needed most: how much of a substance would stay in the blood minutes or hours after administration? Dost thus offered a general theory of the changing level, or concentration, of a given substance in the blood over time, what he called a "blood level curve."[188]

Previous approaches to the issue had failed to coalesce into a "guiding" principle or general law, Dost argued.[189] It was a recognizable sentiment for readers of Teorell, and unsurprisingly, Dost was among them. He cited the kinetics paper throughout *Der Blutspiegel* and credited

Teorell at multiple points for "first" discovering important equations, despite waving his kinetics off partly as "mathematical formalism."[190] Dost instead situated his book in a different academic lineage, connecting himself to another Swede, Erik Widmark, whose research on blood alcohol concentrations Dost claimed as the progenitor for his own theory of blood levels.[191] *Der Blutspiegel* generalized what Widmark had shown for blood alcohol into a broader theory of the metabolism of substances. Dost's introduction thus emphasized the significance of the new "blood level curves," displayed in easily understood graphs and formulas. These mathematical expressions were quantitative, he acknowledged, but Dost also thought they formed part of a deeper, "holistic" methodology that studied the organism as a complete entity, persisting and transforming over time.[192] It was a methodology geared to his eminently practical orientation, allowing a doctor to improve clinical dosing, especially for children.

Although *Der Blutspiegel* is recognized as the text that coined "pharmacokinetics," the term itself appears only halfway through the book. On page 244, Dost explains that elimination of most substances follows a simple exponential pattern, while alcohol is eliminated linearly, the only exception in "all of pharmacokinetics."[193] Betraying no sense of having contributed anything novel with the term, Dost used it sparingly, and "pharmacokinetics" appears in neither the table of contents nor the book's index. Years later, during a discussion group, Dost would inquire with those present where the term had first been used and expressed shock when the answer was his own book.[194] "I suppose it is only rarely that a new expression is created in so modest a way," mused one colleague.[195]

In October 1962, international researchers, including a large cohort from Germany, Switzerland, and the United States, met at the Borstel Research Institute, northeast of Hamburg, to discuss "pharmacokinetics and drug dosage." The meeting was the first international event devoted to the new concept and the latest of Borstel's annual colloquia, covering topics of "acute" interest in medicine, which organizers understood to be united by "methods, principles and standards" with science more broadly. Borstel colloquia were opportunities to "remove the border lines" between such areas of work, and pharmacokinetics, which pharmacologist John Wagner later described as inherently "multinational and multidisciplinary," fit the bill exactly.[196]

The colloquium revealed heterogeneity in methods and style, with papers roughly split between analyses of the pharmacokinetics of two major drug classes: chemotherapeutics and antibiotics. Because participants of an early panel discussion "could not reach agreement" on standard terminology, the resulting edited collection included a bilingual English-German table of pharmacokinetic models and symbols, which featured the models of both Dost and Teorell (the latter could not attend) as well as "suggested" symbols for future equations.[197] Of general agreement, however, was the necessity of imagining the spaces through which drugs move—such as the stomach, intestine, or intracellular fluid—as "compartments."

One of the most significant insights from early pharmacokinetic studies can appear obvious in retrospect: the effect of a drug results not just from its molecular structure, but also from a host of other factors including the manufacturing process, the dosage form, and the amount available to absorption sites. It mattered, to take a contemporaneous example, not just that someone took erythromycin, an antibiotic first isolated from soil in the Philippines in 1952, but also how much of it and how the drug entered their body. This recognition pushed pharmacological research in new directions, and the older term "pharmacodynamics" was gradually redefined in relation to pharmacokinetics: the former referred to a body's response to a drug, the latter to a drug's movement through the body.[198]

Wagner argued in his presentation at Borstel that pharmacokinetics had produced a threshold change in the discipline. Researchers were leaving "classical" pharmacology and entering a brave new world, which he called "biopharmaceutics," the study of the correspondence between "the physical and chemical properties of the drug and its dosage forms."[199] In an article published before the colloquium, Wagner proclaimed this approach "the future of pharmacy."[200] Pharmacokinetics and biopharmaceutics, the latter occasionally described as the former's "daughter," offered the possibility of uniting "biology with physics, and especially with mathematics," promising new dialogue across scientific disciplines about how drugs affect the body.[201] And while dreams of a "mathematics" or "physics" of life stretched back decades, for example to physicist Erwin Schrödinger's influential 1943 lectures, published as *What Is Life?*, pharmacokinetics seemed to put that integration within reach.[202]

Wagner was a historically minded scientist, who read widely in academic history of science and chronicled the history of pharmacokinetics throughout his long career. He was fond of citing scientist and former Harvard president James B. Conant's lecture *On Understanding Science* as offering a basic truth relevant to pharmacology: concepts drive scientific progress.[203] For Wagner, this meant that far too much pharmacological effort was spent on fruitless experimentation when "an hour spent in meditating on what one is going to do is worth more than a year in a laboratory."[204] Of the critical concepts that emerged from the field's intellectual meditations, he considered one of the most important that of imagining the sites of drug distribution and elimination as a series of separate "compartments."

Wagner credited Teorell with being the first compartmentalist, but many, including Italian researcher Giorgio Segre, assigned that responsibility to Gehlen.[205] Neither actually used "compartment" or a comparable term, however. Gehlen's paper from 1933 referred to "volumes" of blood, while Teorell spoke of "distribution" as a series of processes across multiple "boundaries."[206] It is in Teorell's early diagram that the "compartment" vision is most apparent: there the circulatory system exists as a container with paths of ingress and egress. As David Kaiser argues of the dispersal of Richard Feynman's diagrams, which served as both pedagogical and conceptual tools, Teorell's drawn "schema" worked alongside his equations to allow pharmacokineticists to visualize, however abstractly, the minute worlds they studied theoretically.[207] Such diagrams, Stefan Helmreich notes, make complex phenomena amenable to reading and rewriting.[208] "I will use my time to try to do a little bit of drawing," Teorell announced at the end of a later conference, "because without drawings and models we cannot think in this assembly!"[209]

Pharmacokinetics, Dost summarized in 1969, draws its data from observations of concentration curves "inside the compartments of the total organism."[210] In the field's early years, it was the compartment of the circulatory system that took center stage, but already at Borstel, the set of compartments available for analysis was expanding: muscles, organs, even cells.[211] If the varied efforts in pharmacokinetics shared something essential, Dost claimed, it was that the field "thinks" in terms of mathematical models about "the total organism and its compartments." While pharmacokinetics was born in multiple places, it matured through the generalization of the "compartment theory." And before it could think in compartments, pharmacokinetics had to draw in them.

One challenge in pinpointing an origin for the compartment theory, as Segre later noted, is that the concept was used extensively despite "few attempts to define it rigorously." The compartment was an abstract space not just in its theoretical application, with "no physiological counterpart," but also in its formalization.[212] The compartment was almost universal after Borstel, but few paused to consider what assumptions existed within its closed confines. Those who tried to rigorously define a compartment also ran into an eminently practical obstacle: the compartmental vision was an idealization, the "academic euphemism for compromise," according to mathematical biologist Richard Bellman.[213]

A compartment could be as large as the circulatory system or as small as a cell, everything and nothing. Compartmental thinking, in this sense, had been present in epidemiological modeling since at least 1927, when William Ogilvy Kermack and Anderson Gray McKendrick offered a mathematical approach to understanding the movement of epidemics, work Teorell and others appeared unaware of.[214]

Compartment thinking had also appeared in the world of radiobiology. Radioactive isotopes became widely available to biologists in the 1950s as part of an American effort to sell "peaceful" atomic energy to the rest of the world.[215] Radioactive "tracers," useful for a variety of projects in both basic biology and experimental medicine, relied on mathematical models that used compartmental thinking to conceptualize movement, or kinetics, inside biological systems. In an influential article in the *Journal of Applied Physics* in 1948, Charles Sheppard, a Vanderbilt biochemist who moved to Oak Ridge National Laboratory, offered a basic approach to analyzing movement in a theoretical system of *n* compartments.[216] The same year, University of Colorado economist David Hawkins proposed studying economies in the same way: as consisting of "*n* compartments."[217] Scientists at Oak Ridge and other national laboratories such as Brookhaven took up the study of tracer kinetics, analyzing transfers and exchanges between cells and plasma. In a second widely cited paper, Sheppard and a colleague divided paraffin-lined flasks into "compartments" to trace exchanges of potassium between cells and blood plasma.[218] "It is of considerable interest," they wrote, "that human cells and plasma in vitro behave very much like an ideal two-compartment system."[219] Here, multiple kinds of compartment converged: the material, compartmentalized flask; the theoretical compartment; and the vision of cells and plasma as metaphorical compartments. This imbrication of literal and metaphorical compartments, popularized by tracer studies, reappeared in pharmacokinetics.

By mid-1950, while references to compartments and tracers were common in scientific writing, usage was still uneven. When pharmacologist Aldo Rescigno of the Tumor Center in Busto Arsizio, Italy, published equations in 1956 for the movement of tracers in the body, he called the abstract spaces of their motion "components."[220] Segre, who collaborated with Rescigno on an influential textbook about pharmacokinetics and tracers, folded this work into his narrative of the compartment theory's lineage.[221] In this way, disciplinarily and geographically disconnected studies with tracers were drawn into the stories researchers in pharmacokinetics told about how the discipline thought in terms of compartments.

Dost's *Der Blutspiegel* was released in a second edition in 1968, this time under the name *Grundlagen der Pharmakokinetik* (Foundations of pharmacokinetics). "The success of this book," he wrote in the new foreword, "has shown that the need for such an account was present and has not subsided."[222] Students, doctors, and colleagues had begged for a second edition over the years, Dost noted, and the title shift revealed impressive consolidation of the field between editions. Pharmacokinetics, the "word building" that Dost introduced into scientific terminology, was common parlance all over the world by its release.[223] Ruminating on his own impact a year later, Dost reflected that pharmacokinetics had grown "from a curiosity into an applied science" with the "liveliest interest" from pharmaceutical companies.[224]

At Borstel, indeed, most of the authors present were either directly employed by or receiving funding from the pharmaceutical industry. As Teorell had prophesized, pharmacokinetic approaches suggested powerful changes to drug development, making them potentially more effective and cheaper. "Quite often biological screens in animals and man involve only a 'one-point determination' with respect to time," explained Wagner in 1962.[225] Pharmacokinetics was supposed to resolve this, yet Wagner also urged caution: "I would like to emphasize that we still need to study *each drug* and if necessary make specific *in vivo–in vitro* correlations. We are not at the point where we can measure a series of physical constants and make accurate predictions about absorption of a new drug *in vivo*. Let us hope in the future this may be the case."[226] One challenge for researchers, in other words, was that the revolutionary compartment view had not obviated experimental trials; compartments still needed bodies. In the meantime, pharmacokinetic models became ever more complex. By the Schering Workshop on Pharmacokinetics, which followed Borstel in 1969, pharmacologist Edward Garrett could claim that the quantifications in pharmacokinetics were "based on the assumptions that the body, although a complex of many organs, tissues and processes, can be conceived of as a multicompartmental organism."[227] The "general operating rule" in the field, he wrote, was to "to postulate the minimum number of compartments consistent with physiological reality."[228] This preference for minimum-*n* compartment systems also carried its own dangers. Often, Garrett wrote, "parsimony leads to the denial of reality."

Denial of reality was more than possible; it was widely acknowledged. As Teorell pointed out at Schering, the "single force" equations that many researchers had used since his 1937 papers worked, to some extent, but risked dangerously simplifying complicated physiological processes. Treating each compartment as tending toward equilibrium, they ignored "*other* driving forces" involved in drug transport through the body.[229] Research in chemotherapy was also revealing the importance of drug concentrations for the efficacy of treatments. The heavy doses in chemotherapeutic treatment required alternative formulas to account for differential speeds of movement and elimination.

The theoretical movement toward nonlinear, multicompartmental models paralleled the disappearance of physiological imagery from pharmacokinetic papers. Teorell's quasi-representational one-compartment sketch from 1937 seems vitally alive in contrast to the empty boxes that dominated work by 1969. Images of actually existing beings dropped away in favor of formulas, charts, and analogical diagrams. One reason was that the fictive simplification of the body into compartments supported simulations run on analog and digital computers, connecting physiological studies with work in mathematics and physics. The clear mathematical emphasis in pharmacokinetics even made the field an ideal candidate for "early entrance" into Norbert Weiner's cybernetics, Dost thought.[230] The compartment-centered computerization of pharmacokinetics also mirrored concurrent changes in nuclear physics, where simulations of atomic detonations increasingly took the place of direct experimentation with the bomb. As Peter Galison and Joseph Masco have shown, simulation through powerful statistical techniques was central to a postwar realignment of physics and computing.[231] Pharmacokinetics benefited from a similar, yet differently directed "trading zone" between mathematicians, physicists, and biologists.[232]

Analog Dogs, Computerized Kennels

Multicompartment models not only benefitted from computers but seemed to require them. Models were "cumbersome" to resolve otherwise, Charles Sheppard noted.[233] Dost explained in his second edition that "the algorithmic aspect" of pharmacokinetics had taken over the field in recent years because models were "especially [easy] with analog computers," devices that utilized the variation of physical signals to model equations.[234] In the early 1960s, one prominent example of analog computing united the burgeoning field of pharmacokinetics with the growing centrality of the beagle as pharmacological test subject.

From February to March 1961, American newspapers were graced by news from United Press International of a revolutionary invention from the Kalamazoo, Michigan, Upjohn Company. Researchers at Upjohn had devised an "electronic dog," a computer that showed "exactly how medicines work in humans," which could shorten the time required for evaluating new drugs.[235] On reflection, however, the electronic dog was somewhat strange: if the goal was understanding the role of drugs in people, why not design an electronic human? The report was also vague on specifics about the dog and its operation, beyond noting that the dog was an "unusual" computer.

Upjohn's canine simulacrum had been introduced to the scientific world one year earlier as part of a paper on the pharmacokinetics of the antibiotic psicofuranine by Edward Garrett, Richard Thomas, Donald Wallach, and Clayton Alway, members of the company's research laboratories. Thomas and Wallach had studied the drug on dogs using traditional pharmacological methods, while Garrett theorized a way of modeling that data in an analog computer. Alway, later the company's head of research services, constructed the machine itself from a modified Heath Analog Computer, one of the more popular consumer devices of the period, which used variable voltages to represent physical properties. In the "electronic dog," broken up into a series of organ-compartments, voltage represented dosage and allowed the computer to read out curves of drug amount in relation to time. Their dog, argued Upjohn researchers, enabled quantification of processes that were difficult to approach with analytical mathematics.[236]

Did the device work? Garrett presented his dog at the first Borstel colloquium, but the animal remained abstract for many, its utility unclear. One attendee joked, "Dr. Garrett told us about his 'electronic dog'—but only on the blackboard" (Garrett also emphasized it was a "she-dog").[237] After leaving Upjohn to take a job at the University of Florida's School of Pharmacology, Garrett would continue work on his electronic dog for years. Subsequent papers on a modified version even revealed that the device was advanced enough to have a breed: the analog dog was a beagle, the species used for Upjohn's safety testing. Garrett and colleagues later reconfigured the earlier electronic beagle to model the movement of calcium-47, a radioisotope tracer, simulating what some AEC researchers had done to the dogs in reality. "The analog computer is an extraordinary tool for the elucidation of the mechanisms and rates of transformations and compartmental exchanges of drugs in the various organs of the body," one of their papers concluded.[238] At the Schering Workshop in 1969, researchers from the pharmaceutical division of Schering AG lauded Garrett for giving "a new approach to pharmacokinetics by creating the 'electronic dog.'"[239]

The development of computerized dogs had contradictory implications for the expanded regulatory role of living test beagles. On the one hand, Garrett claimed in a profile for the *Pensacola News-Journal* that his electronic beagle might replace living animals while providing better data.[240] On the other hand, computerized and real beagles appeared capable of peaceful coexistence. As Garrett and pharmacologist Howard Lambert explained in 1966, "In an era where both public opinion and federal legislation demand increased proof of drug efficacy and drug safety, the 'art' of drug dosage and formulation design is being supplanted by the sophistications of advanced science and technology," especially computers.[241] As with higher quality animals, computerization represented a way of assuring the public of the safety of Upjohn's products. Yet because the electronic dog was based on studies of multiple beagles, not "pharmacological data gathered from a single dog," it also supported the universalization and abstraction of beagles as consistent and self-same entities.[242] Not many different beagles, but one uniform kind; not multiple computerized beagles, but one electronic dog. This was the Beagle, in theorist Jean Baudrillard's terms, as "third order" simulacrum, superseding reality.[243] Garrett's accomplishment paralleled Marshall's in seeking to realize Andersen's dream of a perfectly standard experimental beagle.

One reason that Garrett built an electronic dog, rather than an electronic man, was the widespread assumption of a dog's simplicity in contrast to seemingly more complicated human bodies. And yet, this abstract beagle remained tied to the breathing, physiological beagle. Newspaper reporting on Garrett's device emphasized that the electronic beagle did not eliminate the necessity of in vivo studies, because "the machine's answers are double-checked . . . in more tests with live animals."[244] Trying new ideas on electronic dogs was, instead, "a prelude to the time-honored and very useful method of trying it on the dog."[245] In this way, although the device appeared to be one of many "alternatives" to animal experimentation, its utility was premised on the continuation of beagle experimentation. The electronic dog complemented real, living dogs, and diagrams of the device even recapitulated the aesthetic form of a beagle. A comment below one image noted that "the 'dog' actually resembles a weird drawing of the real animal," a beagle of lines and triangles rather than flesh and fur.[246] One did not, at the very least, have to sacrifice the computer after its use for studies.

Garrett's electronic dog was a sign of the times, and one of many bred during the period. The multitalented Pennsylvania State University psychologist Howard S. Hoffman, for instance, built his own analog computers to simulate the stimulus response of Pavlov's dogs for students. But the devices also indicated how quickly the times were changing. Analog computers would have a limited lifespan in pharmacokinetics, and by the Schering Workshop in 1969, the devices were already on their way out. Presenters instead extolled the benefits and power of "digital" computers, which had become cheaper and more user-friendly, machines that would drive the next major stage of pharmacokinetic research.[247]

The change was already underway in 1967, when Wagner, also employed by Upjohn, published an article explaining the two principal functions of computers for future pharmacokinetic studies. Analog, digital, or hybrid computers allowed "rapid numerical analysis of data," adding, subtracting, multiplying, and dividing with speeds rarely achievable by hand calculation. Digital computers could also be employed for multiple pharmacokinetic simulations, the "physical embodiment of a mathematical model," which Wagner believed might even recursively drive the development of new mathematical models.[248] From model to computer, as with the analog beagle, researchers could now move from computer to other models.[249] As fond of predicting the future of his field as he was of chronicling its past, Wagner believed the future was digital.

At first, others disagreed: pharmaceutical researchers Horst Röpke and Jürgen Riemann still claimed in 1969 that "an analog computer is much more suitable than a digital computer" for many pharmacokinetic problems. Analog computers, they argued, were "easier to make and change," faster and cheaper.[250] Some of this resistance was undoubtedly connected to accessibility and familiarity: in 1969, computing resources for research were still concentrated in defense laboratories and academic university departments with programs in physics, meteorology, and engineering, leaving limited access to powerful digital computers for many biologists.[251] But at Schering, Richard Bellman argued that the ability of digital computers to store answers for future use in extended sequences of operations, "the way one picks a piece of wood out of a woodbin," had "revolutionized the scientific world." One electronic dog was no longer necessary; digital computers could host and store entire kennels of simulated animals. Bellman, who played a central role in introducing dynamic programming a decade earlier, had become interested in developing these techniques as part of a broader "computational" biology.[252] Yet the first inquiry from his audience, clearly divided about the device's revolutionary capabilities, was for "a very simple example" of feeding a pharmacokinetic problem into a digital computer.[253] "The agreement between a curve and experiments may be very fallacious," Teorell retorted to Bellman. The ultimate question was simple: "Does it mean anything?"[254]

The short answer was yes. The mathematical models and computer simulations produced after 1953, when Dost introduced "pharmacokinetics" to the world, fundamentally transformed much of the basic work in pharmacology and pharmaceutical development. As the editors of journal *Chemotherapy* put it, reflecting on the enduring influence of Dost's *Der Blutspiegel* in 1976: "When clinical pharmacologists or collaborators of a drug authority would be asked to indicate the discipline which influenced most during the last two decades the basic philosophy and the daily practical work in their fields of activity most of them would answer without hesitation: 'Pharmacokinetics and its daughter Biopharmaceutics.'"[255] A trio of contemporary pharmacologists divided the history of pharmacokinetics into periods from 1900 to 1960, 1960 to 1980, and 1980 to 2000, arguing that the emergence of computing techniques from 1960 onward produced an explosion of changes.[256] Summarizing these developments in an article for the *Wall Street Journal* in 1983, reporter Ronald Alsop noted, "Increasingly, scientists . . . are using computers as readily as test tubes and microscopes to search for the next wonder chemical or drug."[257]

But the longer answer was that curves and models could do only so much to direct research or determine safety. The electronic beagle, which Teorell applauded for its "public relations" work for pharmacokinetics, could not displace actual beagles. Pharmacokinetics, like pharmacology generally, was ultimately concerned with living entities.[258] The denial of reality and the generative capacities of computers had limits: pharmacokinetic simulations had to be checked, as many emphasized at Schering, by the grittier work of testing drugs on nonhuman beings. Researchers could not divorce themselves from the continuing centrality of sacrifice studies on living creatures. Beagles remained vital, but a new way of thinking about time would emerge to make sense of the increasingly obvious divergences in the rhythms of drugs across the animal kingdom.

Biological Relativity, Pharmacokinetic Time

One challenge faced by experimental pharmacologists in the twentieth century was the obvious difference in how quickly drugs moved through animals of different sizes, a phenomenon sometimes referred to as "interspecies variance." Studies with nonhuman animals frequently revealed an accelerated elimination of drugs, leading to exaggerated predictions of a treatment's clearance and potentially over- or understatement of its effect and effectiveness. In 1982, University of Connecticut pharmacologist Harold Boxenbaum argued that pharmacology needed not just more compartments and better formulas, but a novel temporal framework. Pharmacology had to finally undergo its own, biological variant of the Einsteinian relativity revolution, beginning to think in terms of physiological, "pharmacokinetic time."[259] The speed of a drug was not universal, but relative to the size of the bodies it moved within.[260]

Boxenbaum was not the first to propose a theory along these lines. In 1931, Alexis Carrel, as famous for his vascular suturing techniques as he was for a fervent belief in the possibility of life's indefinite extension, published a short article in *Science*, "Physiological Time." According to Carrel, it should be obvious to any serious observer that our psychological experience of events in sequences of time based in hours, days, and years in no way exhausts the complexity of physical time. "The value of a year is not identical for short-lived and long-lived animals," he noted.[261] What Carrel called "physiological time" was more uneven than physical time, since it could slow down or accelerate depending on the animal and its aging. It was biologists rather than physicists, Carrel suggested, who should have first discovered the linkage between space and time made famous by Einstein.[262]

Less philosophically than Carrel, Boxenbaum defined "physiological time" broadly construed as "a species dependent unit of chronological time required to complete a species independent physiological event." The notion of pharmacokinetic time was one application of this broader concept.[263] To explain, he drew on an at-hand example: the difference in aging between a dog and a human being. A dog, with an expected lifespan of fourteen years (beagle longevity was now a species-wide assumption), ages at roughly 7.14 percent of its life each year. A human being, on the other hand, ages at 7.14 percent every seven years. One year and seven years are therefore equivalent "physiological times." Similar calculations could be made for the clearance of a drug by any number of different species, factoring in their relative age. Thus, even though Boxenbaum's reflections on the concept occasionally veered in unexpectedly mystical directions, including a history of temporal thinking from Aristotle to Bergson and a peculiar list of time variants in widespread use (government time, Greenwich Mean Time, private time, etc.), "pharmacokinetic" time was intended as part of the practical solution to interspecies variation. Chronological times could be normalized to a "suitable internal barometer or clock": for drugs moving through the bodies of smaller animals, chronological time could be decelerated to match the movement of drugs through larger bodies.[264]

Peter Galison has shown how Einstein's thinking about the relativity of time emerged in part from his experience of coordinated public clocks.[265] Boxenbaum, much later, similarly turned to clocks to make sense of pharmacokinetic time. The natural world, Boxenbaum argued, was shaped by intermingling clocks, from the "systems whose phases are reset or 'entrained' by an external synchronizer," such as sleep-wake cycles, to those that regulate physiological functioning independently from environmental forces, like heart rate.[266] One could understand interspecies scaling, he posited, by imagining that a "small pendulum clock is scaled up to produce a larger one 64 times its size."[267] Against the abstract, reductionist approach of early compartment theories, which distanced themselves from the complexity of living bodies because researchers were "most at ease when the thing they are studying is no longer alive," Boxenbaum imagined reconnecting pharmacokinetics to fleshy, lively things. The goal was a "physiological pharmacokinetics," in contrast to the compartment-oriented "classical pharmacokinetics."[268]

That distinction was somewhat ironic, since Teorell's and Beccari's insights emerged from physiological study, but beginning in the early 1980s, more and more work emphasized how to solve the problem of interspecies variance, developing the very in vivo–in vitro correlations that Wagner deemed essential back in 1961.[269] If drug effects were to be compared across species, using the concept of pharmacokinetic time, there would need to be a sustained study of how the small and larger pendulums were related, how interspecies scaling really worked in practice.

In an ideal world, explained Welsh biochemist Richard Tecwyn Williams in 1972, the testing of drugs, food additives, and other chemicals would rely on a single animal "in which the absorption, distribution, excretion, and the rate and pattern of metabolism of these compounds are similar to those of man." The problem was that such an animal "does not exist."[270] But Williams—a leader in studying how foreign substances were broken down in living bodies, what is sometimes called "xenobiotic" metabolism—argued that properly scaled findings from nonhuman animals could still be useful for human beings. Moving along a ladder of known organisms could enable effective translation of results, even acknowledging the uniqueness of each species. Yet the challenge, argued Robert Dedrick, a researcher in the Biomedical Engineering and Instrumentation Branch of the NIH, was akin to taking a useful chemical reaction produced once and converting it into a consistent product in large quantities. In such cases, it was insufficient to simply multiply the number of times the reaction is produced, because physical and chemical processes scaled in complicated ways.[271]

One could not, in other words, simply take a result in mice and multiply by a standard number, because "the relationship of the pharmacokinetics of any given drug between one species and another" could be "quite straightforward" or "rather obscure."[272] Theoretically, results from a toxicological study in the dog could "scale up" to predict a drug's course in the human body, but variables from pH to drug concentration were still insufficiently supported by experimental data to be adequately simulated by computers. Simulations were apt to produce predictions for the *average* person, rather than highly particular individuals, as a model based on average parameters could be no more precise than how "that individual corresponds to the averages."[273] What was needed was even more biochemical information.

Thus, in studies of interspecies scaling, compartments did not disappear. Rather, computerized models were reimagined to match specific compartments to available physical and chemical data on a particular organism's organs or pathways. Compartments were reconnected to the body (or multiple bodies), and effects were scaled depending on the multiple temporalities of experimental organisms. Organisms that were better described in physiological and histochemical terms, such as rodents and beagles, were therefore easier to develop effective pharmacokinetic models for. This helps explain why, despite serious challenges to the use of beagles in toxicological testing, they remained "almost universally used," as Alan Poole and George B. Leslie noted in their widely used *A Practical Approach to Toxicological Investigations* in 1989.[274] Beagles were an ideal intermediary subject for physiological pharmacokinetics. A host of studies followed and established the staircase from rodent to dog to nonhuman primate to human as the centerpiece of interspecies scaling.[275]

The dispersal of beagles throughout practical pharmacological work and the quiet revolution around pharmacokinetic time offered theoretical and laboratory foundations for opening up human bodies to a host of new pharmaceutical products. Rather than envisioning strict divisions between species when it came to drug use and design, interspecies scaling calculations and computerized models supported a universalizing vision of living bodies as pathways of drug movement, where beagles could stand in for human beings with enough supporting data. The abstraction of "the beagle" from many beagles through the electronic dog, interspecies scaling calculations, and its worldwide commodity form encouraged a view of continuity across each step in the species ladder. Even as researchers recognized the particularity and uniqueness of different creatures, interspecies scaling reinforced a general sense of escalating simplicity: a rat was simpler than a dog which was simpler than a human.

If the decades following the 1980s saw a dramatic expansion of the pharmaceutical industry's marketing of chronic medications, what Joseph Dumit calls a "mass health" model, it was interspecies scaling and a vast testing assemblage that served as critical infrastructure for confident extrapolations between early research and expensive clinical trials.[276] The drug-using body was a temporal as well as a scientific and commercial revolution. Yet this system still relied on the messy unpredictability of living beings, with differences between species and even individuals that researchers considered threats to smooth scaling. "I think people like you in pharmacokinetics may make life more difficult for the toxicologists in [the] future," remarked one researcher at the Schering Workshop in 1969. "You will tell them not to use the dog or rat, experimental subjects with which toxicologists are very familiar. You will tell them instead to use the baboon, marmoset monkey, or Quelt squirrel monkey—or the white elephant, or the alligator in Dr. Garrett's swimming pool!"[277] But rather than yielding a veritable pharmaceutical Noah's Ark of animal models, the advances in pharmacokinetics were effectively reintegrated into conventional pharmacology, sustaining the (beagle) dog's central place. Researchers continued to be told to use beagles, even as many questioned that imperative and new debates raged over just how similar a beagle and a human could possibly be. Some of those debates would come to an explicitly political head in controversies over the proof that cigarettes cause cancer.

Beagle, from the Dogs of the World series for Old Judge Cigarettes. Issued by Goodwin and Company. Commercial color lithograph. 1890. The Jefferson R. Burdick Collection. The Met.

CHAPTER 4

Hot and Smoking Dogs: Cigarettes, Emotions, and Ignorance

"I am writing to you as a representative of the beagle community," begins a letter addressed to Mrs. Jessica Brindle, the PR rep for an unnamed scientific device manufacturer. "We, the beagles, are fed up with ignorant, lazy people automatically associating us with smoking. I am writing to you in your capacity as the public relations officer of an organization that would also probably very much like to eradicate this association, not only because it is a cliché, but also because the world has moved on and I no longer know any beagles who smoke. We are, on the whole, more health-conscious these days."[1] The letter is signed "Maxwell." An open-mouthed beagle, described as "a happy-go-lucky breed with a wonderful disposition," is shown trotting through a field of daisies:. Maxwell, we assume. Appearing in Sally O'Reilly's *The Ambivalents* (2017), a novel inspired by an analysis of laboratory device catalogs, Maxwell's fictional missive raises important questions. How did beagles become so associated with smoking that they might need to complain about it? When did they quit? And why did they ever start smoking in the first place?

This chapter explores how beagles became, first, experimental smokers, and, in the health-conscious latter half of the twentieth century, smoking clichés. While historians have eloquently documented the history of American tobacco over the period that Allan Brandt calls the "cigarette century," this chapter zooms in on one particular nexus of research and the debates it generated: efforts between 1960 and 1975 to construct so-called laboratory *proof*, through studies of dogs, that cigarettes cause cancer.[2] The chapter focuses on work by pathologist Oscar Auerbach and his colleagues at the Veterans Administration Hospital in East Orange, New Jersey, to produce a "smoking beagle," considered by many the first proof of the causal link between cigarettes and cancer in a laboratory setting. Auerbach's research on cigarettes, including epidemiological studies in the late 1950s, served as crucial support for the 1964 Surgeon General's Report on Smoking and Health, the first explicit government acknowledgment of the link between smoking and a wide array of diseases.[3] Auerbach's beagle studies, publicized widely in 1970, were supposed to be even more significant, cementing the connection between smoking and cancer, but that was not to be.[4]

I show here how smoking dogs supported growing antitobacco momentum from the 1950s to the 1970s. Because dogs appeared to develop emphysema and cancer in carefully controlled environments, the studies were powerful rejoinders against industry finger-pointing about alternative cancer risks, such as air pollution, and were more compelling than earlier animal studies that involved painting tar onto the lungs of small creatures. Beagles, it seemed, got cancer just like humans did. But the chapter also shows how Big Tobacco waged an aggressive campaign to construct controversy around Auerbach, his methods, and his beagle studies. Here and in other cases, Robert Proctor notes, the tobacco industry used support for science "as a way to prevent certain kinds of questions from ever becoming 'closed.'"[5]

Such was the case for the beagle studies: industry insiders quickly acknowledged Auerbach's work as clear proof of cigarette carcinogenicity, and Helmut Wakeham, vice president of research and development at Philip Morris, reflected in a 1971 memorandum that "1970 might very properly be called the year of the beagle."[6] Yet the beagle studies failed to fully sway public opinion, in large part because the tobacco industry fought desperately to minimize their impact through intensive surveillance of Auerbach and his staff, interference with the publication of their results, and funding alternative research to keep the controversy open. It is a story that, despite its significance, remains minimally explored in otherwise exhaustive histories of the tobacco industry.[7] Connecting Auerbach's study to a competing project at the Pacific Northwest Laboratory, and building on the previous chapter, I show how beagles remained central to an expanding universe of experimental toxicology, just as dwindling funding pushed earlier beagle research teams into confrontation over the smoking question. Beagles became contested icons: evidence to public health advocates of the dangers of cigarettes, but also proof, according to Big Tobacco, that animal experiments could demonstrate anything—and, therefore, nothing.

In this context, beagle desires and emotions became unexpectedly entangled in debates over their experimental value. This was particularly true with Auerbach's research, because one central industry response to his studies focused on an aspect of the experimental setup that he could do little to change: a beagle dog, analysts claimed, did not smoke *like a human being.* Performatively, the point was relatively obvious, because thumbless beagles could neither hold their own cigarettes nor light up without assistance, but tobacco companies went further: beagles did not smoke like humans because they did not do so for *pleasure*. Lab dogs were *stressed* by being placed in cages and smoking contraptions, rather than puffing away after sex or a stressful day. Beagle studies did not, in other words, recreate the normal experience of smoking; they did not replicate the form of life of American smokers.

Such arguments were not new but represented continuations of a long campaign to portray smoking as an experience of pleasurable, carefree excess. Horace Kornegay, president of the Tobacco Institute, noted in 1971 that cigarette companies profited not from "poison," but "pleasure."[8] Or, as novelist Jonathan Miles later wrote in a review of Brandt's *Cigarette Century*, the "central, vexing paradox of smoking" is that "in return for death, cigarettes give pleasure."[9] Affects such as "pleasure" played key roles in long-raging debates about individual agency versus structural determination—whether smokers made poor *choices* or were *compelled* to do so—but in the 1960s, pleasure became an explicit weapon in epistemological debates around antitobacco research. In response, Auerbach and others described the highly technical assemblages that enabled beagles to smoke in inhalation studies as direct simulations of the human experience of smoking. With the help of surgical interventions and complicated apparatuses, scientists projected ideas about human smoking onto the dogs. With a little help, beagles started to relax and enjoy smoking, even as it killed them.

The chapter explores species projection, then, by revealing how newly constructed technical devices rendered human the decidedly atypical simulation of human smoking, allowing beagles to serve as both physiological and *affective* proxies. I argue that notions of what smoking is, what it looks and feels like, structured beagle experimentation and the controversies that it generated. Researchers not only imagined dogs to feel particular sensations in the presence of tobacco, but some also expressed a surprised fascination in witnessing the beagles' new addiction. In a twist on what Donna Haraway identified in Harry Harlow's infant primate attachment studies as "scientific sadism," this could be called an experimental *voyeurism*: pleasure that emerged from helping dogs to do something they were not supposed to do or enjoy at all.[10] I show how animal affects became chess pieces in the production of, and attempts to dispel, industrially produced ignorance and situate this analysis in the context of broader academic debates about the place of affect in historical studies of laboratory animals. While animal "resistance" has long been highlighted as a form of nonhuman agency, scholars have been more cautious about exploring "pleasure" in laboratory confinement, which I argue here is critical.

Finally, the chapter continues the story about transformations in beagle production and exchange in the United States: while researchers at Davis and Utah managed their own colonies, the utilization of beagle dogs in cigarette toxicity studies was made possible by an expanding laboratory animal industry, explored in the previous chapter, whose goal was efficiently produced, uniform, and widely available research commodities. The disconnect between labs' dogs and commercial breeders even called the beagle's neutrality as a surface for the production of scientific truth into question in the early 1970s, when researchers discovered a new disease in dogs from Auerbach's provider that seemed to mimic the lung cancer he identified in pathological samples. I show how these uncertainties generated important concerns about the universality and utility of the beagle, concerns that were grounded in the continuing lack of basic knowledge—in the *complexity* rather than the *simplicity*—of dogs.

The Beagle Smoked; the Beagle Got Lung Cancer?

On February 6, 1970, the front page of the *New York Times* carried stories about Richard Nixon's new urban policy, set forth during an address in Indianapolis, and described a new "Soviet Satellite Destroyer" believed to be orbiting the earth. Nestled below, at the page's bottom right corner, was a report from science writer Lawrence Altman on the results of a new study connecting cigarettes and lung cancer. "12 Dogs Develop Lung Cancer in Group of 86 Taught to Smoke," the headline noted of research by American Cancer Society (ACS) epidemiologist E. Cuyler Hammond and Veterans Administration (VA) pathologist Oscar Auerbach, presented in a special scientific meeting at New York's Waldorf-Astoria Hotel.[11] Standing before massive projections of tumorous lung tissue, Auerbach argued that the dangers of smoking were now undeniable.

Spread by the wire services of national papers, word of the beagle studies rocketed through dailies across the country. The Tobacco Institute (TI), an industry-supported organization that responded to most major reports of antismoking research, acknowledged being "interested" in the results. The TI would later launch a blitz offensive against the findings, including full-page ads in April attacking their experimental credibility. It was easy to understand the concern: Delos Smith's UPI account appeared in the *Columbia Missourian* under the headline "Scientists Prove Cigarette-Cancer Cause and Effect," while *The Washington Post* version ran in *The Arizona Republic* with an even more dramatic title: "Smoking dogs may be final cancer link."[12] As the title of Walter Sullivan's *New York Times* Week in Review contribution put it: "The Beagle Smoked; the Beagle Got Lung Cancer."[13]

This was not the first time that Auerbach and Hammond received the spotlight for research on smoking and cancer. Nor was it the first time that animal studies had supposedly shown a "causative" link between the two. As Proctor notes, similar research dated back to the beginning of the twentieth century. *Life* magazine had, for instance, devoted multiple pages to a 1953 animal study that seemed to prove the dangers of tobacco "beyond any doubt."[14] But in February 1970, Hammond and Auerbach combined unimpeachable scientific credentials with a strict, almost obsessive, commitment to careful experimentation. Their results appeared to be an existential threat to Big Tobacco, the death knell of the cigarette industry. Auerbach and Hammond had also employed a relatively novel and, to the general public, somewhat unexpected research subject: carefully raised and standardized experimental beagles. The path to this point was anything but obvious.

Oscar Auerbach was described by colleagues and friends as a warm, nearsighted workaholic. He often woke before five and worked until well after ten with a dedication that inspired widespread mythologization. It is said, for instance, that Auerbach once flew from New Jersey to California in the morning to give a talk and then returned on a red-eye the same night so that he could be in his lab the next morning.[15] Born on the first day of January 1905 in New York City, he boasted relatively modest origins. He graduated from New York University in 1925 and New York Medical School in 1929. Informed that training in Europe was his best chance at a research career, he spent 1931–32 as a visiting fellow at the University of Vienna, where he fell in love with pathology. He returned to practice at Seaview Hospital in Staten Island.

Though Auerbach spent most of his later years studying smoking and lung cancer, he specialized initially in tuberculosis. The "white plague"—as it was known when Seaview, "the finest hospital for the treatment of tuberculosis on the continent," first opened—remained a public health menace well into the 1930s.[16] Though mortality rates declined dramatically, from 194 in every one hundred thousand people in 1900, to 40 in every one hundred thousand in 1945, 63,000 people still died of tuberculosis in the United States in 1945 alone.[17] Auerbach became a leading voice in national efforts to eradicate the disease, serving on the Laboratory Subcommittee of the National Tuberculosis Association, which emphasized the development of streptomycin and other antimicrobial agents for tubercle bacilli.[18] Seaview would find a permanent place for itself in the history of tuberculosis when doctor

Edward Robitzek began treating terminally ill patients with isoniazid, an antibiotic that yielded remarkable recoveries, in 1951.[19] Three years later, the work of Auerbach and others was lauded for providing "new hope" in the fight to pharmaceutically control tuberculosis. By 1955, the disease that once ravaged America seemed curable, and many TB researchers moved on to other pulmonary conditions, notably lung cancer.[20]

Following two years in the Navy during World War II and stints at the Halloran and Brooklyn VA Hospitals, Auerbach was appointed chief of the Laboratory Service at the VA Hospital in East Orange, New Jersey, in 1952, a position that gave him flexibility to carry out a large, independent research program.[21] There, his attention gradually turned to cigarettes. Carefully analyzing the lungs of hundreds of tuberculosis patients, he noticed the peculiar frequency of hyperplasia, an enlargement of the lungs that often signals an early stage of cancer, in those belonging to smokers.[22] Inspired by a case of chromium exposure that led to lung cancer, Auerbach began to look specifically for the preliminary stages of the disease in smokers who died of other causes. "If cigarette smoking were a factor," he later explained to a government panel, "we should be able to see these same changes in individuals who die of causes other than lung cancer, and we should see them proportional to the amount of cigarettes which they smoke."[23]

In the mid-1950s, Auerbach teamed up with E. Cuyler Hammond to explore whether a clear correlation could be shown between smoking and cancer.[24] Hammond had collaborated with fellow ACS statistician Daniel Horn on a study of 187,766 white men aged fifty to sixty-nine, living in both rural and urban areas, that connected cigarettes with both cancer and cardiovascular disease in 1955.[25] The link between smoking and premature mortality was proven, they argued, "beyond a reasonable doubt," supporting earlier epidemiological studies by A. Bradford Doll and Richard Hill in England and Ernst Wynder and Evarts Graham of Washington University in St. Louis.[26] Collectively, the studies represented a relatively novel effort to use systematic epidemiology to determine causality in a noninfectious disease.[27] Experimental epidemiology was both "the instrument by which the hazard" of smoking "was proven and the offspring of its proof," in Proctor's words.[28] Yet while this approach built on historical attempts to identify the causes of infectious disease, such as John Snow's groundbreaking nineteenth-century study of the spread of cholera, statistical methods remained difficult to communicate to the lay public. Despite extensive work to shield findings from accusations of "correlation," in contrast to "proof," the tobacco industry reiterated as often as anyone would listen that the connection between smoking and cancer was "merely statistical."[29]

Hammond and Auerbach's collaboration was explicitly designed to rebut those arguments with laboratory confirmation. Auerbach first analyzed nearly twenty thousand lung tissue samples in his laboratory in East Orange, looking for cancer without being able to identify whether samples belonged to light, heavy, or nonsmokers. His results were then checked by Arthur Purdy Stout, an esteemed pathologist from Columbia University, before being sent to Hammond, who alone could match samples with records of individual smokers. The blind study design was intended to counter industry claims that cancer findings came from interested researchers, and this careful approach was emphasized when the data was presented to the public, including in testimony during influential congressional hearings on cigarette advertising in 1957.[30] "Seldom in medical history had such precautions been taken to insure objectivity," claimed a Public Affairs Committee pamphlet about the research.[31]

The study's conclusion was clear: heavier smokers suffered markedly more than nonsmokers. "Cigarette smoking today is the single most important factor in the production of lung cancer in men," summarized Auerbach.[32] The famed New Orleans physician and antismoking advocate Alton Ochsner lauded Auerbach, saying that his was "the most conclusive evidence that has been presented up to the present time."[33] Partly in recognition of these contributions, Auerbach was named senior medical investigator at the VA on January 17, 1960. The early work with Hammond was soon supplemented by studies on pipe and cigar smoke as well as research on the benefits of smoking cessation, all of which earned further public attention. A newspaper cartoon based on his findings depicted a double-barrel shotgun made of cigarettes and lung cancer: "Pre-cancerous lung growths disintegrated and disappeared if a person stopped smoking," it explained.[34] Auerbach gave up smoking and became a staunch public advocate of smoking cessation, speaking at high schools and women's clubs with a clear message: stop as soon as you can.

Despite subsequent restrictions on cigarette advertising, tobacco appeared partially immune to warnings from Auerbach and others. While doctors quit smoking in droves between 1954 and 1960, a period the industry sometimes called "the Cigarette Scare," overall smoking per capita continued to rise in the United States and elsewhere to a peak around 1963.[35] Cigarette companies and their public relations firms held to an insistent mantra: while correlational evidence existed, no laboratory experiment had so far produced cancer in scientific organisms. For experimenters, a careful study in which lung cancer could be displayed inside an animal remained a kind of holy grail, capable of dispelling such protestations: the final piece of the puzzle.

Beginning in 1962, Auerbach's lab turned to dogs to find the missing piece and "definitively demonstrate the cause-and-effect relationship between smoking and lung cancer." The project, "Production of Carcinoma of Tracheobronchial Tree in Dogs," received a $60,000 grant from the American Cancer Society in 1964 to study the effects of cigarette smoke that had been pumped into canine lungs.[36] Intended to be multifaceted, the study had one initial aim to determine the strength of the connection between cigarette smoking and emphysema, recognized as one of the top killers in the United States, with a death rate increasing sixfold between 1953 and 1963.[37] Cigarettes, Auerbach argued in public appearances, were creating a nation of "lung cripples."[38] His animal research was meant to stop that.

Only One Ridiculous Mouse

Oscar Auerbach was not the first to turn to canines in the experimental fight against cigarettes. In November 1955, researcher Edward Rockey began carefully applying tobacco tar to the bronchial mucosa of dogs at the New York University–Bellevue Medical Center. Initial funding came from the Northwoods Sanatorium, a tuberculosis treatment facility in Saranac Lake, New York, where one of his coauthors, Edgar Mayer, was a consulting physician.[39] A native of Hungary who retained his distinctive accent, Rockey moved to New York following service as a MASH medic in Korea and a post at the Murphy Army Hospital in Waltham, Massachusetts, where he developed expertise in thoracic surgery.[40] In New York, he hoped to add "experimental support" of his own to the "impressive evidence" offered by epidemiological studies such as Auerbach and Hammond's.[41]

The ideal animal experiment, Rockey noted, "would be one that most closely simulated human exposure" by means of "some unique smoking device."[42] This was not, of course, the only possible approach: numerous researchers would instead place animals inside sealed containers that forced them to inhale smoke over time. A team in Florida, for instance, attempted to induce emphysema in retired racing greyhounds by putting them in wooden inhalation chambers that mimicked their racetrack starting boxes.[43] But, in an era prior to widespread concern about environmental exposure, Rockey and many others felt that the perfect experiment would simulate the individual's experience of smoking a cigarette. A dog need not necessarily hold its own smokes, but a hypothetical experimental setup should mimic that activity to some degree. The problem was that no such device existed.

Wynder, Graham, and Adele Croninger of Washington University in St. Louis had received considerable attention in 1953 for painting tar onto the lungs of mice. Their research generated pronounced anxiety within the tobacco industry. But tar-painting experiments were also criticized as a far cry from the "authentic" experience of human smoking.[44] Without an ideal smoking device, Rockey thought that the larger lungs of dogs might at least allow for application of tar directly onto the bronchial mucosa, the connective tissues in the body's airways, to produce an effect closer to what happened in a human smoker's lungs. To do so, he drew on a new technique developed with Mayer, starting in 1954, called tracheal fenestration, which involved creating a surgical opening in the trachea with small, "door-like skin valves."[45] Mayer and colleague Israel Rappaport originally envisioned it as an improvement on conventional tracheostomy, which left a painful hole in the patient's neck, but Rockey saw other possibilities as they practiced the technique on dogs. "The fenestrated dogs can easily be catheterized or bronchoscoped through this opening," which "makes them excellent subjects for experimental studies," he explained.[46] When eight painted dogs showed symptoms of incipient cancer, Rockey and his collaborators felt they had found something significant. But because tracheal fenestration was itself relatively novel, they could not rule it out as a factor in the pulmonary damage.

Another challenge also reared its head: continued funding for the research was in short supply. The grant from Northwoods was enough to cover operating expenses until September 1956, but a longer study would require another source of support. Rockey turned to the Tobacco Industry Research Committee (TIRC), formed in 1953 by the major American tobacco companies with a supposedly objective pledge to assist research in tobacco and health, as that potential sponsor, requesting $7,375 to keep the project afloat until March 1957. The extra time would allow him to move the dogs to new facilities at the New York Medical College and keep them alive longer before their sacrifice to await further pathological changes. With additional funding, Rockey explained to Robert Hockett, TIRC associate scientific director, "we can comfortably make plans for a more adequate broadening of the study."[47] Hockett thought highly of Rockey, "who has done nearly all the work single-handed," and his tracheal fenestration technique, which Hockett described as "several cuts above" comparable research. But he cautioned that Rockey "has been influenced considerably, I should judge, by Auerbach's work" in a letter to the TIRC Scientific Advisory Board, now chaired by a recently retired C. C. Little.[48] "He is, however, a conscientious and careful experimenter," Hockett added, in an apparent slight at Auerbach.[49]

Hockett encouraged Rockey's project but cautioned against broadening it "into new channels."[50] In one early edition of Rockey's proposal, the claim that one of the "ultimate goals" was to "produce cancer of the lung" is crossed out and replaced with "observe whether" cancer of the lung "will be produced."[51] The switch to passive voice fit the TIRC's modus operandi: funding a diverse array of research that might serve as an alibi for public-spirited investigation in legal cases and as a deflection from cause-and-effect language about cigarettes and cancer. Rockey's final proposal included the altered phrasing, a compromise that he may have deemed worthwhile for saving the dogs and ten months of work, but the TIRC denied his proposal in August 1956 nonetheless, citing several other ongoing studies "of a similar character" whose "course of development [needed to] be ascertained more fully."[52] That September, Rockey informed Hockett of additional support from Northwoods to temporarily keep the dogs and the study alive.[53]

Two years later, Rockey found firmer support, expanding the study to "a more adequate scale" thanks to a grant of $62,333 from the American Cancer Society, which emphasized the value of his "painless" technique.[54] From an original eight dogs, 101 dogs divided into two groups now received tracheal fenestrations and either "smoke condensate" or a control "rubbing manipulation."[55] Although earlier publications included minimal justification for choosing dogs, this time Rockey and coauthors noted that dogs "were selected primarily because [they are] not prone to develop spontaneous tumors of the lung," a claim supported not by the ongoing atomic studies, which Rockey appeared unaware of, but by research from the Royal Veterinary College in London.[56] Dogs were thus framed as blank slates for research purposes and more reliable experimental technologies than rodents. In 1958, Robert DuPuis, vice president for research at Philip Morris, urged Paul Kotin, an industry-friendly pathologist at the University of Southern California, to consider Rockey's new technique, "which would certainly be more meaningful than mouse-skin experiments."[57]

In the summer of 1958, Rockey's first detailed results appeared in the journal *Cancer*.[58] Hockett wrote to request reprints of the article, encouraging Rockey to investigate "synergism" between tobacco smoke and "powerful chemical carcinogens," because Hockett was "still among those skeptical whether tobacco smoke has any significant direct carcinogenic action on human tissues."[59] The suggestion seemed a transparent attempt to redirect Rockey's research focus, and it went unheeded. The next year, Leonard Zahn of industry PR firm Hill and Knowlton received word from freelance journalist Fred Hodgson that Rockey had produced "squamous carcinoma" in the lungs of his dogs and would submit the results to a major journal within months.[60] Tobacco's nightmare appeared imminent.

After a delay, the research finally appeared in 1962. Rockey's team reported on two years of comprehensive analysis that showed that a small number of dogs receiving the smoke condensate experienced "proliferative changes of the bronchial mucosa, including carcinoma in situ and invasive carcinoma."[61] Smoke condensate produced cancer, in other words, appearing to confirm what Auerbach and Hammond had shown in their epidemiological work in humans. Rockey was confident enough in his findings to proclaim at the annual American Medical Association (AMA) meeting that year in Chicago that cigarette advertising "through any medium" should be banned entirely.[62] The dog studies added, according to Rockey and his collaborators, "still another link" in a growing pile of "overwhelming" evidence.[63]

As usual, the skeptics queued up. Carl Seltzer, an independent researcher who leveraged a marginal affiliation with Harvard's Peabody Museum and later Harvard's School of Public Health to support the tobacco line on cancer, wrote Hockett in January 1963 offering his initial criticisms and soliciting other opinions.[64] A few weeks later, Hockett reached out to Israel Rappaport, one of Rockey's original collaborators and a dedicated, decades-long skeptic who served as a confidential consultant to the industry. Rappaport was quick to indicate his distaste for both Rockey and the fenestration technique and argued that the animal experiments were poorly structured. "Reading the conclusions one is indeed amazed at the meagre results," he wrote. "Pariuntior montes et peperit ridiculus mus," Rappaport summarized, misquoting a famous line of Horace: "Mountains will go into labor, and a silly little mouse will be born."[65]

Yet T. D. Day, head of the TIRC-funded research unit at Harrogate in the United Kingdom, took a very different view, noting the "valuable information" that condensate could produce "true bronchial cancer in an animal which is not usually susceptible to this disease." Cautioning that Rockey was reporting only on work in progress, he allowed that "there can be no doubt that figure 10 represents a true epidermoid cancer" and that "the technique of tracheal fenestration appears to be a very good one."[66] Rockey and colleagues continued studying the production of lung carcinomas in canines with smoke condensate, publishing an overview of their progress in December 1962 in the *Journal of the American Medical Association* (*JAMA*). From November 1955 to December 31, 1961, 325 dogs had served as either experimental or control animals.[67] Of the 182 dogs that were not eliminated from the study for various reasons, 127 were male and 55 were female. The emphasis on male dogs here and elsewhere implied a focus on the threat to male smokers.

Dogs, the authors reemphasized, made for good experimental animals because "spontaneous bronchogenic carcinoma is not prone to develop," a point attributed this time to pathologist R. M. Mulligan of the University of Colorado School of Medicine, which demonstrated the shifting nature of canine expertise.[68] This gave "added weight" to the findings of the study, because carcinoma in situ and invasive carcinoma "were produced only in the dogs treated with cigarette smoke condensate."[69] Russell Weller, a doctor at the Memorial Hospital of Chester County in Pennsylvania, urged Hockett to take the article seriously, even as he criticized Rockey's experimental technique and his decision to use dogs "of unknown age, breed, and history."[70]

By the mid-1960s, dogs were increasingly dominant animals in tobacco studies, and the National Cancer Institute (NCI) affirmed them as an ideal species for research on cancer generally. Mulligan's work, for example, was produced with funding from the NCI. One year after Rockey's study appeared in *JAMA*, two researchers at the UCSF School of Medicine used nineteen "normal" dogs in a study of bronchial metaplasia to argue that the condition might be reversible.[71] That research, like Rockey's and later Auerbach's, received funding from the American Cancer Society. Concurrent attention from toxicologists and drug researchers, as shown in the previous chapter, supported the broader emphasis on dogs as the ideal animals with which to explore cancer risks. Some internal tobacco sources agreed, and a Council for Tobacco Research (CTR, the new name for the TIRC) memo in advance of a January 1969 meeting noted that dogs "will prove to be the best and most convenient large laboratory animal" for future studies, unexpectedly citing the past experiences of Stockard, Scott, and Fuller.[72] Day hesitated about using dogs at Harrogate, but largely because "necessary numbers" of them "might arouse anti-vivisection prejudices" in England.[73]

Despite his apparent successes, Rockey still desired what he had first dreamed of: a "simulation" of smoking in dogs. Drawing on grant funding from the US Public Health Service and the ACS, his team turned its attention to constructing a device that could expose the canine bronchial tree to cigarette smoke without the need for tar painting or smoke condensate.[74] At first, they used a small glass tube shaped like the letter Y that could be inserted, with a metal adapter and rubber tubing, into the opening created by tracheal fenestration. One side of the Y was covered with a paper valve while the other side held a lit cigarette so that a dog's inhalation brought in cigarette smoke and exhalations sent it out through the valve. On May 28, 1963, a dog "smoked" a cigarette for the first time under controlled conditions.

The smoking technology was nonetheless a work in progress: the metal adapter was eliminated; later, the Y shape was replaced by a straight glass tube holding the cigarette, which could be removed for the dogs to exhale smoke. To test out a nonfenestration approach, the team tried inserting a cigarette into one nostril, compressing the other, and then releasing the nostril for exhalation. By 1966, they had applied the "straight" tube and "nostril smoking" techniques on thirty-three dogs "with success."[75] Rockey's animals were not always particularly eager participants in the experiments, however: technicians had to grasp some by the snout to hold them still during smoking and manually compress the chests of others in order to produce deep inhalation. Yet for Rockey, perfecting the simulation was more important than an animal's participation or pleasure.

Toward the end of July 1964, Rockey demonstrated his techniques to two members of Oscar Auerbach's team, surgeon William Cahan and lead technician David Kirman.[76] Cahan had independently experimented with dogs as models of smoking while employed at the Sloan-Kettering Cancer Center in New York, but lacked funding to carry out extensive research and asked whether he could move his study to the East Orange VA in October 1962, where Auerbach would oversee future progress.[77] Rockey himself fell out of the spotlight of smoking research as funding for his experiments slowed and he was forced to conclude work entirely in 1967.[78] As his techniques were taken up by other researchers, he reluctantly turned his attention back to therapeutic applications for tracheal fenestration. In later years, he delighted in a small orchard on the terraces of his East Side penthouse, where dinner guests were treated to nectarines plucked straight from the tree.[79]

How to Make a Beagle Smoke, Voluntarily

By February 1963, Kirman, a VA research assistant who previously worked with animals at the Rockefeller Institute, had developed an improved approach to canine smoking designed to skirt criticisms of previous experiments as "unnatural, unrealistic, and alien to the actual human experience" of smoking.[80] The device needed to bypass a dog's nasal filtration, it was decided, since human smokers typically do not smoke through their noses. Building on Rockey's original approach, the apparatus used a surgical tracheostomy and specialized collar with a tube opening attached to the lab's "smoking machine." Composed of "two sets of belted pulleys and a drive shaft mounted in roller bearings," the smoking machine could be controlled by researchers via a foot pedal that pumped smoke, an optional step when dogs refused to inhale by themselves. Pump-induced smoking was contrasted with what researchers called "voluntary smoking."

A list of challenges from a 1964 progress report makes clear just how difficult it was to make the system work: "Initially there were problems concerning tracheostomy procedure, type of tracheostomy tube and type collar to retain it in situ, type connector from cigarette smoking machine to tracheostomy tube, restraint of animals during cigarette smoking procedure, and type cigarette smoking apparatus."[81] Originally made of silver, the tracheostomy tubes were later fabricated from plastic and then Teflon to avoid continuous fracturing and irritation to the animals. The tube question, Auerbach summarized, was "baffling." The dog collars that kept the tubes stable were originally constructed from two pieces of leather, with clamps securing them in place, but the clamps continually came undone and caused the collars to fall off. Kirman resolved this by producing a "ball and slot" device attached to thicker leather. The resulting collar offered particularly striking evidence of the ways in which quotidian technologies of dog ownership were reconfigured for experimental purposes. Then came the challenge of keeping the animals themselves stationary: an initial canvas wrapper, so tight that it encouraged resistance from the dogs, was replaced by a looser leather harness attached to a new smoking table. By giving the dogs additional room, Kirman found, they resisted less actively. The "major problems" in the early phases of the experiment, summed up Auerbach, "were mechanical in nature." The group anticipated sharing results in five years, but solving the technical issues was a prerequisite: "Before worthwhile progress could be made in actual smoking of animals, each challenging problem had to be met."

The problems, however, were not entirely mechanical. The device needed to not only assist dogs in receiving cigarette smoke, but also generate the conditions for "voluntary smoking." Kirman was happy to report that after a few months of smoking training, most beagles "learned to inhale voluntarily, much like their human counterparts."[82] Yet how would one know whether dogs "chose" to smoke when attached to a machine? The key appeared to be in analyzing their behavior. Did the dogs appear excited about connecting to the devices, as a human might be about smoking a first cigarette of the morning? Did they seem to "enjoy" it? While the phrase "voluntary smoking" appears innocuous, it invites a focus for analysis different from that taken in previous histories of experimental animals and technology, where animal desire and affect are often absent.[83] Traditionally, the central question has been that of animal pain, whether as a rhetorical point for organizations such as People for the Ethical Treatment of Animals or as a concern for producing reliable data among scientists.[84]

Historical studies of nonhuman organisms in science have been framed by a focus on "resistance," moments in laboratory life when creatures thwart procedures or refuse participation. If a scientific experiment fails because a nonhuman refused to play its expected role, scholars have taken this as a sign of activity and agency.[85] The approach traces back to Michel Callon's analysis of the scallops and fishermen of St. Brieuc Bay: one might not know precisely what scallops want (if they "want" at all), but their occasional obstinacy in the face of human plans is indexed as a sign of multispecies agency.[86] This negative theory of agency appears throughout early science studies texts, partly popularized by Bruno Latour's description of nonhuman "actants" as "entities that hesitate, quake, and induce perplexity."[87]

The negative theory makes for a convenient methodological approach, since disruptions are omnipresent even in researchers' own accounts. Irene Pepperberg's studies of the intelligence of African grey parrots, for instance, offer countless examples of picky birds making for tricky laboratory companions.[88] Refusals are important, historian Amanda Rees notes, because assignations of "agency" through resistance have allowed scholars to accord significance to marginalized actors who appear only glancingly in written records.[89] Opposing portrayals of animals as passive entities, such accounts often equate "agency" and "resistance," allowing quasi-ethical arguments to slip through the backdoor, even—perhaps particularly—when authors avoid frontal assaults against animal experimentation.[90] That animals might, instead, enjoy participating in research that wrongs them appears to be a dangerous acknowledgment, and pleasure is neither as obviously agential nor as heartwarming as resistance to anthropocentric projects is. Yet, as Justyna Włodarczyk notes, understanding the history of human-canine relations requires historicizing not just resistance, but also obedience and cooperation.[91]

Dogs did defy the smoking machines, and their inhalations were never entirely voluntary, despite assurances from researchers. Rockey, for his part, insisted that dogs disliked smoking as a rule, and attachment to the machine meant that canine smoking could only ever be a technical feat. But the beagles' participation cannot be reduced to resistance. Instead, smoking dogs revealed an interweaving of pleasure and pain, passivity and activity. After all, pleasure coupled with resistance and pain was precisely what Auerbach *and* his scientific adversaries looked for in the beagles.[92] Locating stress within a study was a way of accusing experimentalists of failure, while finding voluntary smoking was proof that the experiment transcended its technical limitations. Devices like the smoking machine needed to be constructed, technically and representationally, as generators of particular experiences of smoking. Dogs had to be not only physiological proxies, but proxies of a performative and emotional experience. When Auerbach and his team looked at the dogs, they saw them as human analogues in different ways: as children, as first-time smokers, and as addicts. These forms of species projection were important, we will see, in the epistemological fight over the smoking inhalation studies.

The Rockey-Cahan-Kirman smoking machine was both a mechanical and an imaginative technology; yet a further aspect of the experimental setup that received less explicit comment was equally critical: it used a new kind of dog. Unlike the generic strays acquired by Rockey and others, Cahan and Kirman explained that the beagle breed had been selected for Auerbach's studies because "the species is relatively large and tractable" with "a bronchial epithelium and lung parenchyma closely resembling that of a human being." The beagles used in East Orange were one- or two-year old males from pedigreed stock, registered with the American Kennel Club and bred, the authors noted, "specifically for laboratory purposes."[93] This decision was clearly intended to respond to criticisms of Rockey's mixed-breed dogs: their "haphazard selection," Rappaport had suggested to Hockett, might have "effects on the poor results."[94] Where exactly the East Orange beagles came from was left unspoken in publications, save vague allusions to a "reputable" breeder in upstate New York, but as we will see toward the end of the chapter, the breeder turned out to be none other than Marshall Farms.

The first results from Auerbach's group, supported by Hammond and the American Cancer Society, arrived earlier than anticipated, in a paper and news conference at the annual meeting of the American Medical Association in Chicago in June 1966.[95] An Associated Press account appeared in newspapers across the country, announcing "Cigarette Smoking Gives Dogs Disease."[96] While Rockey's findings had garnered mild attention, interest was acutely piqued by Auerbach and Hammond's new study given the duo's name recognition. Emphysema, newspapers stressed, had been "produced experimentally" in cigarette-smoking dogs.[97] The Tobacco Institute was quick to respond that the experiment was "not at all comparable to human smoking conditions," although Auerbach asserted that his beagles loved listening to stereo music, particularly rock-and-roll, as they smoked.[98] What could be more authentic than that? A joke from the syndicated humor column Senator Soaper Says quipped in September, "Experiments show smoking is bad for dogs. Also, it is difficult to train them to use the ashtray."[99]

Hill and Knowlton, the firm that aided multiple big tobacco companies in developing a prosmoking, research-centric response to earlier animal studies, reported internally that coverage was "fairly extensive throughout the country," although thankfully little in magazines, radio, or television.[100] In one sign of Auerbach's wide impact, an editorial in the "Church News" section of the *Salt Lake City Tribune* praised success in teaching beagles to smoke as "further evidence of the divine inspiration which came to the Prophet Joseph Smith" when he said "tobacco is not for the body, neither for the belly, and is not good for man."[101]

Despite assurances from Hill and Knowlton, Big Tobacco was agitated. The combination of a novel research animal, a well-funded laboratory, and qualified researchers made the task of refutation imposing. A later, confidential Philip Morris memo revealed the threat Auerbach's emphysematous beagles represented, calling them "one of the most talked-about contributions to the etiology and pathology of pulmonary emphysema since 1963."[102] In turn, the industry carefully monitored Auerbach's activities, eagle-eyed for further updates on the project. Reports took a variety of forms: in July, Hiram Langston, a professor of surgery at the University of Illinois College of Medicine who later testified on behalf of the industry, reported to a tobacco-connected lawyer that he was unimpressed when he saw "our friend" Auerbach "stumble through" his "travesty" of a paper.[103] A confidential November memorandum from Hill and Knowlton contained detailed notes from two of Auerbach's public presentations, noting minor changes to the emphysema paper since its initial presentation.[104] Highlighting such inconsistencies, collected by an extensive surveillance network, was a crucial weapon against smoking researchers, which the industry repeatedly deployed against Auerbach.

One reason for the almost obsessive attention on Auerbach was his team's success in framing their technique. In dozens of national newspaper articles, Auerbach and colleagues emphasized how similar the behavior of experimental beagles was to typical smokers, relying on a rhetorical humanization of the beagle's experience of smoking. It was noted that "the dogs initially behaved like a child smoking his first cigarette": they salivated, coughed, teared up, and became nauseous. But after a short period, "the dogs showed signs of enjoying smoking—as evidenced by tail wagging and voluntarily jumping into the smoking box."[105] Beagles transitioned, over the course of the study, from children to adults, nonsmokers to experienced addicts. They became, Auerbach told the *Sunday Oregonian*, "addicted, just like human beings."[106] They ran toward the box when it was smoking time and were "short of breath" when playing, proportional to their cigarette usage. One dog showed signs of withdrawal after being removed from smoking machines because of an illness, while another would bite researchers if they distracted him from smoking.[107]

In this way, the smoking apparatus was analogized to well-known and recognizable (for the nearly half of Americans who smoked) human experiences with cigarettes. The emphasis on canine characteristics, such as tail wagging and jumping, also seemed to suggest, precisely because of the species differences, just how similar to human smoking the dogs' smoking was. Species projection and anthropomorphism were mainstream: Americans, trained to see themselves in dogs and dogs in themselves, could recognize their own forms of tail wagging. Newspaper cartoons inspired by the research supported this sense by repeatedly portraying dogs smoking in quasi-human ways: "Let me have a puff!" a beagle-like hound shouts, in a cartoon for the *Houston Post* by Bud Bentley, to a friend smoking happily in front of their doghouse. "I'd rather fight than switch!" barks an angry dog in front of a trashcan in a cartoon by Ben Wicks, referencing a popular Tareyton cigarette advertising campaign.[108] Dogs trained to smoke, or at least to performatively hold cigarettes, had been subjects of occasional popular bemusement long before Auerbach's research.[109] And in 1976, a widely published newspaper snippet about Tiny, a small terrier from Franklin, Indiana, even showed a dog doing what Auerbach's beagles could not: smoking directly from a lit cigarette. "Like some humans we know," the *Philadelphia Daily News* version noted, "Tiny begs and whines whenever he wants a smoke."[110]

The doggification of smoking worked against the industry's insistence that cigarettes were an individual's elevated choice. Philadelphia physician and antismoking researcher William Weiss, who visited Auerbach's lab in 1969, described the powerfully alienating experience of witnessing the smoking beagles:

> As we entered a ward, the dogs raised a great din. . . . A handler opened a cage door and a beagle jumped gleefully to the floor, trotted immediately to the cubicle, hopped into the chair and let the handler attach the tubing to a tracheostomy in his neck. The dog began to salivate in anticipation as the cigarette was lit. . . . We realized with something of a shock that this animal was enjoying the whole procedure. The scene was ridiculous. Here was a dog doing what most people think of as a sophisticated, peculiarly human accomplishment—smoking a cigarette. I had the feeling man was being degraded.[111]

To see dogs, creatures generally considered to be less subjectively motivated and cognitively complex, puffing happily away seemed to eviscerate the image of smoking as a voluntary, pleasurable act. In Auerbach's lab, on the contrary, it emerged as Pavlovian conditioning and learned helplessness, whether for dogs or for degraded men. Although the reasoning was less than explicit, that the experimental dogs were all male, like many of the researchers, appeared significant: the study suggested a fundamental critique of the imagined pleasure and agency involved with smoking by America's powerful and learned men.

When the emphysema study was finally published in January 1967 in *JAMA*, humanizing descriptions of the beagles were retained.[112] More than throwaway anecdotes, they were key elements in the lab's strategy to defend the validity of its experimental setup. Dogs not only got emphysema like humans at a pathological level, and Auerbach emphasized that the lungs were indistinguishable, but they also *became* smokers, experiencing and performing elements of that conformation of life. The technical setup, far from creating a stressful experience, appeared to replicate an eminently human one. It re-created smoking as a conformation and produced a seemingly higher truth than had prior research, even if one could nitpick differences in how smoke was delivered.

Such compelling animal tests had been the goal of antismoking researchers since long before Rockey. While the tobacco industry was thrown into disarray by the first tar-painting studies with mice, it had responded with an organized public relations campaign to refute any similar work.[113] Although specifics shifted to fit individual cases, the retort boiled down to a consistent assertion: the experiment in question did not replicate actual human smoking. Evidence needed to be human to be convincing. For that reason, when Auerbach testified before the Senate Committee on Commerce back in 1965 during a hearing on cigarette labeling, he relied on large images comparing the lungs of human smokers and nonsmokers. His testimony was among the first in congressional history to use images in this way, so impressing Chairman Warren Magnuson that the latter quipped toward the end of the proceedings, “We are reminded that one picture is worth a thousand words,” and requested that Auerbach leave the photos for later perusal.[114]

Yet committee members also worried that Auerbach's reliance on data from human studies stood at odds with prevailing cancer research methodologies, which emphasized the reproduction of cancers in experimental animals. “I know of no better experiment than the one that we have in the human being,” Auerbach retorted. “The cry cannot be raised that animals are not human beings.”[115] Senator Ross Bass of Tennessee nevertheless prodded him about the failure of animal experiments. “There is no test that shows that smoking produces cancer in animals,” Bass noted, implicitly referring to “Koch's postulates,” a set of four criteria for establishing the causal relationship between a disease and its pathogen offered by Robert Koch and Friedrich Loeffler in 1884, which included inducing disease in a laboratory organism.[116] Auerbach's response was revealing: "There has been no successful way up to now of getting a large animal that you can smoke long enough, comparable to the human being, so that you can develop the cancer. There are several experiments that are now underway. I think they may turn out to be the answer to your question."[117] Without fully tipping his hand, Auerbach was referring to the ongoing beagle studies. Despite his own insistence that experimental proof was unnecessary given a wealth of other evidence, Auerbach recognized the persuasive power that such a positive demonstration might have. As Richard Kluger put it, Auerbach and Hammond "knew full well the graphic value of successfully inducing lung cancer in animals with smoking as the only variable factor."[118]

Ultimately, the controversy concerned two forms of experimentation and two logics of persuasion. For Auerbach, human studies were ideal because they obviated the messiness of translating results between species. He largely agreed with the dozen or so New Jersey residents who wrote Senator Harrison Williams in 1967 to criticize the beagle studies for cruelly proving something that everyone already knew.[119] Yet for many like Bass, there could be nothing clearer than an animal in a controlled environment going from not having to having a particular condition. The repeated failure of animal experiments to meet this standard was, as Proctor has shown, a key element in the tobacco industry's battle against experimental epidemiology: "The companies put forward this argument in the 1950s to dismiss evidence of hazards; they put forth similar arguments today to exculpate their actions in legal retrospect."[120] The insistence on experimental demonstration was ultimately about a deeper connection between animal testing and truth in the public understanding of science. Like microscopes or beakers, animals were understood to be vital experimental tools. If the polio vaccine was tested in animals, and if other cancer researchers were using them, why had tobacco researchers failed to generate the final proof?

Letting Smoking Dogs Lie

As much interest as the beagle emphysema research generated, the results came from a relatively small study: ten smoking dogs and ten controls. Originally, the VA team saw it as a "preliminary" test to determine the maximum number of cigarettes that dogs could smoke before becoming ill. But after beagles appeared to tolerate a heavier smoking load than predicted, it was decided to analyze the results.[121] The initial findings were then followed up by histopathological analysis of changes in the bronchial tubes in December 1967 and electron microscopy analysis in December 1968.[122] The miraculous smoking beagles of New Jersey appeared poised to reveal further truths about the danger of cigarettes. When Auerbach found what seemed to be the first invasive lung tumor in one of his beagles, he went to Boston to show his microscopic slides to Shields Warren, now an emeritus professor at Harvard, who agreed that he was onto something profoundly important.[123]

Yet Auerbach was not the only investigator using beagles to study smoking. One month after his emphysema findings appeared in newspapers, Bill Bair, now head of the Inhalation Toxicology Program at the Pacific Northwest Laboratory (PNL), wrote to TI president George V. Allen to report his "concern" over reports that some of Auerbach's beagles had died while smoking.[124] In his own population of beagles smoking twenty cigarettes per day, Bair reported no evidence of marked health effects. The smoking technology at PNL, he went on, which used masks rather than tracheal openings, "more nearly simulates human cigarette smoking than Dr. Auerbach's." An expanded study focused on cigarettes was a logical extension of existing capabilities, so Bair ended the letter by wondering about "possible sponsors of such a research program."[125]

Supporting Big Tobacco was not the original reason for PNL's beagle studies. Beagles started smoking in Washington, instead, because America's nuclear planners wanted to know why uranium miners were dying. The US Public Health Service had announced in 1967 that miners were suffering alarming rates of lung cancer, roughly ten times higher than the general population.[126] An almost identical announcement had come in 1961, along with "controls" to "curb the rate" of disease—these evidently failed.[127] Although epidemiological evidence pointed to the role of uranium ore dust, most miners were also smokers, and researchers hoped to disentangle the effects of tobacco, uranium dust, and the radon daughters floating in the mines. Whether uranium inhalation caused cancer, and at what levels, would also be relevant for safety standards at future nuclear installations and for members of the Air Force transporting nuclear weapons. To study the issue, technicians fabricated masks, trained beagles to sit in glass boxes, and started lighting them up.[128]

The inhalation research emerged directly from and utilized many of the apparatuses designed for studies of plutonium inhalation in beagles, as well as even earlier PNL studies with rats. The rat experiments had been simple: old Coca-Cola bottles were converted into small inhalation chambers and a rubber stopper replaced the bottom of the bottles. Rats were "totally comfortable and could breathe easily," Bair recalled, but rodents generated major challenges for inhalation research.[129] Rats appeared to filter particulate matter differently than humans did, and some rodent strains showed high incidence of spontaneous neoplasms, which made drawing clear connections between inhalation and cancer difficult. These reasons pushed atomic researchers to study inhalation in dogs, just as they had compelled Hammond and Auerbach to choose beagles.

When Bair wrote to the Tobacco Institute, he explained, "Our limited study of cigarette smoking was specifically related to our Atomic Energy Commission–sponsored research on inhaled radioactive aerosols."[130] Yet the formerly munificent pockets of the AEC had partially dried up after 1964, when Hanford Laboratories, formerly administered by General Electric, was transferred to the management of Ohio-based contractor Battelle Memorial Institute. The change, eventually resulting in Hanford's renaming to the Pacific Northwest Laboratory, came amid government cutbacks in funding for the AEC's plutonium reactors and fears of economic stagnation in the region around Hanford.[131] After losing the major inhalation contract to Lovelace, Bair's team needed funding, and their beagle expertise made the inhalation toxicology group an appealing partner for tobacco industry actors concerned about the influence of Auerbach's research. Bair might also have heard about a large grant given by England's Tobacco Research Council a few years earlier to Battelle's German outpost in Frankfurt am Main to study methods for exposing animals to measurable quantities of smoke.[132]

Hill and Knowlton vice president Edward DeHart, whom Bair knew from his university days at Ohio Wesleyan, wrote a confidential letter to industry advisers on July 21, 1966 about the proposal, attaching prepublication research and requesting "any comments you may have on this matter."[133] Eight days later, a meeting of counsel in Washington, DC, singled out Bair's proposal as a "Possible Special Project." Hockett and Charles Kensler, an industry-friendly pharmacologist, were "impressed with the results to date and apparently think it desirable to consider a grant." An initial meeting between Bair, Hockett, DeHardt, and in-house council Alexander Holtzman was scheduled for August to discuss how the project might proceed. Auerbach's beagles had forced a 180 on dog research by the group that turned down Rockey's proposal years before.

An airline strike kept Bair from visiting the Council for Tobacco Research in New York, but by November his inhalation toxicology group submitted a proposal for a one-year study of dogs inhaling cigarette smoke.[134] The project would make use of PNL's radioactive aerosol setup, which taught beagles to sit in a small box and inhale the aerosols directly through a mask that pumped in cigarette smoke. The study would also attempt to outdo Auerbach's experiments on crucial grounds: by hewing even more closely, they claimed, to a direct simulation of "human smoking." And just as Auerbach's team had humanized the beagle's experience of smoking, so PNL technician Vanis Daniels remembered that the dogs "were just like humans," and would "chain smoke, if you allowed them to."[135] The difference with Auerbach's project, however, was clear: "Inhalation of cigarette smoke by a limited number of animals (12 dogs) using our technique to simulate human smoking is not expected to cause biological effects," a proposal noted.[136] The expectation was to find nothing.

In the mid-1960s, beagles were thus enrolled simultaneously in two contradictory studies with immense public significance, both with support from architects of the AEC's atomic dog research. Where one expanded on previous animal research with cigarettes, the other represented the alter-life of Cold War radiobiological studies on beagles. In the latter, beagles were both implicitly and explicitly "agnotological," in that their goal was to *produce ignorance*: the PNL studies would not only find "no biological effects" (a self-evident absurdity), but also generate confusion about the state of research. Beagles were useful research animals precisely because previous experience suggested that nothing would happen when the dogs inhaled tobacco smoke through the lab's specialized masks. "It is also expected that the results here may tend to refute some of Auerbach's work," explained David Hardy, a partner in law firm Shook, Hardy, Ottman, Mitchell, and Bacon, to Philip Grant of cigarette company Lorillard.[137]

Yet funding dog studies was still a risk, because even failed research might end up supporting Auerbach. Early internal praise of Rockey's fenestration technique had suggested exactly this. A memorandum from the American Tobacco Company in March 1966 thus noted, "Any work undertaken in an area relating to effects of smoking on the respiratory mucosa could well open the Company to charges of accepting Auerbach's premises."[138] Auerbach and Bair, as well as industry strategists, were technically in agreement that beagles represented a good experimental animal. Looking for the emphysematous changes that Auerbach had found might hazard unexpected successes. "It is not intended, nor expected," Hardy nevertheless affirmed, "that there is anything about the test that would permit complete extrapolation of results on dogs to human beings." The tests could not but fail.

What the industry desired was to maintain the appearance of a commitment to research while doing little of the sort. In the best case, the PNL project might disprove Auerbach's findings, but even in the worst case, it would demonstrate nothing. This, as scholars have shown, was one of the central tactics of the Tobacco Research Council and its allied scientists: funding research that kept the question open and that maintained the possibility of denial; studies, in other words, that generated doubt.[139]

The Campaign against Smoking Beagles

Despite the auspicious arrival of Bair's beagles, Auerbach remained acutely concerning to the tobacco industry. Informants for the CTR monitored his public presentations, noting that by late 1967 he was frequently mentioning "pre-cancerous conditions" in beagles.[140] Emphysema had been bad, but producing cancer, especially given the disease's symbolic significance, would be a disaster. A confidential memo from February 1967 revealed plans for a formal public campaign against Auerbach, including a suggestion that Hill and Knowlton's Len Zahn get "one of the tobacco journals" to have a staff correspondent write about Auerbach's many "contradictions and outright damn lies."[141] Zahn, Allan Brandt has shown, was lauded for his skill in planting such stories, which offered just enough scientific cover to magnify suspicion about findings without drawing unwanted attention.[142] The proposal, however, was met with caution amid concerns that it might descend into libel. Instead, the Tobacco Institute sent a special communication to *JAMA* in May 1967, commenting on a presentation of Auerbach's in Dallas and raising concerns about his methodology, partly in order to establish doubt before he submitted findings to the journal.[143]

The dreaded findings were finally announced on February 5, 1970, at the ACS press conference at the Waldorf-Astoria.[144] Hammond spoke first, emphasizing that a team effort had decisively shown, first, that filtered cigarettes were less dangerous than unfiltered cigarettes and, second, that cancer had been experimentally produced in the dogs. Auerbach went next, stressing technician David Kirman's "mechanical ingenuity" in creating the amazing smoking machine—rare public recognition for Kirman, who painstakingly tinkered with smoking devices and cared for the dogs.[145] The results, Auerbach continued, had been confirmed by esteemed scientists, including Warren and Ray Yesner, a Yale lung cancer specialist and chairman of the VA lung cancer chemotherapy research group. Auerbach then walked the assembled reporters through a series of slides covering the findings, zooming in on and detailing concerning images, much as he had done during his congressional testimony.

The cancer study was more experimentally robust than the emphysema work, with ninety-six experimental dogs compared to ten. The ACS press release, carried in hundreds of newspapers, did not mince words in extolling its importance: "For the first time, scientists have produced lung cancer in a significantly large experimental animal as a result of heavy cigarette smoking."[146] Chet Huntley reported on the results for *NBC Evening News* that night, and Walter Cronkite did the same for *CBS Evening News*, announcing the first "direct cause-effect link between cigarettes and [cancer in] higher animals made to inhale."[147] In this way, and in the voluminous press coverage that followed, Auerbach's smoking beagles made their way into the homes of everyday Americans. Harvard cardiac surgeon Dwight Harken, who would play a crucial role in assembling academic support for legal activist group Action on Smoking and Health (ASH), wrote Auerbach on February 10 with congratulations "on your genuine breakthrough in the search for cause and [effect] relationship between smoking and lung cancer."[148] Rockey, for his part, would remain bitter that few stories mentioned his earlier work.[149] In April, Steven Spencer summarized in *Reader's Digest*, "Never again can the tobacco industry hide behind its most-used defense—that no one has induced cancer in laboratory animals with cigarette smoke."[150]

Internally, many tobacco researchers agreed, acutely aware of what the results meant. Brian Davis, a researcher at the industry's Harrogate facility, wrote of Auerbach and Hammond's method in March that its only justification was the obvious one: "it works."[151] Davis did not "share the optimism" that further studies would disprove the findings, pointing out that the "invasive tumors" reported by Auerbach and Hammond "should be the end-point of a test for carcinogenicity." In April, V. D. Tughan, a researcher with British tobacco multinational Gallaher Limited, explained to his managing director that Auerbach's research was "undoubtedly a step forward," even compared to Battelle's sophisticated methods. Tughan continued: "We believe that the Auerbach work proves beyond reasonable doubt that fresh whole cigarette smoke is carcinogenic to dog lungs and therefore it is highly likely that it is carcinogenic to human lungs. . . . We have to bear in mind that the anatomy of the dog is relatively close to human anatomy and the type of tumour found in the dog was the same type as found in heavy smokers."[152] The beagle results were not even isolated findings, Tughan added, but appeared to confirm results from Walter Dontenwill in Germany and from researchers at Harrogate. "It therefore seems to us that it is more than coincidence," he concluded, "that experimental evidence is building up in this direction from several independent research organisations, each of which is of a very high calibre."[153]

Such admissions were part of a broader internal understanding about the cigarette's dangers that was barely reflected in corporate behavior or advertising. Yet responses to the beagle research made something else additionally clear. The industry's rhetorical commitment to the supposed value of animal experiments as a gold standard left it vulnerable to exactly the results that Auerbach and Hammond presented. After Rockey's findings were publicized, many industry-funded researchers had operated within a paradigm in which dogs, and specifically beagles, were high points of experimental practice. The anatomy of the dog was "relatively close" to human anatomy, as Tughan had noted. Research over multiple decades had placed dogs, and particularly beagles, as prime translational subjects, and Auerbach's cancerous beagles represented potential catastrophe precisely because they offered proof in the terms that the industry had long demanded: a viable, large experimental animal model smoking "just like" people did.

This was not, however, something that would be acknowledged publicly. Tobacco companies were already reeling from the successful efforts of a young lawyer, John Banzhaf, to apply the fairness doctrine to cigarette advertising in the mid-1960s and require balanced presentation of antismoking messaging.[154] Any further attacks on cigarettes seemed poised to spell doom for the industry as a whole: indeed, tobacco stocks fell across the board within a day of the Auerbach and Hammond announcement, with shares in Philip Morris down 2.5 points and shares in R. J. Reynolds dropping most precipitously, by 3.3 points.[155] To stem the bleeding, the Tobacco Institute issued an immediate statement on February 5. It was "naturally intensely interested" in Auerbach's results, even though such studies had mistreated dogs by forcing smoke "through holes cut in their throats." Such concern for the ethics of animal experimentation was, at minimum, hypocritical, given that the industry had asked for demonstrative animal experiments and was actively supporting a beagle-smoking study of its own. But the comments also reflect, in retrospect, an incipient movement among American corporate actors to draw on animal rights rhetoric in service of a critique of regulation. "It is impossible," the TI statement continued, "to draw a meaningful parallel between human smoking and dogs subjected to these most stressful laboratory conditions."[156] Human smokers did not smoke through holes in their necks or in strange laboratory conditions. They smoked of their own free will and for pleasure.

Initially, the stress argument appeared promising. Zahn informed William Kloepfer Jr., the TI's head of PR, of video that Auerbach had shown the November prior while presenting about his dogs. "The animals' eyes were bloodshot and bulging," Zahn emphasized, and they were "clearly under extreme stress." This was important, he continued, because it "belies the Hammond-Auerbach claim that the dogs learned to enjoy, and look forward to, the cigarettes."[157] In direct contrast to the framing of dogs as delighted human smokers, the industry recognized value in establishing imaginative distance between dogs and men. In the coming months, tobacco would oscillate in its reasons for why the beagles were not models of "real" smoking, but efforts to paint the dogs as hapless victims of violent surgical methods remained central. The dogs did not represent the form of life of smoking but, instead, species rejection: they were so different they could not possibly be compared with humans. Yet Auerbach saw the role of the beagles' "sacrifice" very differently: "People think nothing of sending a dog out at midnight to protect his master, and if killed, we practically bury him with medals and honors," he explained. "Here, in our experiments, we're sending dogs out to help the human race."[158]

Resistance to the findings could not focus solely on the research itself, however. It was the men themselves—and especially Auerbach, whose antismoking zeal was well known—that initially represented the most plausible targets. Thus, from an initially scattershot response, an all-encompassing program of denial slowly emerged, spanning multiple rhetorical modes and every available form of media. In hindsight, the industry's response represents one of the most astonishing and comprehensive efforts to resist and sabotage a single scientific study in history.

Because Hammond and Auerbach's findings had been discussed but not yet published, the first goal was understanding just what claims they were actually making. The Tobacco Institute's Fred Panzer wrote Kloepfer in February with "significant points" that he had "elicited" from Doug Gasner of *Time*, who was present for the ACS press conference. For one, Panzer questioned whether Auerbach's laboratory had been evaluated by the American Association for the Accreditation of Laboratory Animal Care.[159] Separately, Zahn wondered whether "some independent pathologist" could be found to write a letter to the *New York Times* criticizing the experiments. Whether they had located their man, or another had found them, within a few days Amherst entomologist Lincoln Brower wrote a letter to the editors of the *Times* describing Auerbach's work as a "misuse" of science for "vested interests."[160] Although Brower claimed no prosmoking bias, citing his lack of stock ownership, his letter would become a centerpiece in tobacco reports and presentations in the following months, evidence that a "professor of biology" rejected Auerbach's findings in an "unsolicited opinion."[161] As the industry's favorite excerpt argued, "All this experiment proves is: Whether or not smoking causes lung cancer in beagles is as inconclusive now as it was before the experiment was carried out."[162]

Although Brower's letter appeared disinterested, its origin remains mysterious. According to one preserved letter, Edward Cooke Jr., who served as legal counsel for the industry, met with him shortly after it was published, but no public record survives from the meeting.[163] Brower wrote a second prominent letter later that same year, again raising questions about claims of carcinogenicity, but this time to the editor of *Science* and this time about sodium cyclamate, an artificial sweetener that remains banned in the United States. Brower's questioning of the cyclamate study appeared, ironically, in the same issue that Robert Bazell reported on industry efforts to "discount cancer studies" such as Auerbach and Hammond's.[164] Brower's publications on carcinoma begin and end suddenly in 1970, which is especially suggestive because they were far afield from his primary area of expertise: the biology and conservation of monarch butterflies.

Brower's letter was useful, but the industry soon focused on a different avenue of attack: the failure of Hammond and Auerbach's paper to appear in a major scientific journal. If the results were as important and accurate as their authors claimed, the industry reasoned, then surely they should be published quickly in a flagship journal, such as *JAMA* or the *New England Journal of Medicine*. As long as the results were not published, there would also remain at least a vague justification for assertions that the beagle results had not been adequately evaluated by qualified specialists. The industry's plan, developed partly by Alexander Holtzman, would be for Joseph Cullman III, chairman of Philip Morris and the TI's Executive Committee, to write directly to the head of the ACS requesting a review of the findings by independent scientists. Holtzman attributed the idea to an anonymous pathologist who "is an editor of a cancer journal to which [Auerbach and Hammond's] paper will in all likelihood be submitted." While discussing the results, this pathologist argued that the ACS was "guilty of improper conduct" in by sharing the findings before an actual journal article was published and suggested the industry "call upon the ACS to make available for examination the lung sections on which the finding of lung cancer is based." The proposed panel, including the anonymous individual, would be "unlikely to confirm Auerbach's findings."[165] This piece of the tobacco project of denial, then, had, it would seem, origins within the scientific community.

Tobacco Institute president Earle Clements thought the idea not only had "considerable merit" but represented one of the only opportunities to reshape public opinion about the research and defuse its impact on pending cigarette legislation and antismoking petitions.[166] Therefore, on February 27, Cullman wrote to William Lewis, chairman of the ACS, requesting permission for "a thorough evaluation of the experiment and its results by a panel of independent scientists," indicating a willingness to "bear all costs" relating to the review.[167] Lewis did not respond immediately. On March 2, Kloepfer announced the letter to TI staff and explained his decision "to withhold publicity" on the proposal temporarily, out of fear that the "scientists who have tentatively agreed to participate" might resist if the proposal appeared to be a "publicity stunt" by Big Tobacco.[168]

Ten days later, Lewis finally answered. He acknowledged Cullman's letter but roundly refused the offer: "Your appraisal and suggestions," he retorted, "are based on a summary which was prepared for the edification of science writers." Lewis anticipated that results would be published soon, making this "meticulous work" that met "the highest traditions and protocol of scientific investigations" available to the public in due time. A review panel was thus unnecessary, because conventional journal peer review would accomplish the same thing, and he directed the TI to review Hammond's and Auerbach's papers to "satisfy any scientific or other questions regarding their findings."[169] The letter was brief, evidence of a relative lack of concern about the industry's objections. In 1970, the ACS had few reasons to believe that the TI concern for scientific integrity was sincere.

Undeterred, Cullman wrote back on March 20. He was "greatly disappointed" (advisers insisted on his expressing "disappointment") by the unwillingness of the ACS to "permit the impartial review" requested by the Tobacco Institute. He reiterated the industry's interest in evaluating the evidence "immediately" and insisted that waiting for publication was inadequate to resolve concerns about "proper interpretation of the pathological material." The letter ended with a threat: if the ACS would not agree to this proposal, "we plan to use every means at our disposal to see to it that the medical and lay public are made aware of our respective positions in this matter."[170] At a meeting of the TI Communications Committee three days earlier, plans were already in place to release the letters to media outlets "with a covering background statement" as soon as possible.[171]

Lewis waited about a month to reply. In April, he again refused review by a panel of scientists "chosen by the Tobacco Institute," a proposal that "is without precedent in the scientific community." As Banzhaf later put the situation in a letter to Harken, "No reasonable principle of justice or of scientific inquiry would suggest that the evidence against a murderer be evaluated by a jury of his own selection."[172] Nevertheless, the "iron curtain" had been drawn down, as one internal industry document put it, and the planned campaign against the ACS commenced. Philip Morris vice president James Bowling reported to American Tobacco president Robert Heimann that scientific advisers thought "this issue would be a favorable one on which to make a stand since they feel we have a 95% chance of discrediting the beagle work with the scientific community."[173] Those were difficult odds to resist.

Readers of major American newspapers were greeted, on May 1, with a bold advertisement: "The Tobacco Institute believes the American public is entitled to complete, authenticated information about cigarette smoking and health," it announced. "The American Cancer Society does not seem to agree."[174] The ads included reprinted copies of the letters sent by Cullman and Lewis, first announced by Cullman with great fanfare during a news conference at the Overseas Press Club shortly before they ran in print. Eight days after the appearance of the ad, the *New York Times* published an editorial that supported the industry's position, characterizing the ACS refusal as "an error of judgment" because it threw "needless doubt on the already overwhelming case against cigarette smoking."[175] On June 17, five days after Surgeon General Jesse L. Steinfeld informed the ACS he was satisfied with the quality of the dog studies, the industry mailed a printed collection of the correspondence to physicians and relevant science writers across the United States. Mailing directly to doctors was necessary because "most major medical periodicals had refused the advertisement."[176]

The unstated irony of accusations of secrecy against Auerbach was that he had long opened his lab to visits from specialists of every stripe. One reason that tobacco researcher Brian Davis believed Auerbach's research, for instance, was that he had received a tour from Auerbach and Kirman in 1967. Although he raised criticisms of their method in a report on the visit, Davis summarized, "This appears to be a very good project of Auerbach's."[177] After announcing the beagle findings, Auerbach promised to allow any interested researchers to see his photographs and slides for themselves. Harken was one of the first to visit after the news conference, so that he could cite firsthand knowledge of Auerbach's method in upcoming legal testimony against the industry. Afterward, Harken reflected that Auerbach had applied "something like Koch's postulate to smoking" and provided the "'closing link' in the chain of evidence" against cigarettes. "To you, the public, and indeed mankind, must remain eternally grateful," he wrote.[178]

On June 3, 1970, Auerbach and Hammond wrote to Sheldon Sommers, the CTR's research director, extending a similar invitation.[179] In it, Auerbach noted coyly that he had already shown the histological slides to "some of our mutual friends." Sommers was on vacation in New Zealand when the invitation came, but Holtzman quickly telegrammed to inform him of the offer, explaining how Sommers should respond: "Would greatly appreciate your cabling him about as follows. Have been informed by my office that you have invited me to review your data on smoking beagles. Believe that such review should be conducted by panel of independent experts as proposed by Tobacco Institute to American Cancer Society."[180] Sommers sent Auerbach a nearly identical message and proposed further discussion upon his return on June 28.[181] He never visited the laboratory.

The reasons were obvious: delaying the visit allowed the industry's political theater to continue. Sommers would even contend for years after that he received neither an invitation nor permission to evaluate the pathological slides. That account hardly squared with descriptions of Auerbach, who had, according to British toxicologist Francis Roe, "a stream of visitors and has developed a fixed routine for dealing with them under the motto 'I have nothing to hide, I show everything.'"[182] By refusing to meet Auerbach, the industry could continue to insist in public that he, Hammond, and the ACS were disallowing fair evaluation from their independent panel.

Tasting blood and savoring their success, the industry decided to ramp up the media blitz. The March 1970 issue of the industry magazine *Tobacco Reporter* carried an unflattering account titled, "Dr. Auerbach's Smoking Beagles," which nitpicked at the results using many of the same lines of attack that appeared in letters between industry lawyers and public relations gurus.[183] By October, members of the US Congress would receive a letter about the smoking beagles from Horace Kornegay, a former representative from North Carolina who became president of the Tobacco Institute that year. The beagles were "one of the greatest scientific hoaxes of our time," he wrote.[184] An industry strategy document from May 1970 explained that its audience was threefold: "scientists, legislators and laymen" concerned about cigarettes and health; "investors, suppliers, employees and businessmen" with ties to or interest in the industry; and "the broad cross section of the concerned public whose attitude to, and opinion on, the cigarette-health controversy vitally affect the welfare of the industry."[185] The media budget for that month alone reached nearly $400,000, enough to fund replication of Auerbach and Hammond's study.

Print publications were only one front in the conflict. In July 1970, New York media company 1492 Pictures sent a draft script proposal for an audio-visual presentation rebutting the smoking beagles to Philip Morris. "As the show starts, the sound is that of hunting music with the sound of the beagle pack. Then a shot of beagles running is seen on center screen with individual shots frozen on the side screen."[186] An inquisition-style dialogue follows where two men debate the significance of twelve dead beagles; hardly the "proper subject of an inquest," they agree. After retelling the story of the TI press conference and its subsequent fallout, the men tediously discuss Koch's postulate, which they assert the beagle studies have failed to meet, and debate the strange selection of dogs, which they chalk up to pricing and availability. Though resisting the findings of the dog studies, the dialogue nevertheless demonstrates once again the perceived importance of animal experimentation in producing truth: in this case, it was just the wrong animals.

The script was met with little enthusiasm, so 1492 suggested a multimedia presentation instead, which could be given by a single individual, such as Kornegay, to reporters at the next major media briefing. The final proposal, presented in 1971, differed dramatically from the inquisition dialogue. This time, a speaker, seated at a table or behind a lectern, begins by directly addressing the audience. "I guess none of us needs to be reminded that we're living in a highly institutionalized culture," he begins. "I work for a large institution"—a nod to his tobacco connections—"and most of you work for large institutions." The man proceeds to reflect on the ways that institutions create myths, including one that "reality shattered": that of the smoking dogs. The beagle sounds of the first script had disappeared; so had the corny drama. The point was not to look backward to beagling. Instead, the "feeling" that should be expressed, 1492 noted, is that of the TI "not simply as advocate, but also as teacher."[187] Rather than attacking Hammond, Auerbach, *and* the ACS, their proposal advocated to "attack the ACS as the main villain."

Publishing Interference

The attacks were a success, and the beagle study had fizzled. Kloepfer reported in June 1970 that press and scientific reaction "has been substantially favorable."[188] Yet an obvious question appeared: why had Hammond and Auerbach not immediately published their study to avoid the slow-burning controversy? According to a letter from Holtzman, recounting his conversation with Lauren Ackerman, an editor of *Cancer* who received substantial funding from the industry during his career and possibly the anonymous pathologist referenced earlier, Auerbach decided not to send the manuscript to *Cancer* out of concern for how his slides would be evaluated.[189]

Instead, his first choice for publication was the *New England Journal of Medicine* (*NEJM*), then and now one of the most prestigious medical journals, so Auerbach submitted the paper to editor Franz Ingelfinger on February 13, 1970, a few days before the ACS presentation.[190] He was informed of the paper's acceptance on February 20, a secretarial mistake, but heard seven days later that the paper would actually be rejected because of *NEJM* policy not to publish results that received prior publicity. An article about Auerbach and Hammond's research in *Medical Tribune* had included both numerical data and direct quotations from the manuscript, triggering the decision, which Ingelfinger called "agonizing." As long as the policy existed, he sighed, "we must apply it to the Hammonds and Auerbachs as well as to the Smiths and Joneses."[191] On May 22, 1970, *Medical World News*, a magazine that routinely touted industry lines on smoking, announced the study's *NEJM* rejection. Auerbach's sense of betrayal was likely exacerbated when *NEJM* became the sole medical or scientific publication to accept the Tobacco Institute's advertisement, sparking a flurry of letters to the editor, one of them complaining that "people who might otherwise stop smoking may die as a result."[192]

Auerbach turned instead to *JAMA*, where the emphysema paper had appeared, and asked for the submission to be handled by a friend, senior editor Therese Southgate.[193] Without his knowledge, however, the tobacco industry was carefully monitoring the paper's movement. On June 3, Holtzman reported that the editor of *Tobacco Reporter* had contacted Hugh Hussey, who became editor of *JAMA* that January, and learned about a two-week delay in the review of Auerbach and Hammond's work. Holtzman added, "as a matter of confidential information," that he knew one of the doctors serving as a reviewer.[194] On June 22, one day before Auerbach and Hammond were set to present their work at the AMA's annual meeting, Philip Morris's James Bowling and the CTR's J. Morrison Brady visited the hotel room of AMA president Gerald Dorman and found him busily watching a Mets game with his wife. Brady had served under Dorman during World War II, so the industry men were warmly received: Dorman's wife offered them a drink. Bowling and Brady handed over copies of the Cullman-Lewis correspondence, requesting that Dorman read them and publicly release the *JAMA* peer-review reports. Dorman, in turn, relayed the information directly to Hussey and told the two tobacco reps that *JAMA* "unanimously" rejected the manuscript. Although Hussey refused to give out the names of his reviewers, a violation of the blindness of the peer-review process, he affirmed that he would "tell the press if asked" about the fate of the paper.[195] Hussey would pass away, a little over ten years later, from lung cancer, a disease that Auerbach knew was typically associated with smoking.

Unaware of the backroom dealings involving their research, Auerbach and Hammond brought their results to a press conference the next day, this time taking questions from the audience. The purpose of the study, Hammond explained, had not been to produce cancer in dogs, which human studies had already demonstrated, but to use the animals as a model to develop less dangerous filtered cigarettes. Hammond, a long-term pipe smoker, insisted on this framing even though Auerbach, more obstinate in his resistance to smoking, appeared displeased with the explanation, always understanding the dog studies as providing the cancer proof. Nevertheless, Hammond continued, their experiments had generated two "invasive bronchial carcinomas" and twelve "invasive" tumors. Their declining to refer to these tumors as cancers, he noted, was due to a difference in pathological opinion about "where cancer begins," not a shift in their findings.[196]

Auerbach and Hammond's *JAMA* submission—which Therese Southgate jokingly called "the most famous unpublished manuscript in the Western world!"—was finally returned with a request for revisions and new image reproductions in July.[197] While preparing that altered version, Auerbach spoke with his colleague Katharine Boucot, editor of *Archives of Environmental Health,* who informed him that she would be delighted to consider it for *Archives* instead. The two ultimately decided to send their article to Boucot. As they explained to Southgate later in July 1970, "To put it mildly, the situation is complicated!"[198] Auerbach brought new and improved slides to members of the National Cancer Institute's Tobacco Working Group in November, and four of the scientists present agreed that he had undoubtedly showed them cancer. "If you saw the same kind of cells in the human lung, you would remove the lung," the NCI's John Berg remarked.[199] At the meeting, Auerbach apologized for his earlier slides, which he admitted were not of a high enough quality, and informed everyone that the long-awaited publication would come in December.[200] The industry, which would have learned the same news via a report in the *Winston-Salem Journal* on September 16—if not earlier—quickly moved to discredit Boucot as "a long-time campaigner against smoking."[201] Auerbach sat on the journal's editorial board, and both he and Boucot had previously received ACS funding, which the Tobacco Institute considered severe demerits.

Earlier in August, Len Zahn encouraged one of his press partners, the *Washington Star*'s James Kilpatrick, to write a story chronicling the paper's rejection. "You have given me a swell column, but forgive me if I test this limb before I crawl to the end of it," Kilpatrick responded, inquiring how he might prove the rejection.[202] On September 15, the *Washington Evening Star* ran Kilpatrick's column under the headline, "What Became of 'Landmark' Cancer Study?"[203] Fred Panzer wrote Zahn two days later forwarding information about the upcoming publication and pointed out Boucot's purported conflicts of interest, noting it might "be very valuable for Kilpatrick as rebuttal material."[204] Kilpatrick, an avowed smoker, followed up with a further article, but his rebuttal was marred by basic mistakes, including misnaming the *Archives of Environmental Health.*[205]

In December, the paper was finally published. The fight, however, was hardly over. "As soon as possible, we would encourage the formation of a panel of experts, ideally headed by Dr. Sommers, with the express assignment of preparing a critique of the study," wrote J. V. Blalock, of tobacco company Brown and Williamson, to colleagues a month earlier.[206] Hardly the disinterested "teachers" that their PR campaign had suggested, the industry was carefully planning a takedown before Auerbach and Hammond's publication was even released. Any response would need to be rapid, Blalock reasoned, so "the scientific panel should plan its methodology now—assembling all that is now known about the study, making specific assignments for analyzing the paper, devising means of getting an advance copy of the published paper, an overall scheme for bringing together all the various parts and writing the critique." The scheme, in this case, could hardly have been more explicit. Blalock hoped that as much attention as possible could be given to this "highly important project, because it *can* be anticipated."

Neither Crusader, nor Judge

The fight against Auerbach and Hammond's findings continued long after publication. But by drawing a cloud over their work in advance, the industry avoided the need for a campaign at the scale of the previous year's. Still, in 1972, after the National Cancer Institute, pressured by the Surgeon General, planned a follow-up project to confirm Auerbach and Hammond's results and further investigate the cardiovascular effects of smoking, industry representatives encouraged Gio Batta Gori, the head of the NCI's Smoking and Health Program, to abandon the project before it had even begun. Gori, who later served as a full-time industry consultant, was sympathetic to some tobacco talking points, including recommendations to use the Battelle-style smoking masks, but denied their requests for cessation of research. "We are not looking for cancer effects," he assured them, acknowledging that Auerbach might make "other interpretations."[207]

Make other interpretations he would, but the industry had largely survived the storm augured by initial news of the smoking beagles. Auerbach would spend multiple years overseeing an NCI-supported beagle study of high- and low-nicotine cigarettes, but the project ended early and inconclusively in 1978, and little analysis was ever published.[208] After initial hype, Bair's PNL beagle research had minimal impact. The industry demurred from funding the large, long-term studies Bair hoped to carry out, and the findings that were produced failed to generate the counterpoint initially imagined. Publications from PNL were slowed by the need for CTR representatives to review them, occasionally eliminating phrases that suggested the possibility of cancer findings. But another reason for circumspection about Bair's beagles was that the industry's early suspicions proved correct: a 1977 report noted "functional and morphological responses of the respiratory system to smoke exposure in beagle dogs" that were "similar to those found in humans," including damage to the trachea and impairment of bacterial suppression.[209] Researchers from PNL were consulted on the NCI's follow-up beagle studies, contributing their technical know-how, but their work was eventually transferred to Borriston Research Laboratories, in Maryland, and the NCI would stick to a modified version of Auerbach's smoking method.

More threatening to tobacco profits in the long run was a new political coalition, still in its nascency when the ACS press conference took place in 1970: the nonsmokers' rights movement. As historian Sarah Milov has argued, citizen-led antismoking campaigns were responsible for many of the largest victories undermining Big Tobacco's control over American life in the second half of the century, from efforts to eliminate smoking aboard airplanes and in restaurants to the growing concern over cigarette smoke as "indoor air pollution."[210] The incipient "right" to not breathe tobacco smoke in public, combined with increasing scientific concerns over "secondhand smoke," would dominate activism and research going forward.

In subsequent years, dogs appeared in smoking research less frequently as test subjects than as pets, subjected to "environmental" tobacco smoke. Companion dogs were, in this view, possible "sentinels" of indoor cancer threats.[211] In the 1990s, researchers at Colorado State University found, for instance, an elevated risk of lung and nasal cancers among dogs living with smokers.[212] Such smoking studies with pet dogs foreshadowed the shift of scientific attention to volunteered, "companion" dogs in research, as chapter 5 explores. "Smoking Problem Also Hounds Dogs," went one newspaper headline.[213] It was an apt descriptor for the second half of the twentieth century: by the time Maxwell and his kin ceased smoking, they were joining an overwhelming trend in the United States, becoming just some of the cigarette's many victims.

When industry-connected researcher Francis Roe visited Auerbach in East Orange on April 18, 1972, he found a man who could only be described as "super dedicated." Roe was startled by how early Auerbach had arrived to the lab that day, around four in the morning, and by Auerbach's claim to have taken no vacations in fourteen years. "[He] has a department of some 30 souls[;] 80% of these are negro. I met a Dr. Jewel (a white man) and one other qualified person (a white female). Otherwise, Dr. Auerbach sits in resplendent white isolation like a sphinx on a pyramid of black staff. If the Good Lord should take him the whole edifice would surely crumble and disappear."[214] Roe's racist depiction failed to recognize that Auerbach's black staff were utterly essential to his work, part of a long-standing tendency to discount the scientific contributions of black laboratory technicians.[215] Staff members had provided the care and management necessary to keep the beagles alive and useful. But the account of his lab nevertheless hinted that Auerbach's was a long, solitary quest. Attempting to summarize his visit, Roe admitted that Auerbach "is sincere" if "dedicated to a fanatic degree."[216] Auerbach made rounds of the dogs every morning and personally supervised their autopsies.[217] The man who showed him around the VA hospital had done undeniably impressive work, including analysis of thousands of slides every week of the year, yet he had "not trained a successor at East Orange," Roe concluded. "His work and research are likely to die with him."[218]

The industry battle against Oscar Auerbach's research remains one of the most stunning campaigns to discredit a scientific study. The two-part paper in *Archives of Environmental Health* received less than a seventh of the number of citations that earlier *NEJM* articles had and was engaged on a more limited basis than much of the rest of their research.[219] In November 1970, Auerbach described the industry's efforts as a "scurrilous attack" on the validity of his work, which the *Oklahoman* described as "probably . . . the most widely publicized study of the effects of smoking" since the Surgeon General's Report in 1964.[220] Yet scurrilous barely touched the surface of the industry's struggle, much of which Auerbach only partially understood.

While tobacco publications went to great lengths to portray him as a hopelessly deluded fanatic, Auerbach saw his own position quite differently: "I'm not a crusader and I'm not a judge. I'm interested in the scientific aspects," he noted. Seeing him in his thick glasses and often walking rapidly about, "you[']d think he was going to a fire," remembered pathologist Raymond Yesner.[221] From the beagle controversy, Auerbach had nevertheless learned a lesson about overzealousness and the relationship between media and science: "I would never present a paper to the press in the same way again . . . and I would consult first with the editor of the journal in which I expected to publish the paper."[222] His interest carried him a long way, but the gratitude of all mankind that Harken foresaw never quite came. Auerbach continued teaching until a week before his death, at ninety-two, in 1997, and is remembered by an obituary as "the man who helped indict smoking."[223]

On Cancers and Lungworms

One of the less predictable results of VA smoking research was in highlighting the beagles themselves. After the successful rhetorical and scientific work of establishing beagles as a viable proxy for cigarette inhalation studies, the NCI's Smoking and Health Program confirmed the dogs to be key models for future research in the mid-1970s. Whether in Washington, New Jersey, or England, the animal of choice for many high-quality inhalation studies would be the beagle. As the response to Auerbach's research revealed, British tobacco researchers were as eager to use the breed as their American counterparts. This helps to explain the other, ultimately more iconic smoking beagle story of the era.

In 1975, British journalist Mary Beith went undercover at the Imperial Chemical Industries (ICI) laboratory in Macclesfield, a market town in Cheshire, England. During her brief, seven-day employment, Beith was able to capture a photograph of beagles attached to PNL-type inhalation masks. Published on the front page of the English tabloid *People* on January 26, 1975, Beith's photo inspired outrage in the country and renewed attention from animal rights protestors.[224] Chairman Jack Callard of the ICI argued, in defense, that "use of beagles to test non-tobacco smoking material could help cut down on 50,000 annual British deaths blamed on smoking."[225] Beith's image remains a touchstone: years later, in 2009, when the soft drink Lucozade, formerly owned by GlaxoSmithKline, was connected to a failed bomb plot, PR agent Mark Borkowski noted that the parent company might see it as a "beagle-smoking" moment, something so controversial it requires crisis meetings.[226]

Although not initially the subject of popular indignation, the anonymous breeder of Auerbach's beagles would receive unexpected attention as well. In 1973, Robert Hirth and Girard Hottendorf, two members of the research division of pharmaceutical company Bristol Laboratories in Syracuse, New York, announced a surprising finding: an enzootic lungworm infection, previously considered uncommon in canines, appeared in test beagles originating from five commercial production facilities. Although additional sources were included to trace how widespread the infection was, Bristol had originally purchased their dogs to evaluate the safety of "25 pharmaceutical compounds over a 6-year period" from a "commercial breeder in the northeastern United States."[227] They did not name that breeder, but a follow-up article on "spontaneous" diseases suggests a majority of the beagles came from Gilman Marshall's "Marshall Research Animals" in North Rose, New York.[228]

When the lungworm discovery was first presented in 1969 to a conference of toxicologists, its significance was framed squarely within the realm of pharmaceutical testing: "It is important," Hirth and Hottendorf stressed, "that the lesions not be interpreted as drug-induced changes, especially in chronic toxicity studies."[229] A cynical eyebrow might be raised at two pharmaceutical toxicologists cautioning against interpreting pathological changes as the result of drug activity. However, in their

1973 paper, the authors amplified the potential importance of the discovery: "Canine pulmonary lesions that were reported to be neoplastic and experimentally produced are similar to some of the changes present in our lungworm-infected dogs."[230] Their example of such a possible misinterpretation was Auerbach and Hammond's study. The new lungworm species, *Filaroides hirthi*, named in Hirth's honor, now reared its microscopic tail as a possible alternative cause for the results Auerbach and Hammond had attributed to smoking in their 1970 publication, with all the drama that had transpired.[231]

The tobacco industry, attentive to any and all avenues of critique, jumped on the findings. British researcher S. R. Evelyn sent Hirth and Hottendorf's research to Brown and Williamson's Wally Hughes in November 1974, and Hughes forwarded the study ("published in an obscure European journal") to William Gardner, of the CTR, and David Hardy in November 1974. Hughes noted that "Auerbach got his beagles from the same place," and wondered if "this could have influenced his findings."[232] Shook, Hardy, and Bacon's Donald Hoel passed the study on to Kloepfer at the TI in December.[233] Holtzman then wrote various individuals within Philip Morris in February 1975, attaching the paper and explaining its finding that lungworms could mimic neoplastic lesions. "You may remember that in 1970 Drs. Auerbach and Hammond published a paper on a study they conducted using commercially bred Beagle dogs," he noted. "The present paper . . . suggests that the lesions referred to in the Auerbach and Hammond experiment could have been produced by the lungworm rather than by cigarette smoke."[234]

The rapid movement of the study across tobacco companies, their legal partners, and associated scientists on multiple continents is evidence, simultaneously, of how quickly Auerbach and Hammond's work had passed from view as well as of the lingering danger their findings represented.[235] The potential implications of the lungworm were clear to industry insiders: tumors in dogs were not caused by cigarettes but by infections, further proof of Auerbach's supposed carelessness. It was a line of argumentation that also fit well with the industry's long-standing support, as Robin Wolfe Scheffler has shown, for "viral" and alternative explanations of cancer.[236] Auerbach had in fact been asked in 1970 whether his tumors might be caused by something akin to jaagsiekte, a contagious lung disease in sheep.[237]

Because they went so far as to suggest that Auerbach and Hammond's important cancer findings were misinterpretations of a canine veterinary ailment, Hirth and Hottendorf could not go unanswered. "No clinical evidence of lungworm infestation has been found in any of the dogs of our study," the East Orange lab responded in 1974 in the journal

Experimental and Molecular Pathology. After a careful second search of the slides found no histological evidence of a lungworm infestation, the team reasserted that the changes should "properly be ascribed to exposure to cigarette smoke."[238] In September 1975, however, Auerbach reported to members of the Tobacco Working Group that his team had found lungworms in a number of deceased dogs participating in his follow-up study.[239] The suggestion that beagle dogs might be poor models for lung cancer studies because of a strange lungworm was unanticipated and difficult to decisively disprove. Unsurprisingly, the tobacco industry's supporters and legal counsel continued to assert that Auerbach's tumors were unrecognized lungworm damage until well into the 1990s.[240] A deposition for *United States of America v. Philip Morris USA Inc.* in 2001 also featured extensive discussion of Auerbach's work and suggestive reference to the lungworm paper.[241]

Further investigations of *Filaroides hirthi,* carried out for two decades after, offered a parasitic trace of the spread of beagles across continents. In 1976, scientists at Japan's Institute of Public Health and at the University of Tokyo reported extensive lungworm traces in experimental beagles imported from an unnamed American breeder.[242] By 1983, Australian researchers had identified lungworms in beagle colonies at both the Institute of Medical and Veterinary Science in Adelaide and the University of Melbourne's Veterinary Clinical Centre. These populations were derived from a colony at the Nicholas Institute in Sherbrooke, which was itself founded from imported American beagles.[243] The first case in England was reported in 1985, and batches of laboratory beagles imported to Israel from England tested positive for lungworm by 1991.[244] In 1994, scientists at German chemical company BASF reported lungworm in their dogs as well, imported over time from both England and the United States, and argued that the entire colony was likely infected.[245] Lungworm would be found in beagles in France, the Czech Republic, Italy, Turkey, and more.[246]

Because the parasites appeared in dogs from all five major American beagle-breeding companies and at laboratories around the world, contamination represented a broader threat to the beagle's place as a stable basis for scientific truth. While debates over whether beagles "naturally" developed lung cancers were frequent in responses to the East Orange studies, that beagle dogs were good tools for studying numerous problems was largely accepted—by Auerbach, PNL, and members of the industry. Yet because the lungworm seemed to mimic the changes induced by drugs and carcinogens, it raised questions about a significant quantity of previous toxicological research. Could beagles remain *the* laboratory dog if results from a wide variety of safety studies were misinterpretations of basic endemic infections?

The lungworm question highlighted once more the basic challenge of determining whether beagles were "good" proxies for understanding human bodies and their ailments. As the AEC researchers had found, the answer was not easily reduced to a binary yes or no. Andersen's *Beagle as an Experimental Dog* appeared the same year as Auerbach and Hammond's smoking beagle study was first published; and as proxies not simply for basic physiology but also behaviors and emotions, smoking beagles revealed both the publicly persuasive nature of dog findings and the ways in which laboratory environments shaped those results. The affective beagle was a powerful tool, but it also raised questions about whether laboratories were a good place for dogs to be at all. The persistent difficulties in raising healthy beagles, and the even knottier question of what exactly a "normal" dog was, continued to plague researchers even as beagles became conspicuous norms of experimental life. Those challenges could be taken up into the engines of agnotological contest in contradictory forms: beagle studies might be, simultaneously, and depending on one's perspective, the experimental gold standard or fundamentally misleading.

The lungworm scare thus revealed the experimental implications of shifting systems of production and exchange for laboratory organisms. Beagles were increasingly products of a consolidating international industry rather than locally produced tools. If dogs were infected at the AEC colonies, it would have largely been the fault of the scientists and veterinarians themselves. But now that researchers were primarily buying beagles from large producers, framed explicitly as market commodities, researchers no longer controlled the conditions of a beagle's reproduction. Instead, they had to trust that producers were sending them good tools. In 1979, *The NIH Record* included a short story on efforts to combat Hirth's parasite, not so surprisingly lead by a Cornell veterinary researcher, Jay Georgi, whose team was studying the antiparasitic albendazole as a weapon against lungworm. Georgi posited that the parasite may have initially spread to beagles from animals such as skunks or weasels, like the ferrets Marshall raised, rendering lungworm a kind of biological memory of Marshall's history.[247]

Smoking remains one of, if not the, largest cause of premature death in the developed world.[248] As Americans continue to steadily quit their daily cigarettes, demographers expect that life expectancy will rise in the United States, even with growth in other potentially offsetting factors.[249] Yet as the world's population grows older, living longer than ever before, many worry about a new public health challenge: a predicted rise in dementia. In order to understand that challenge, researchers would once again turn to beagles and other elderly dogs, as the role and value of experimental canines shifted close to the dawn of the new millennium.

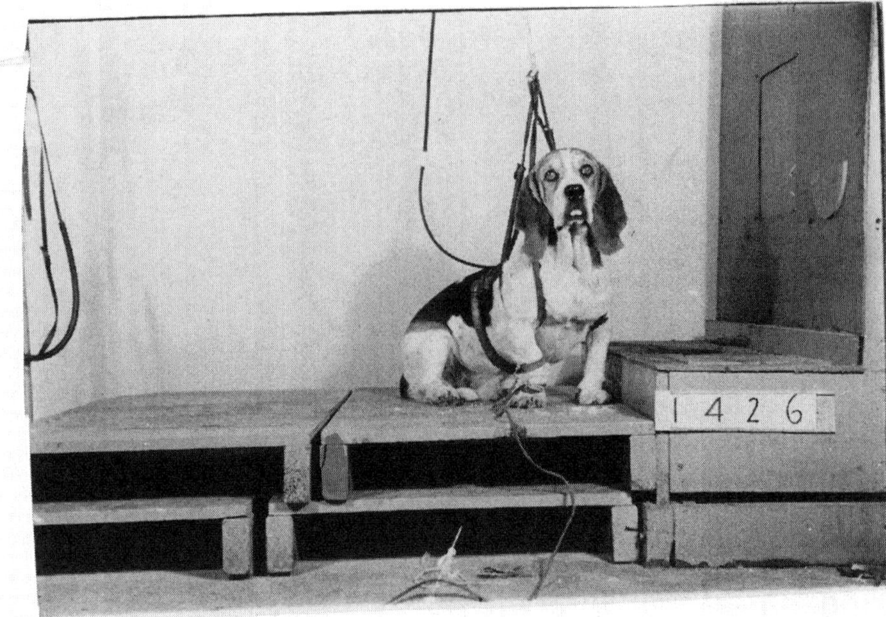

A dog sitting on an experimental platform like those used at Stockard's lab. Courtesy of Hargrett Rare Book and Manuscript Library / University of Georgia Libraries.

A lab dog absconding into an experimental box during studies by William T. James. Courtesy of Hargrett Rare Book and Manuscript Library / University of Georgia Libraries.

CHAPTER 5

The Age of Dogs: Cognitive Dysfunction, Canine Pharmaceuticals, and Companion Science

How long can a dog live? In 2023, the oldest dog in recorded memory, an exemplary representative of the Portuguese rafeiro do Alentejo breed named Bobi, passed away after more than thirty-one years.[1] According to the traditional "dog years" calculation, Bobi was nearly 217 in human terms. In the early days of the Utah and Davis beagle projects, researchers estimated a far more limited average life expectancy for their little hounds, between nine and twelve years. Many of the dogs, however, outlived the most optimistic expectations, surviving to the ripe old age of eighteen. Even though recent reevaluations of the notion of "dog years" would place the geriatric beagles only at the century mark, their longevity was impressive, and most who live with dogs would cherish nearly two decades with their beloved companions.

Recently, biotech researchers and gerontologists have turned their attention to fulfilling the wish for more years with our dogs. Canine anti-aging medications, they argue, might allow dogs to live longer than ever before. Rather than a record-book entry, Bobi could be the norm. In November 2023, one company, known for predictably doggy reasons as Loyal, announced word from the US Food and Drug Administration that its flagship life-extension drug carried "a reasonable expectation of effectiveness."[2] That seemingly tepid appraisal was welcome news for Loyal, however, because it potentially set the stage for "expanded conditional approval," a regulatory designation that allows certain drugs to reach the market before large clinical trials are concluded if they treat "serious or life-threatening conditions" or respond to "unmet animal or human health needs" that are challenging to study.[3] Loyal supporters cheered, and Lee Sung Jin, creator of the Netflix series *Beef*, encouraged the FDA to fast-track the drug at the 2024 Emmy Awards.[4]

The Loyal drug, known only as LOY-001, is an extended-release implant designed for large dogs that seeks to modulate a hormone called insulin-like growth factor-1 (IGF-1). A priority target of aging researchers since the early 2000s, studies of smaller mammals such as mice have shown that down-regulation of IGF-1 can prolong an organism's life. The exact role of IGF-1 in aging remains inadequately understood, however. While it seems to be involved in protecting neurological and cardiovascular health, IGF-1 also appears linked to the presence of certain cancers.[5] In turn, Loyal's claims about their drug are rather limited: chief executive Celine Halioua hopes her company will be able to promise dog owners "at least one year of healthy life span extension."[6] Even that remains to be proven. But if it one day is, Loyal could have a blockbuster drug on their hands, selling the promise of longevity to dog owners across the globe and, according to their website, "giving our best friends more time."[7]

This chapter traces shifting understandings of the capabilities of aging canines amid growing utilization of laboratory dogs as *cognitive* proxies: useful subjects for studying human cognition and disease. It connects these changes to the emergence of the international market for canine pharmaceuticals; unsurprisingly, beagles would play a key role. The chapter begins, however, in an unexpected place. From the late 1980s to the present, a network of researchers at the University of Toronto and the University of California, Irvine, sought to translate the degenerative disease known as Alzheimer's across the species divide by arguing that old beagles and other dogs suffered from something called "cognitive dysfunction syndrome" (CDS). By studying CDS in beagles, they argued, scientists might better understand the mechanisms of dementia in humans.

Alzheimer's is typically diagnosed through symptoms such as memory loss and behavioral change, but identifying "cognitive dysfunction" in dogs required complicated interpretations of canine behavior. After an initial focus on pinpointing parallel biomarkers in dog and human brains, such as beta-amyloid plaques, beagle researchers turned to the "lifestyle" factors that many now understand to be central in Alzheimer's, such as exercise and diet. In doing so, the research also suggested the possibility that a dog's environment, its life shared with humans, was in fact triggering cognitive deterioration.[8] In this view, canine cognitive dysfunction syndrome was a "biological imprint of human actions," written into a longer, coevolutionary history.[9]

Using dogs to study cognitive decline was relatively radical at the time, however, and the scientists initially found scant support for the idea that dogs even possessed complex cognition. For multiple centuries, the prevailing sentiment was that *Canis familiaris* thought minimally, if at all, learning only through rudimentary trial and error. Dogs were loyal, but not brilliant. While canine cognition is now the subject of extensive research and continuously extolled by pet books and popular science reporting, the techniques and devices relevant for studying canine brains were late arrivals in the twentieth century. Beagle studies of dementia laid some of the foundations for recent discoveries about Fido's brain, just as they were also significant for understanding Alzheimer's itself.

Equally important in the turn to beagles was its preeminent initial justification: canine pharmaceuticals. Many still disagree on whether a dog can get Alzheimer's, a debate that is both semantic and scientific, but it is widely accepted that dogs acquire CDS. Yet the concept of CDS was initially introduced, in part, for the marketing of a new pharmaceutical aimed at geriatric dogs: Anipryl. First released in 1997 by Deprenyl Animal Health, Anipryl became one of the first successful dog-focused pharmaceuticals. Its origin as a possible treatment for Parkinson's disease in humans brings us to the second major theme of this chapter: the cross-cutting intersections between human and pet (or companion-animal) pharmaceuticals.[10]

Already valued at over a billion dollars in 1997, the market for animal drugs has grown explosively, just like pet ownership, and companies both large and small have directed their expanding animal divisions to develop drugs for our pets. Loyal is just one of the more high-profile recent entrants. By tracing the origins of CDS and Anipryl, the chapter reveals the feedback loops between the production of canine medications and research on human diseases. Pet medications were both logical extensions of the broader use of dogs to test pharmaceutical safety and vehicles for salvaging value from failed clinical trials.[11]

Returning once more to life extension in dogs, the chapter ends by exploring a reversal in the fate of laboratory beagles. Old dogs, once perceived as relatively useless lab organisms because they do not represent healthy young people, are now the focus of intensive study, whether for drugs like Loyal's or for new cancer treatments. What I call the *environmenticity* of dogs, the sense that they are defined by sharing an environment with people, has emerged as a crucial laboratory benefit. Some of the first elderly experimental dogs were beagles, survivors of the old atomic projects, but such specialty-bred beagles are increasingly being replaced by a cheaper alternative: volunteered pets. Here, the "salvage" model of using strays and former pets in research, which purpose-bred beagles once replaced, has reemerged in a new guise: *companion science*. A new Loyal study called STAY, announced in 2024, thus hopes to enroll over one thousand volunteered pet dogs in the coming years. Combined with revelations from studying dogs as cognitive proxies, including that they are intelligent and thoughtful beings, companion science may signal the end of the laboratory beagle.

A Series of Miracles

When Bill Milgram was a young psychology professor at the University of Toronto in the final years of the 1980s, he liked to play competitive, high-stakes games of rubber bridge. Over the pauses between bids, Milgram gradually befriended a wealthy opponent named Morton Shulman. Shulman, the former chief coroner of Ontario and an inspiration for the Canadian TV drama *Wojek*, had struck unexpected gold through stock market speculation years earlier. His talk show, *The Shulman File*, made him a national media personality, but in 1982 Shulman was diagnosed with Parkinson's, which turned his attention and his substantial fortune to the search for treatments and possibly even a cure.[12] In their conversations at the card table, Shulman told Milgram about a new drug that he was taking, one with impressive claims of efficacy. Its name was selegiline, known chemically as L-deprenyl.

First synthesized in the early 1960s under the name "E-250" by researcher Elizabeth Miller, working under Zoltan Eczeri at Budapest's Chinoin Pharmaceutical Company, deprenyl was originally envisioned as an antihypertensive agent.[13] Early tests on rats by Chinoin scientific adviser Joseph Knoll and his colleagues at the University Medical School in Budapest, however, suggested E-250 might also be an antidepressant, what Knoll called a "psychic energizer."[14] The improbable survivor of Auschwitz, Buchenwald, and Dachau, Knoll spent the postwar years studying acquired drives in animals, compelled to make scientific sense of the vagaries of human emotion he experienced in the Nazi concentration camps.[15] Since 1960, with effervescent enthusiasm, he had analyzed how amphetamine and methamphetamine transformed animal behavior, and E-250, part of the same chemical family as amphetamine, appeared to hold a number of valuable properties.[16]

Knoll's early research on E-250, published in Hungarian in 1964, received little attention.[17] An English-language article followed in 1965 with an additional claim that deprenyl inhibited monoamine oxidase, an enzyme that controls the function of neurotransmitters.[18] Monoamine oxidase inhibitors (MAOIs) were the first clinical antidepressants, but high-profile examples such as iproniazid, a tuberculosis drug that turned out to elevate mood as well, fell off the market when patients who consumed aged cheese, red wine, or chocolate experienced severe hypertensive episodes.[19] Knoll's childhood friend Ervin Varga, who worked at a nearby psychiatric clinic, had administered E-250 to volunteers in

1965 and found it avoided this "cheese effect," but neither could explain what differentiated the compound. Company leaders at Chinoin, one of the nationalized firms that sustained Hungary's position as a prime drug developer for and exporter to the Soviet Union, were wary of selling a dangerous medicine.[20]

Then, in 1968, J. P. Johnston of British chemical company May and Baker demonstrated that monoamine oxidase was actually a "binary system" of two subtypes, MAO-A and MAO-B.[21] In subsequent research, a drug known as clorgyline was found to selectively inhibit MAO-A.[22] Knoll and collaborators confirmed that L-deprenyl (or selegiline), an isomer of their original compound, selectively inhibited MAO-B, presenting the findings at a 1971 conference in Sardinia. But Chinoin's continuing concern was matched this time by international skepticism that researchers behind the Iron Curtain had solved the "cheese effect."[23]

Around the same time, Walther Birkmayer and Oleh Hornykiewicz, from the Ludwig Boltzmann Neurochemistry Institute in Austria, were looking for a drug to complement their levodopa treatment regimen for Parkinson's. Because the disease was known to stem from dopamine deficiencies in the brain, injections of levodopa (or L-DOPA), the biological precursor to dopamine, appeared to be a promising treatment. Initial results were miraculous, including bed-ridden patients standing up to walk once more, yet they came with life-threatening risks. Birkmayer and Hornykiewicz tried combining L-DOPA with a MAO inhibitor, but the latter's side effects left them stumped—until Moussa Youdim, an Israeli biochemist working at Oxford, arrived for an invited lecture in Vienna in October 1974 to discuss the MAO system.[24]

By chance, Youdim had met Birkmayer's colleague Peter Riederer during Riederer's visit to England the year before, and the men had discussed connections between Parkinson's and MAOIs. During his lecture in Vienna, Youdim listed selegiline as a drug of potential interest and, over a wine-enriched dinner following the talk, discussed using E-250 with L-DOPA. Birkmayer agreed to try it out, and initial studies confirmed that the combination eliminated many problems from earlier treatment regimes. After results appeared in the *Lancet* in 1977, selegiline became a key component of Parkinson's treatment.[25] The initial report made no reference to Knoll's work with selegiline.[26]

By the time Shulman was swapping tricks with Milgram in Toronto, he was taking selegiline and extolling its benefits. A few years earlier, finding that Chinoin's drug was widely used in Europe but unavailable in Canada, Shulman convinced a friend, the Hungarian-born trust fund manager Andrew Sarlos, to bring back samples of the drug.[27] Upon his taking it, Shulman's symptoms—"shaking, drooling, I couldn't walk or talk"—rapidly cleared. When he prescribed the treatment to two of his own Parkinson's patients, it worked for them as well. "It was like being God," he remembered in 1991.[28]

Convinced of selegiline's otherworldly power, Shulman moved to acquire the company that held distribution rights in the Western Hemisphere: New Jersey–based Somerset Pharmaceuticals, founded in the early 1980s by researcher Donald Buyske, a biochemist and former park ranger. Buyske had zealously sought access to selegiline and other Hungarian pharmaceuticals, first on behalf of Warner-Lambert and then S. C. Johnson and Son, going so far as lobbying President Jimmy Carter to return to Hungary a relic, the Holy Crown of Hungary, taken into American custody during World War II, to improve his chances.[29] Yet Somerset, spun off from S. C. Johnson's research division, had mixed success with Buyske's hard-won prize: the FDA, unmoved by suggestive European findings, requested additional clinical data before approving what Somerset planned to call "Eldepryl." The ensuing slow approval process seriously strained the company's finances.[30]

Shulman appeared at exactly the right moment. Unable to straightforwardly buy Somerset, he purchased a significant percentage of the company and rights to market and sell Eldepryl in Canada. Shulman liked to do things on his own, so he founded a new pharmaceutical concern, Deprenyl Research Limited (DRL), for the project.[31] Beginning operations not long after Black Monday, the global stock market crash of October 1987, DRL's start was shaky, but Shulman sold nearly two million shares to the public, including many close friends, which financed further operations. The price of DRL stock, initially at $3 in 1988, rose to $17 by 1989, and Eldepryl was soon available to Parkinson's patients across Canada.[32] In 1989, the US FDA finally approved Eldepryl, just as Somerset was being jointly acquired by Mylan Pharmaceuticals and generic drug manufacturer Bolar.[33]

During their bridge games, Shulman told Milgram about additional research from Knoll, work that showed that selegiline had not only antidepressant functionality, but the ability to extend longevity. Published in 1989, Knoll's study had claimed that rats taking selegiline lived not just longer, but beyond their expected biological limit—and by months, not just days or weeks.[34] The findings were particularly attractive to an aging Shulman, as was the study's suggestion that selegiline produced extended reproductive activity: Shulman's speeches from the period often included allusions to the "aphrodisiac side-effects" that sustained his vibrant personal life.[35] Because Milgram had extensive research experience with rats, long the favorite animal model for physiological psychology, he was an ideal candidate to test Knoll's findings. Despite some skepticism about the promised longevity, Milgram informed Shulman that nobody would believe it unless the studies were replicated, and he offered to repeat them if Shulman would foot the bill.

After several hundred thousand dollars and countless hours of bored technicians examining aging rats for signs of renewed sexual vigor, Milgram and colleagues including Gwen Ivy found only minor improvements in lifespan and cognitive processing from selegiline.[36] "Clearly," he joked later, "the Hungarian rats are different from the rats in North America."[37] Concluding that selegiline was hardly the miracle drug Knoll's work suggested, Milgram noted nevertheless that a comparison of his results with ongoing tests in Parkinson's patients supported early administration of the drug to maximize its benefits. Sensing the end of their relationship, he relayed the disappointing results to Shulman.

Undeterred, Shulman reasoned that a drug that failed to extend life in humans could still help other animals. The extended life and improved processing that Milgram found in rats might even appear in dogs and cats, too. So Shulman turned to Ralston Purina, the St. Louis–based animal products conglomerate, with an option to use Eldepryl in pet foods and other products, but their interest was limited.[38] In response, Shulman formed a subsidiary of DRL called Deprenyl Animal Health, Inc. (DAHI) in July 1990.[39] Headquartered first in Kansas City, Missouri—closer to more extensive veterinary expertise—DAHI stated in its initial public offering that its goal was to "develop and implement the regulatory steps required to obtain marketing approval in the United States and Canada of L-deprenyl as a prescriptive drug for use in dogs and later in cats."[40] The aim, in short, was selling selegiline to pet owners.

Yet one immediate challenge was that all supporting research for DAHI's product came from rats. A wary investment analyst cautioned in 1991 that the same effects might "not occur with other animals" or be "significant enough to attract pet owners' attention."[41] With only three employees, including its president, David Stevens, DAHI planned to rely on collaborations with academic researchers such as Milgram to produce the missing evidence.[42] But more significantly still, the original assumption that what was true for rats must be true for dogs reflected, rather than resolved, a significant gap in existing knowledge. Did dogs experience cognitive decline? How would we know, and what would it look like?

The Intelligence of Dogs

The DAHI plan went against not just some common sense of the time, but centuries of thinking about canine intelligence. Dogs were known to be faithful, certainly, since before Plato. But few people saw them as bearers of higher cognitive abilities. As classicist Cristiana Franco notes, trustworthiness and controllability, rather than intellect, were the key canine traits in Greece and the later Roman Empire.[43] The French priest and philosopher Nicolas Malebranche later summed up centuries of popular, Descartian wisdom when he wrote that animals "eat without pleasure, cry without pain, grow without knowing it; they desire nothing, fear nothing, know nothing."[44] While challenged in a variety of ways, the kernel of that view held long after its initial appearance.

It was Charles Darwin, a lifelong dog obsessive, who most prominently disputed the Descartian understanding, arguing that man's best friends were moral, perhaps even consciously reasoning creatures. Remembered for a close connection with his fox terrier Polly, Darwin had insisted in *The Descent of Man* that dogs shared many of the most apparently human emotions, everything from jealousy and excitement to ennui.[45] That they additionally had the capacity for memory was, to Darwin, self-evident: "I had a dog who was savage and averse to all strangers," he recalled, "and I purposely tried his memory after an absence of five years and two days. I went near the stable where he lived, and shouted to him in my old manner; he shewed no joy, but instantly followed me out walking, and obeyed me, exactly as if I had parted with him only half an hour before."[46] For Darwin and many followers, a clear continuum existed between the intelligence of dogs and that of human beings. That this belief grew alongside the establishment of modern dog breeds was no coincidence: the world of wildly varying dogs was vital food for Darwin's thought and inspiration for his theory of natural selection.[47]

But this magnanimous view of canine intelligence foundered on the rocks of early experimental psychology. In 1892, the British zoologist Conwy Lloyd Morgan formulated his "canon" concerning animal intelligence, that no animal activity should "be interpreted as the outcome of the exercise of a higher psychical faculty, if it can be fairly interpreted as the outcome of one which stands lower on the psychological scale."[48] Morgan's fox terrier, Tony, might be able to open the garden gate latch, but such ingenuity demonstrated simple learning rather than virtuosic scheming.[49] Framed explicitly against the work of Darwin's protégé George John Romanes, who considered animals capable of abstract thought, Morgan's canon was supported by Oxford Sanskrit scholar Friedrich Max Müller's popular argument that language and reason were the sole possession of human beings.[50] Romanes, for his part, celebrated the "high intelligence" of dogs, which he placed at least on par with that of apes.[51]

Although Morgan cautioned against interpreting his canon as a prohibition on acknowledging higher cognition in nonhuman animals, it was influentially taught in this way, shaping comparative psychology for nearly a century. Believing that Morgan had relied too heavily on limited, anecdotal evidence, psychologist Edward Thorndike began building "puzzle boxes" that dogs and other animals could learn to solve while he studied at Harvard in the late 1890s. Thorndike charted solution times over successive attempts in order to quantify animal intelligence and provided some of the first documentation of "learning curves," which became a term of art in the 1920s.[52] While he acknowledged the presence of trial-and-error learning in dogs, Thorndike went little further than Morgan.[53] "A dog may very well growl in his sleep without any idea of a hostile dog," he noted, in response to the suggestion that dogs possessed memory or "imagination."[54]

Horsley Gantt, an American acolyte of Pavlov, carried out wide-ranging research with canines during the early and mid-twentieth century. Gantt and his followers studied individual psychic differences in dogs as well as the effect of traumatic events, known at the time as "nervous breakdowns," but they were less interested in higher reasoning or complex memory than in experimenting with conditioned responses.[55] This focus on "conditioning" was dominant for America's Pavlovians, including Howard Liddell at Cornell and a group of Gantt's followers at the University of Arkansas, who studied nervous breakdowns in a lab colony of pointer dogs. Two sets of experiments, however, began to push in other directions.

First was the behavior project at Jackson. Little's and Gregg's initial interest in demonstrating that "intelligence" was inherited, as we saw in chapter 1, already suggested that dogs might be useful cognitive proxies, if in a eugenic, analogical mode. Scott and Fuller's research began to sketch a more multifaceted view of canine cognition by breaking down a dog's intelligence into capacities shaped by emotion and motivation. There were no differences in pure intelligence between breeds, they argued: a beagle might be better at certain tasks than a spaniel, but that was due to motivation, not because they were "smarter."[56] Yet Scott and Fuller, focused on questions of behavior and genetics, continued to place dogs lower in the hierarchy of intelligence. Fuller argued in 1962, for instance, that dogs were "well adapted to many programs of psychological research," with a caveat about tests that "require the superior learning capacity of primates."[57]

In 1952, Ogden Lindsley, a graduate student of psychologist B. F. Skinner, built the first canine operant conditioning chamber, what many know as a "Skinner box," to explore how radiation affected learning for an Atomic Energy Commission–sponsored project.[58] Lindsley's first chamber was tested on nearly sixty-five beagles, kept in small internal colonies for the work.[59] One survivor, Hunter, named for the psychologist Walter Hunter, went on to become the Skinner family pet, pulling a cart at their summer home on Monhegan Island.[60] Lindsley's project served as his Harvard dissertation and quietly marked the beagle's arrival as a cognitive proxy in experimental psychology, but he soon moved on to studying schizophrenia in human patients.[61] In 1953, however, three other Harvard-connected psychologists, Richard Solomon, Leon Kamin, and Lyman Wynne, showed that a mixed-breed dog's conditioned response to traumatic electric shock—in this case, jumping over a barrier into a safer compartment—could be gradually extinguished with countervailing procedures.[62] Building on this work in the late 1960s, Martin Seligman and collaborators would demonstrate that dogs will give up jumping away from a shock if they learn that such efforts are futile, the foundation for Seligman's influential concept of "learned helplessness."[63] While at Cornell in 1970, Seligman and Dennis Groves found that persistent learned helplessness was particularly pronounced in cage-raised beagles, which were selected for their local availability and known ancestry.[64] In much of this work, researchers saw a clear analogy between canine and human learning.

Yet despite this, a broader disinterest in the specificities of canine memory and cognition prevailed. Scott reflected in 1985, not without a degree of self-aggrandizement, "Since '65, nothing better has ever been done on dog research."[65] One reason for the absence of dogs in experimental psychology and psychiatry was that neurophysiologists and their neuroscientific heirs preferred to work with rats, nonhuman primates, and—if a pet animal was to be included—cats. Although reasons for the feline predilection varied, one decisive precedent was set with Sir Victor Horsley and Robert Clarke's "stereotaxic instrument." The device, constructed in 1905 and publicized in 1908, allowed precise location of brain sections and would eventually inspire a multidecade wave of "lesion studies" that explored how small brain incisions impaired cognitive functioning.

Writing about the apparatus, Horsley and Clarke explained their choice not to enlarge it to accommodate dogs: "In almost every respect the cat's brain is superior to the dog's for elementary neurological purposes; the nerve tracts are better marked, the size of the encephalon is more convenient for serial sections, and, most important of all, cats' heads are of much more uniform size and shape than those of dogs; in fact, the endless variations in the size and shape of dogs' heads make them unsuitable for a research involving accurate cranio-encephalic topography."[66] Horsley, often remembered for a photograph with his terrier Scamp, was not averse to research on dogs, but the wide variation in modern breeds that early eugenic researchers saw as a benefit was here a serious defect.[67] When the stereotaxic apparatus was reconstructed after the First World War, initial prototypes, built in 1931 and 1934, were again designed for cats and nonhuman primates. The earliest brain "atlases," which mapped brain structure to coordinates, featured rhesus macaques and cats, and stereotaxic instruments were successfully deployed in the decades that followed with rats and various primates.

"For some strange reason," however, "nearly everyone has avoided the dog, physiologists' favorite experimental subject," neuroanatomist William Windle remarked in 1960.[68] For Windle, the explanation was simple: the "spade work for investigations had been carried out in other species."[69] The 1950s, however, marked a turning point for dogs and neurophysiology. Alongside the beagle's rise to popular adulation, a number of methods for using dogs in stereotaxic studies were introduced. In 1960, a stereotaxic atlas focused explicitly on beagles was published by Robert K. S. Lim and Robert Moffit of medical research company Miles Laboratories and Chan-Nao Liu of the University of Pennsylvania.[70] In his review of their book, University of Chicago neurophysiologist Kao Liang Chow voiced hope that it might remedy American psychology's "neglect of the dog" in contrast to attention to the dog in countries such as the Soviet Union.[71] Indeed, one of the earliest stereotaxic devices designed explicitly for dogs was produced in the Laboratory of Physiology at the Polish Academy of Sciences in Warsaw in the late 1950s. Introducing the device, Władysław Traczyk explained that dogs were "the most convenient object" for chronic experiments.[72]

The authors of the beagle atlas also justified their decision to work with dogs on the grounds of ease and availability. The researchers were "situated in locations where dogs were more readily available than cats," a widespread problem given difficulties in purpose-breeding cats for laboratories.[73] They had simply chosen a breed with medium-sized snouts and adjusted the popular Lab-Tronics "cat headholder" to fit them. In 1960, University of Michigan pharmacologist Edward Domino shared his own version of a canine head holder that was also designed for beagles.[74] Beagles were now officially on the map for experimental psychology.

This research found an unexpected public spotlight in 1962 after the Associated Press reported on Ohio Democratic senator Stephen Young's criticism of NIH funding policies. Young was dubious about studies of ungulates and "really upset about spending more tax money on such projects as 'the stereotactic atlas of the beagle brain.'"[75] Angry letters to the editor flooded in to defend the projects against Young's apparent ignorance. Biophysicist Edward MacNichol Jr., for instance, wrote acidly to the *Baltimore Sun* that any neurophysiologist worth her salt could confirm that much of what was known about brain function came from such work.[76] America's Central Intelligence Agency in fact agreed, carefully monitoring Soviet publications on stereotaxic methods and electrical brain stimulation as part of a broader interest in controlling human behavior.[77]

Canine atlases were thus a significant advance, but they failed to displace broader disinterest in dogs and their cognition. As Stanley Coren noted wryly in his popular 1994 book, *The Intelligence of Dogs*, "When I first did my training in psychology, the belief was quite strong that dogs (and all other nonhuman animals) did not have consciousness. We were assured, for instance, that a beagle is not a conscious, thinking creature with self-awareness and emotional feelings but rather a beagle-shaped bag of reflexes, automatic responses, and genetic programming."[78] This was the Gantt-Scott-Fuller view in unflattering caricature. But Coren's book, which emerged just before the "dog paper boom" of the new millennium and propagated the notion that border collies are the smartest dogs, followed by poodles and others, relied on surveys of dog obedience judges, because laboratory tests of a sufficient number of dogs seemed implausible.[79] As a result, canine "intelligence" became firmly yoked to "obedience" and "utility" to human beings.

Creative studies by Michael W. Fox, emerging from postdoctoral work with Scott and Fuller, had already argued for the need to study canine behavior in the context of brain development in the 1960s and 1970s. Fox had, coincidentally, visited the Davis beagle colony in 1968 to explore how environmental changes affected behavior and emotion in the dogs.[80] But few saw the need to take Fox's insights into general experimental psychology. A rare exception was a study of object permanence in cats and dogs by Estrella Triana and Robert Pasnak at George Mason University, published in 1981, which showed that dogs could master a number of tasks used to study object permanence in infants.[81] The authors, nevertheless, worked with only a small number of dogs and seemed mostly interested in their feline results. Later in the decade, Harry and Martha Frank started to explore differences in cognitive ability between wolves and domestic dogs, thinking explicitly in terms of "information processing," but most work along these lines still interpreted canine psychology within a broader "behavioral" umbrella that placed the inner, conscious life of dogs inside a black box.[82] By the last decade of the millennium, when Milgram and his doctoral student Elizabeth Head began to look at what effects selegiline might have on a dog's memory, many still debated whether the species possessed procedural memory, the processing associated with motor skills, let alone episodic memory, the ability to mentally time travel. Setting off into barren lands, the chill of disinterest still blew.

It is therefore ironic that, at roughly the same time, psychologist François Doré and his student Sylvain Gagnon at the University of Laval in Quebec were beginning to extend Triana and Pasnak's initial foray into canine intelligence with a broad program exploring whether Jean Piaget's stages of sensorimotor development appeared in dogs. Systematically studying a number of breeds, including a large group of terriers, Gagnon and Doré found that dogs were frequently able to solve complicated "invisible displacement" tasks, in which an object is surreptitiously switched from one container to another. Doing so, they argued, implied a canine ability to "mentally reconstruct the trajectory" of an object that dogs did not perceive, something cats routinely failed to do.[83] Yet despite the relative proximity of Toronto and Laval, the two research groups, looking at canine cognition on opposite ends of a dog's life, initially had little to say to each other.

Old Dogs, New Schticks

The first Toronto study population was a small internal colony of twenty-two beagles, purchased from Marshall Farms, with five mixed-breed dogs from a Canadian animal supplier.[84] Head and Milgram developed a modified "field test" based on Fox's experiments, using a computer to track the movement of dogs around their lab space after administration of selegiline. Their goal, at first, was to replicate selegiline's mood- and activity-enhancing benefits in rats, but they found little to get excited about: dogs were walking, sniffing, and urinating in new places, but there was minimal revelation. The next year, however, in collaboration with veterinarians at DAHI and neuropsychiatrists at the University of Saskatchewan, they published more substantial findings concerning behavioral changes and improvements to "cognitive function."[85] Some of their old dogs, originally incapable of certain experimental tasks, performed dramatically better after a dose of selegiline. The study supported parallel findings from the NIH Geriatric Psychopharmacology Unit that both short- and long-term selegiline treatment of human Alzheimer's patients produced minor improvements in free recall.[86]

The Toronto team found few studies of canine cognition while designing their protocols and drew instead on a fragmentary, half-century line of inquiry into biological markers of aging in dogs. That trail was not particularly warm: German researchers working in the 1950s had noticed parallels in brain pathology between old dogs and old people, and Allen Andersen had proposed Davis beagles as ideal models of aging in the

1970s.[87] But by the 1980s, scientists disagreed about the extent to which beagle brains displayed the biological changes characteristic of Alzheimer's, particularly the amyloid-beta plaques that were becoming central to research after a paradigmatic paper, published in 1992, attributed Alzheimer's to an "amyloid cascade."[88] Milgram, Head, and two other psychologists note that "in an extensive search of the literature, we were not able to find any experimental studies on cognitive function and aging in which the dog was used, although the dog is possibly the most widely used species in other areas of biomedical research."[89] The few citations to canine studies in their first published paper, from 1992, refer mostly to research decades earlier on canine responses to amphetamines.[90]

Following the halting but positive progress from Toronto, DAHI signed an agreement with the Inhalation Toxicology Research Institute at Lovelace in 1991 to study aging in beagles at the Lovelace colony.[91] In this way, the dogs, surviving control animals from earlier AEC studies, became living connections between the beagle studies of the past and those of the present. In 1992, the Toronto-DAHI collaboration also brought in a new set of partners: neuropathologist Carl Cotman, of the University of California, Irvine, whom Milgram knew professionally, and a number of Cotman's students and colleagues. The Irvine group had coincidentally begun studying dogs as models for Alzheimer's around the same time, building on initial research by Michael Russell at Davis and specialists from the Sanders-Brown Center on Aging at the University of Kentucky, who used archived beagle brain samples to explore familial inheritance of Alzheimer's.[92]

In 1993, Cotman and colleagues at Irvine published a more extensive study exploring amyloid deposition in beagle brains. The dogs, raised "in a controlled setting, maintained in good health," and with "available medical records," were some of the final survivors of the multidecade radiobiology project at Davis, which formally ended that year.[93] The beagles, who either died spontaneously or were euthanized at late stages of terminal illness, had a median age of sixteen, placing them alongside the Lovelace dogs as some of the oldest living exemplars of the breed. Descendants of survivors of the nuclear catastrophe that befell many of their brethren, some had Alzheimer's-like neuropathology akin to that of many humans in a similar position. Seen from one vantage point, the beagle model of Alzheimer's was yet another spinoff of Cold War radiobiology. From a second angle, however, all the debates about "normal" aging in dogs at the atomic beagle colonies had been aiming toward precisely this kind of study. Here was a summation of nearly a century of biological inquiry: from fears of abnormal development and premature mortality, to worries about ever-longer lives.

A prominent collaborative paper from Irvine, Toronto, and DAHI argued in 1995 that beagles could serve as models for "age-related cognitive decline," an umbrella term connecting everything from minor dementia to Alzheimer's.[94] Acknowledging that there was neither a formal definition for, nor a standard approach to, studying cognitive deterioration in dogs, they cited anecdotal evidence of learning deficits, confusion, reduced vitality, and related symptoms, arguing that these were "equivalent" to the *Diagnostic and Statistical Manual of Mental Disorders* (*DSM*) criteria for dementia. The maxim that old dogs struggle to learn new tricks now carried a potential veterinary diagnosis, what DAHI researchers called canine "cognitive dysfunction syndrome" (CDS).[95] Because the beagles appeared to age more similarly to humans than rats did, were cheaper to house than primates, and produced age-related pathological changes faster, the group argued that "elderly dogs with cognitive dysfunction provide an excellent spontaneously occurring animal model of dementia, Alzheimer's disease, and other neurodegenerative disorders."[96] The term *spontaneous* was important and signaled a shift in the use value of laboratory beagles: from their value as standard, blank slates for experimentation, increasing attention would be paid to the "natural" development of parallel disease conditions.

The 1995 paper was a moment of convergence, between laboratories as well as academic and business concerns, but it cloaked underlying tensions. The scholarly side of the research collaboration had been interested from the very beginning in developing tests and metrics that would allow dogs to serve as experimental models for studying human cognitive decline. Researchers such as Head wanted to know whether aging dogs might reveal aspects of the progression and treatment of dementia that rats and primates did not. The DAHI researchers, on the other hand, were squarely focused on establishing a scientific foundation in order to sell selegiline to veterinarians and pet owners. Indeed, both parties acknowledged, at least to each other, that CDS was designed as a marketing term.[97] In a literal sense, the condition did not exist before 1992.

The company's first drug, trademarked that year under the name "Anipryl," was selegiline by another name and dosage. But what exactly Anipryl did was initially up in the air. A pilot study in 1992 at Kansas State University's veterinary school suggested the drug might treat Cushing's disease in dogs, a condition resulting from overproduction of cortisol in the adrenal glands occasionally described as the dog version of Parkinson's.[98] Anipryl was thus initially, in name and function, what DAHI's name suggested: L-deprenyl for animals. The "Ani" prefix merely signaled the drug's crossing of the species border. But in July 1992, the company took out a second patent for use of Anipryl as a treatment for "cognitive dysfunction" in dogs, following the vague suggestions of research at Toronto. "The diagnosis and therapy of cognitive dysfunction is relatively new in veterinary medicine," noted a report on the patent, which was an understatement: CDS existed as a diagnosis in part so that Anipryl could treat it.[99] To build capacity for the drug's production, whatever diseases it ended up targeting ("maintaining weight during mammalian aging" was added later in 1992), DAHI purchased equity in Phoenix Scientific, a manufacturer based in St. Joseph, Missouri, with experience in generic animal pharmaceuticals.[100] In September, the company received its patent for treating Cushing's but reported extensive losses, spooking investors.

The expanded effort to sell Anipryl nevertheless broadcast a new understanding of canine cognition and aging to a wider audience. What the Toronto researchers saw as an exciting avenue for research into Alzheimer's rapidly became a novel way for pet owners to conceptualize and act on changes they occasionally noticed in their beloved dogs. As a short, syndicated article explained in October 1992, "Many old dogs suffer from a dementia that is similar to that in humans: they no longer recognize familiar people, wander aimlessly and often lose control of their bladder functions without any apparent physical cause." A cartoon ran alongside the piece in Santa Fe's *New Mexican* depicting a confused and unkempt dog standing on hind legs, holding a street map in his or her right hand.[101]

The cartoon makes explicit the analogical work behind CDS. The dog is clearly anthropomorphic: in his or her map reading, standing on hind legs, and human-like front arms. The comparison is highlighted by a worried feline sitting on a nearby fire hydrant, portrayed very much like a conventional cat. Recycling folk wisdom about the surprising intelligence of dogs and the similarity of theirs to our own, the cartoon also indexed a significant reversal in the neuroscientific fortunes of dogs, who were now far more familiarly cerebral than cats, the old favorites of neurophysiology. Here species projection helped make cognitive dysfunction syndrome sensible to the general public: a dog with CDS is just like a very forgetful man. This dog is your grandfather, trying to find his way in a new city. And yet, a certain ambiguity remained about the human-dog comparison, just as it had in Auerbach's description of the smoking beagles, because the forgetful, humanized old dog stands atop an exceedingly chewable bone.

With the term "cognitive dysfunction syndrome" now out in the open, DAHI worked quickly to establish CDS as common sense. A 1993 press release described the condition as "a syndrome frequently encountered in pet dogs which is recognized by many veterinarians and pet owners."[102] The next year, a newspaper article noted that pets were living longer than ever and that vets were noticing behaviors similar to those in people with Alzheimer's. "In vet-speak," the article continued, "it's called cognitive dysfunction syndrome."[103] In two years, the diagnosis had gone from "relatively new" to "frequently encountered," from novelty to established "vet-speak." On the one hand, CDS was a clear-cut example of "disease mongering," the "selling of sickness that widens the boundaries of illness and grows the markets for those who sell and deliver treatments"—but this time directed at pets and their owners.[104] Many vets who spoke the condition into existence were connected to DAHI, and Head recalled many Toronto-area practitioners resisting the idea of CDS at first. They suspected, not without reason, that it was a pharmaceutical trojan horse.[105]

Although DAHI initially planned a large internal study of Anipryl, the company instead began encouraging pet owners in the United States and Canada to volunteer their dogs for treatment. Veterinarian Ron Worb was "looking for a few good dogs" to participate in one such study in 1994.[106] Partly a marketing scheme, the request also reflected a growing trend of research with "volunteered" dogs and hinted at the shaky scientific basis of early claims about Anipryl. The studies at Kansas State that found cognitive benefits in aging dogs, for instance, included only three experimental subjects—nowhere close to FDA requirements for testing a drug for the human market.[107] But volunteering for Anipryl treatments would allow both "dog and owner to lead more normal, rewarding and healthy lives."[108] This was veterinary drug testing as citizen science, but it also drew dogs into wider discourses around pharmaceutical use by modern consumers: just as human subjects were envisioned as actors attempting to maximize their "health" and "quality of life" through safe drugs and lifestyle interventions, a *normal* dog was necessarily a responsible user of pharmaceuticals, according to CDS sales pitches.[109]

The Market in Canine Illness

Anipryl was slow to make its way to market, but Shulman's subsidiary nonetheless found itself at the leading edge of a new area of pharmaceutical research: drugs for dogs. Traders on Toronto's Bay Street, the Canadian counterpart of Wall Street, had referred to Deprenyl Animal Health from the beginning by a shorter moniker, "Deprenyl Dog."[110] While Canadian investors were wary of involvement because of Shulman's controversial personality, Americans saw opportunity in Anipryl's promise to mine value from a new market of absentminded pups. "The approval process could take several years, but might be worth the wait for adventuresome investors," noted Richard Wholey of Chicago investment firm Wayne Hummer and Company in 1991.[111]

Adventure was there, and not without risk: few chronic-use drugs marketed squarely at domestic dogs existed in the early 1990s. Heartgard-30, Merck's monthly preventive ivermectin-based pill for heartworm in dogs, appeared in 1987, but its marketing focused initially on vets.[112] Advertising for Program, however, the first long-term flea and tick medication, shifted the focus squarely to consumers, promising a "quantum leap in flea control."[113] Beginning its two-year approval process in 1993, Program, which worked by inhibiting reproduction in fleas that consumed blood carrying the compound, was initially controversial even within the company that developed it, Ciba-Geigy Animal Health. Because Program prevented fleas rather than killing existing ones, researchers thought dog owners would be largely uninterested: owners still wanted quick fixes, not preventive care.[114] Nevertheless, the Swiss Ciba-Geigy, which was invested in expanding its presence in the American pet market, embarked on a "big-bang" advertising push to explain why a dog owner might pay for Program in the first place: Murray, the border collie star of *TV's Mad about You*, appeared at Ciba's first press conference to endorse the drug.[115] Program was like "birth control for fleas," ad copy asserted, while the company's internal slogan regarding the drug was "Think big."[116]

When Anipryl began its regulatory journey, therefore, there were only a few examples of commercial success for drugs for pets. Approaches to regulating them also varied dramatically between countries, leaving a potential regulatory maze. For this reason, DAHI was selected in 1991 to participate in a binational review aimed at harmonizing American and Canadian veterinary medicine regulation, a collaborative project between the FDA and Health and Welfare Canada's veterinary division.[117] Anipryl's regulatory journey would serve as a test case for future pet pharmaceuticals headed for the world market.

Cross-border harmony was slow to arrive. As Shulman stepped down from the board of DRL in May 1993 under pressure about his leadership, Anipryl was still slogging toward regulatory approval. Following his brief departure, Shulman reappeared in August with a major investment offer to regain control of the company but withdrew it just as quickly. In preparation for Anipryl's approval, DAHI inked an agreement with Hoechst Veterinär, a subsidiary of the company through which Siegfried Kamphans helped introduce beagles to Germany, to market and develop Anipryl for the European market. But reporting large losses, running short on funding, and floating on a one-time loan from its parent company, DAHI needed a victory soon to stave off collapse.

Early in 1995, Draxis Health, the new name of Deprenyl Research Limited, ratcheted up its equity holdings in DAHI. Company insiders whispered that government approval was finally around the corner for Anipryl. A letter of intent was signed with Pfizer to distribute and market Anipryl, but Pfizer grew distracted acquiring SmithKline Beecham's animal health business, and the deal fell apart. Draxis formally filed with the US Food and Drug Administration in September 1995 for use of Anipryl to treat dogs suffering from Cushing's, a population estimated around 170,000 in North America. The drug received approval in Canada one month later.[118] By New Year's Eve, Draxis took Pfizer's place as Anipryl's proposed North American distributor, and the drug began arriving to Canadian consumers in the spring of 1996.[119]

Canadian approval was welcome news, but the United States and its massive pet population were the big prize: analysts predicted that stock prices would double on news of American approval. The FDA, however, requested more clinical trial data, which slowed the process.[120] Even more than its approval for Cushing's, it was Anipryl's indication for "cognitive dysfunction" that represented the true motherlode. Covering a larger and more ambiguous set of symptoms, CDS affected more than ten times the number of dogs with Cushing's. Papers were filed with Canadian regulators in June 1996 to treat this "Alzheimer's-like disease" in dogs, one supposedly affecting as many as 150,000 Canadian dogs each year.[121] DAHI hoped to follow with approval in America in 1997, where CDS supposedly afflicted at least 1.5 million dogs in any given year. But there was virtually no meaningful research on the condition's veterinary burden, and these estimates were ultimately just calculations of elderly dogs.

Anipryl finally hit the American market in 1997. Prices for Draxis stock rocketed upward. DAHI dusted off its earlier distribution agreement with Pfizer Animal Health, who had successfully released canine arthritis treatment Rimadyl earlier that year and saw Anipryl's obvious value. Like Rimadyl, but unlike many earlier veterinary medicines, Anipryl was an everyday pill. It demanded impressive spending from owners: nearly three dollars per day, one thousand dollars per year. This was far more than Program, whose once-a-month regimen was already contentious at its debut. The two much-hyped arrivals, Anipryl and Rimadyl, marked a dramatic transformation in the pet medicine market and the very idea of companion medicine.

That change, however, was analogous to changes in human pharmaceuticals over the previous decades, as firms turned aggressively to marketing drugs for chronic, long-term conditions with vast, ill-defined patient markets.[122] Antidepressants or antianxiety medications were lifetime commitments; their users, lifetime customers. Pet drugs promised something similar, particularly because the size of the market was tricky to determine. Statistics on pet ownership or canine disease were less exhaustively maintained than those concerning human illness, and most articles about Anipryl referred only to disease "estimates." Conservative approximations placed the number of dogs who could benefit from Anipryl in Canada around 150,000, but others simply listed the country's population of "old dogs": around 700,000, all of whom could be imagined, at least potentially, to need the drug.[123] In the United States, the number of such dogs was calculated to be 7.3 million in 1999.[124] As with human Alzheimer's patients, the population of longer-living, old dogs appeared poised for further growth, not least thanks to a "whole generation of aging boomers who are deeply attached to their old dogs."[125]

This recognition was itself novel. Anipryl, guessed Martin Barkin, the aptly named chairman of DAHI, was "probably a $100-million a year drug." Yet it was only recently, he noted, that the size of the animal market seemed "sufficiently rewarding to justify the risks."[126] Where most veterinary pharmaceuticals had focused on farm animals, the modern animal pharm grasped onto companion animals as a new frontier for profit. Companion drugs also typically reached the market more quickly than those for livestock, because no data about residues needed to be provided to regulators.[127] Although Novartis's Clomicalm, a canine antianxiety medication released just after Anipryl, received much of the initial press—because of the humor of "chill pills for fido" and echoes of a 1976 *Saturday Night Live* skit about "puppy uppers" and "doggy downers" (featuring a prominent beagle)—Anipryl was at the leading edge: "You rarely see anything better than this, especially with animal products," noted a Boston financial analyst about DAHI's deal with Pfizer.[128]

"When something's just not right with your older dog," began a half-page Anipryl spread in 1998, "*it may not be natural aging.*"[129] The ad's language indexed a remarkable transformation in the understanding and discourse surrounding senility in dogs. What Andersen and the researchers at Davis had struggled desperately to decipher—the boundary between what was "normal" and "abnormal" in a beagle's aging—was now quotidian knowledge, observable by any thoughtful pet owner. "If your dog is experiencing one or more of the following signs," including limited interaction, loss of house training, or changes in activity level, "he may be suffering from CDS rather than simply the effects of aging," noted Pfizer's advertisement.[130]

Anipryl marketing left a fuzzy boundary between the apparently inevitable loss of functionality due to normal aging and potentially reversible losses due to abnormal aging, shifting the ground about what aging owners should consider natural. Here was real-world application of Lawrence Cohen's insight that "aging" is not a strictly objective condition but one of many natural orders useful for projects of ideological worlding.[131] With Anipryl, dogs were formally integrated into the marketing of a Western conception of aging. A host of behavioral signs, traditionally accepted as eminently natural, could now be reinterpreted by pet owners and vets alike as justifications for a prescription.[132]

A Dog-Too Pipeline

The advertising incitement to "ask your doctor" about treatments successfully spurred many patients to seek out novel drugs, and ads for Anipryl almost universally included the directive to "consult your veterinarian today."[133] Partly aimed at discouraging the uncontrolled use of human drugs on companion animals, the recommendation had the undeniable further effect of hailing a generation of pet owners into pharmaceuticalized companionship.[134] "I knew that if we got our message out, people would flock to veterinarians," explained John Payne, the vice president of animal health at Bayer, echoing the era's advertising common sense.[135] It was no longer enough to properly feed and exercise your dog; one also needed to know which drugs they required. This responsibilizing tactic was implemented through items like the 1999 "CDS Signs Quiz," which asked owners to check for the condition, because "the signs of CDS will most often be seen only at home and not during a routine veterinary exam."[136] Here a paradox emerged, and expertise was inverted: pet owners needed to consult a veterinarian even though they were themselves uniquely capable of interpreting signs correctly. This paradox reflected a lingering tension in veterinary practice between trained expertise and lay contestation from owners. Pet pharmaceutical companies are now investing heavily in genetic testing and AI image recognition diagnostic tools partly in order to sidestep these thorny questions about expertise.[137]

Extensive marketing behind canine pharmaceuticals such as Anipryl embraced and magnified the popular rhetoric of "bonding" and "companionship." Advertisers and corporate executives referred extensively in the late 1990s to the ever-expanding American commitment to dogs, one that appeared as recession-proof as bars and liquor stores. While "animals" might not obviously deserve costly drugs, "companions" did, earning what Haraway has called the "right (obligation) to health."[138] Companies embraced this framing after research showed that pet owners would pay whatever it took to guarantee a companion one or two more years—long before Loyal could promise it to them. And even before Anipryl's approval, studies found that many owners were already using the drug off-label to treat CDS. "There is a tremendous human-animal bonding relationship established, especially with senior dogs," noted Pfizer veterinarian Edward Kanara, who argued that Anipryl could "restore" that bond's deterioration with age.[139] Corporate biocapitalism thus accommodated the companion species outlook, a process that should provoke caution about the central role that care and companionship have taken in academic theorization of multispecies worlding. As Haraway summarizes, "Rights to health and family-making practices are heavily capitalized and stratified, for dogs as well as for their humans."[140] Fewer drugs targeted cats, on the other hand, because the industry found that cat owners were more likely to leave them outside and less likely to pay for medicine.[141]

But drugs for dogs were intriguing not just because of the market in emotional investment. Canine pharmaceuticals also intersected with the drug development process itself in unique ways. On the one hand, despite DAHI's various hurdles, approval was typically easier to reach with nonhuman pharmaceuticals than with blockbuster human drugs. Where the latter might require rodent, beagle, and human studies, a drug destined entirely for dogs meant that companies could skip the human clinical trial, a major, costly stage in the process that typically entailed additional years of testing and approval. Clinical trials for pet pharmaceuticals were different in form, and many relied on collected reports from veterinarians, based on the assumption that they were the "most reliable sources of test subjects and the most sensitive measures of an experimental drug's effectiveness."[142] By succeeding first in dogs, companies could also test the market for future human drug possibilities, as for instance with Fort Dodge's successful canine Lyme vaccine, LymeVax, which came out in 1992 and inspired many predictions of a natural human follow-up.[143]

Conversely, drugs that succeeded in humans were often already partially proven in dogs, making canine variants of human pharmaceuticals a kind of alternative "me-too" pipeline: the dog-too drug. Pet pharmaceuticals did not command the massive research budgets of human drugs, so "what we do is piggyback on a lot of the research done in human medicine," noted Guy Teddit of Novartis.[144] As veterinarian Barbara Simpson optimistically put it, "Animals have been used so long to test drugs for humans. Now, tests in humans are guaranteeing the safety of drugs for animals."[145] More cynically, psychopharmaceuticals for companion animals represented an excellent way to extract surplus value from previously marketed drugs, with Valium already given to cats and Prozac surreptitiously prescribed to dogs.[146] The dog-too approach was the origin for both Anipryl and Clomicalm, a pet-approved version of the Novartis OCD drug Anafranil. Puppy uppers and doggie downers differed from human medications mostly in dosing and labeling, plus a meat element added to make some palatable.[147]

Because a large number of drug candidates failed in their final human trials but succeeded in canines—deprenyl had minimal effects on human cognition—pet pharmaceuticals allowed corporations to recoup a degree of lost investment. Pfizer's Rimadyl was originally destined to treat arthritis for humans, but the drug did not last in the human body long enough to have its desired effect. It did last, however, within the slower metabolism and pharmacokinetic time of dogs. And, because the system for removing drugs from the pet market was less consistently enforced, a dangerous treatment could remain available to consumers in the face of countervailing evidence. Such was arguably the case for Rimadyl, which appeared to cause life-threatening side effects in some dogs, but nevertheless retained its FDA approval. In response to an outcry from enraged pet owners in 2000, the agency simply urged veterinarians to take precaution in articulating Rimadyl's risks.[148]

Anipryl was unique, however, because it treated something more nebulous than arthritis. Cognitive dysfunction syndrome, which was always ambiguous, gradually supplanted the earlier veterinary notion of "old dog's disease," which referred to any number of suspect transformations in an aging pet. If your dog was slower, less affectionate, having trouble finding things, or displaying a number of other symptoms, he or she might suffer from CDS. Many owners found that Anipryl did resolve some concerns, but like other psychopharmaceuticals marketed to canines, the drug benefited from a relative lack of understanding of canine health. Because dogs could not exactly speak their ailments into existence, they were even more open to projective interpretation than were people. With patents for use against Cushing's, CDS, weight loss, and numerous other conditions, Anipryl was positioned as a kind of canine cure-all. Given failures to develop anything similar for human beings, the framing naturally invites suspicion.

Nonetheless, drugs for dogs were also a safer investment than many human treatments. They had nearly guaranteed returns and certainly promised a larger market than the flatlining one for livestock medications. In turn, companies spent heavily to sell them: Pfizer funneled at least ten million dollars into Anipryl's launch, including "print, direct mail and broadcast and cable TV ads."[149] The Minnesota advertising firm Colle and Mcvoy, who oversaw the introduction of Rimadyl, produced an extensive campaign backing Anipryl. Early commercials featured a sad-looking dog interspersed with joyful human-dog bonding moments and informed viewers that Anipryl was "giving old dogs a new lease on life."[150] Where pharmaceutical companies spent seventeen million dollars on advertising for pet drugs in 1994, the collective investment had jumped to nearly $97 million dollars only three years later. With a total US market estimated around $2.6 billion dollars, growing between 10 and 20 percent annually, there were reasons to invest in dogs. When Pfizer spun off its animal health division as "Zoetis" in 2013, the company's initial public offering was the largest from an American company since Facebook.[151] In 2024, the size of the companion-animal health market was valued around $22.7 billion and expected to reach $61.6 billion by 2033.[152]

Following safe money was good business, but many saw the growth of the pet market as a reflection of industry tendencies to deprioritize research into treatments for some of the most dangerous diseases affecting large swaths of the world's human population, such as malaria, tuberculosis, and acute lower-respiratory infections. As Ken Silverstein noted in the *Nation*, "Only 1 percent of all new medicines brought to market by multinational pharmaceutical companies between 1975 and 1997 were designed specifically to treat tropical diseases," just thirteen of 1,223 medications.[153] Yet in 1999, the twenty largest drug companies spent around $215 million on pet pharmaceutical research and development, a number that was far smaller than the equivalent for human research but still a vast sum.[154] With the companion animal market "exploding," Silverstein pointed to a paradox: "The same companies that are indifferent to malaria are enormously troubled by the plight of dysfunctional First World pets."[155] Such concern for pets was matched by concern for the plight of wasted research.

Finding Alzheimer's in the Canine Brain

In November 1996, Draxis acquired the remaining shares in DAHI, closing the circle that spun it off from its predecessor. By the time of Anipryl's FDA approval, many key players in the early history were less than thrilled about how their story had concluded. Knoll filed a lawsuit against Draxis in 1998, alleging a violation of his right to profit from the use of selegiline in animals, a clause in an early agreement with Shulman. Shulman supported Knoll's claim, alleging that he was also duped into selling off stock in the company shortly after being ousted as chairman. Even Milgram claimed he "got burned horribly" when twenty thousand options in DAHI were "taken away without his knowledge."[156] Knoll eventually won access to share options, but Shulman took his battle against Draxis further by fomenting a proxy battle over the company's board. He was eventually defeated.[157]

As their involvement with Anipryl lessened, Milgram and Head continued to believe that their explorations of cognitive aging in dogs might offer fundamental breakthroughs for the study of Alzheimer's. In particular, they thought beagles represented a model for the study of the disease with unique benefits, including natural development of the disease and brain pathology that was remarkably consistent with that found in human beings. Mice were also popular Alzheimer's research subjects, and a report on the first transgenic mouse model of Alzheimer's appeared in 1995, but a standardized canine model for Alzheimer's could conceivably allow scientists to track the disease's natural progression over months or years in a way that genetically modified rodents did not.[158] Canine brains were also gyrencephalic, or folded, like human brains, a factor associated with expanded cognitive function; rodents, by contrast, show almost no gyrification.[159] "We believe that the aged canine will prove valuable in bridging the gap between transgenic models and man," the Toronto-Irvine collaboration explained in 1996, extending the long-standing discourse of dogs as a species "bridge" between mice and men.[160] Back in 1992, Milgram and Head had also noted an added benefit of dogs: the animals shared a common environment with humans and might be "subject to similar environmental stresses."[161] If Alzheimer's was due not just to the appearance of amyloid plaques, but also to triggers in the environment—what Margaret Lock differentiates as the "localization" and "entanglement" theories—dog studies might reveal the operation of these factors in dementia while still supporting efforts to localize brain pathology.[162]

The Toronto team had worked with mixed breeds and volunteered dogs, but gradually they settled on beagles. Access was key: major commercial breeders, such as Marshall Farms, sold beagles, and the few surviving lab colonies, such as those at Davis and Lovelace, had beagles too. But beagles were also better than other plausible canine candidates because of their small size. For earlier researchers, the breed's size had principally been an economic factor limiting food costs, but now it also came with the promise of extended life. Dog experts were long aware of significant breed differences in aging—small dogs tended to live much longer than large dogs—but such variation was relatively unexplored, scientifically, at the time. In her dissertation, submitted in 1997, Elizabeth Head summarized that "beagle dogs and pound source dogs show differential aging effects in terms of behavior. Breed and task interact such that aged beagles are impaired on tasks that aged pound source dogs show no deficits and vice versa."[163] The comment suggested value in studying a multiplicity of breeds, but accounting for crossbreed susceptibility with a mixed study population was too expensive to undertake on limited research budgets.[164] Beagles were, once again, good enough.

Throughout the 1990s, much of the Toronto and Irvine beagle work centered on determining whether dogs actually displayed the biomarkers of Alzheimer's. The answer was unclear. Back in 1911, the Polish neurologist Teofil Simchowicz, supervised by Alois Alzheimer, had studied brains from a twelve-year-old and a seventeen-year-old dog and found no evidence of senile plaques.[165] Three years later, Spanish neurologist Gonzalo Rodríguez Lafora, a disciple of the renowned neuroanatomist Santiago Ramón y Cajal, looked at the brain of a fifteen-year-old basset hound and largely confirmed Simchowicz's findings.[166] Only in the 1950s did psychiatrist Anton von Braunmühl, at the Haar Mental Hospital near Munich, finally identify plaques similar to those in Alzheimer's in older dogs, studying brain samples from twenty individuals.[167] The larger number and older average age of his dogs was likely crucial. Building on these findings, a cohort of researchers at the Albert Einstein College of Medicine, led by pathologist Henryk Wiśniewski, had argued in 1970 that the neuropathology in old dogs mirrored human cases, later proposing use of dogs as a model for understanding the disease.[168]

Still, understanding of how canine brains changed with age was limited, and existing findings, contested. The chapter of Andersen's book on "beagle geriatrics," once predicted by Betsy Stover, had yet to be written.[169] Looking at material from over one hundred beagles and mixed-breed dogs, the Irvine-Toronto-DAHI collaboration was able to identify extensive evidence of early-stage amyloid plaques but little sign of the later-stage plaques that seemed especially important in severe dementia.[170] This, they argued optimistically, positioned dogs as useful models for understanding the initial stages of plaque formation and thus the initial onset of dementia—a species bridge *and* an aging bridge. On the other hand, there was widespread agreement that dogs did not possess the neurofibrillary tangles that also appeared significant for the disease's progression, meaning that studying them could illuminate only some of the vital processes.[171] There were, thus, reasons for skepticism about using dogs to study the characteristic pathological changes.

Another challenge lay in determining whether living dogs could actually be diagnosed with "dementia" or "Alzheimer's." While the fourth *Diagnostic and Statistical Manual of Mental Disorders* (*DSM-IV*), published in 1994, gave defined diagnostic criteria for dementia, they were difficult to translate to dogs. The *DSM-IV* noted, for instance, that dementia would "cause significant impairment in social or occupational functioning and represent a significant decline from a previous level of functioning."[172] In service dogs, these changes might be more or less clear, but what was occupational impairment for a dog without a job? What was a decline in function for animals who are notoriously quirky? An older dog might be less interested in food, sleeping differently, or more frightened, but those changes could be signs of a number of things. Many pet owners might also struggle to monitor changes in their older dogs' behavior closely enough to even notice. "Consult your vet" was one answer, but finding an objective baseline for research was trickier.

To screen for cognitive dysfunction in dogs, Head took Stanley Coren's dog intelligence test, a new diagnostic tool, and developed a "Canine Mini Mental Exam."[173] Questions included whether dogs ate sloppily or exhibited poor grooming habits—answered by their owners or caretakers—and queried changes in their social interaction with humans. While each question related canine behavior to analogous symptoms of Alzheimer's in humans, the beagle researchers also saw the diversity of canine aging symptoms as its own small benefit, a "spectrum of behavioral problems" that was "similar to that observed in people with dementia of the Alzheimer's type classified at stages 4–6 on the global deterioration scale."[174] Although the general similarity of beagles was important, the diversity of individuals was here valuable for research.

The "Mini" exam was a rapid screening device, but more complicated tools were needed to see cognitive decline in action. Head adapted the Wisconsin General Test Apparatus (WGTA), a laboratory artifact first introduced by Harry Harlow and John Bromer to study learning in primates in the 1930s, to accommodate dogs. The WGTA, which "brought to the study of primate learning capabilities what Henry Ford's assembly line brought to manufacturing," was a "prolific and reliable" scientific platform, endlessly reconfigured and redeployed by researchers after its invention.[175] In Head's version, a multiple-box contraption allowed for stimulus trays with food-smeared objects to be presented to dogs while blocking experimenters from view, limiting the role of human influence and allowing Toronto researchers to test large numbers of dogs quickly.[176]

Beagles were pretrained with a series of tasks and then run through a battery of cognitive tests. Their discrimination learning, or ability to differentiate similar stimuli, was tested. So was reversal learning, their ability to disengage from previously learned behavior. A delayed-nonmatching-to-sample test assessed their memory of previous stimuli. Just as a device designed for primates was reequipped for dogs, behavioral measures that were age sensitive in nonhuman primates were translated to make sense of the dogs' performance. When it came to behavior, finding Alzheimer's in dogs presumed a fundamental comparability between signs of aging in dogs and in nonhuman primates, such as macaques, whose cognition was more comprehensively studied. This dog-primate-human chain of comparisons was essential, because a successful model needed to express parallel, if not precisely identical, behaviors. As cognitive proxies, then, beagles required additional mediation. For Alzheimer's, a complicated multispecies ladder now ran from mice to dogs to nonhuman primates to people.

The similarity of dogs and macaques was complicated, however. While elderly primates typically learned visual discrimination tasks more slowly than their younger peers, older dogs were often incapable of acquiring competency in them at all. Head and others hypothesized that physiological differences might be a factor, since canine brains seemed to dedicate less real estate to processing object attributes. In turn, they designed an altered version of the nonmatching-to-sample task that emphasized the spatial location of a test object rather than visual recognition of it. In the new version, dogs performed extraordinarily well.[177] A follow-up study revealed meaningful differences between aged and younger dogs, including deficits in spatial learning and working memory.[178]

Building on this initial differentiation of species capacity, the lab began to wonder whether focusing on visualization was part of the problem. Canine vision remains poorly understood, but research in the 1990s began to break down a number of older myths: dogs do not see only in black and white, for example, as optometrist Gordon Lynn Walls influentially argued in the 1940s, but possess photoreceptor cones in their eyes that allow them to see at least grays, blues, and yellows. Nevertheless, evidence strongly suggests limitations in canine visual acuity compared with that of humans: a dog is unlikely to see something as clearly from as far away as you or I can.[179] The legendary canine olfactory capabilities, on the other hand, seemed to offer an alternative pathway to studying age-related deterioration. In a task that required young and aged dogs to discriminate between two identical objects, one that smelled like vanilla and the other like aftershave, many aged dogs who were "nonimpaired" on visual tasks demonstrated significant deficits in olfactory differentiation. Visual tasks, it seemed, were inadequate to diagnose minor aging changes in dogs.

Like dogs' vision, the canine sense of smell had long puzzled scientists. A dog's nose is filled to the brim with sensory neurons, hundreds of millions of which allow for minute differentiation between stimuli and for the ability to detect food or predators from significant distances, a talent already investigated extensively by the 1950s.[180] But how and why this worked, as well as what it meant for canine experience in the world, were far less clear. Despite decades of research into "artificial" or simulated canine noses, for instance, dogs remain the contraband detectors of choice at airports and commercial docks. Smell seems essential to a dog's very sense of identity, and research suggests that dogs may differentiate self and other primarily through smell.[181] If Alzheimer's is, in part, a dissolution of the self, the loss of olfactory discrimination might mean the same thing for a dog.

Much of this, however, remained mysterious during the early stages of Alzheimer's research with beagles. After years of studies, the cognitive tasks designed by Head and others were refined to the point that they appeared to meaningfully distinguish between animals with "normal" levels of decline and those experiencing moderate or severe cognitive decline. At the annual meeting of the Society for Neuroscience in 1993, Head and colleagues argued that tests of spatial memory provided "a measure of age-dependent cognitive deterioration," while tests of spatial learning could be predictors of abnormal amyloid accumulation.[182] The tests also appeared highly sensitive to treatment effects, including interventions such as, unsurprisingly, Anipryl.

After finishing her doctorate in 1997, Head left to join Cotman at Irvine's Institute for Brain Aging and Dementia, the beginning of a pipeline that would take many of Milgram's students to California's sunnier shores for postdoctoral positions. The Irvine-Toronto collaboration, with dogs from Lovelace and occasional support from the University of Kentucky's Sanders-Brown Center on Aging, where Head went in 2009, continued to work on clarifying age-dependent decline in the cognitive capacities of beagles. That the tests worked relatively well in beagles, however, also suggested the possibility of applying them to other populations with parallel linguistic or visual limitations—cats, for instance, or pigs. But one of the groups that researchers turned their attention to was surprising: in the early 2000s, members of the Irvine Alzheimer's group began to look into a noted but underanalyzed connection between Alzheimer's and Down syndrome, the genetic disorder linked to a third copy of chromosome 21. In doing so, their work raised difficult questions about the level of "similarity" between human and canine cognition and the role of beagles as cognitive proxies.

Troubling Comparisons

As far back as the late 1940s, those who studied dementia noted surprising parallels between postmortem brain analyses of individuals with Alzheimer's pathology and those with Down syndrome. People with Down syndrome are statistically very likely to develop dementia after the age of forty, which is far earlier than the general population. But this high incidence of dementia was poorly understood prior to the 1980s, partly because of a limited life expectancy for those with Down syndrome that began to seriously improve only over the last thirty years.[183] Amid the late 1980s and early 1990s boom in brain pathology studies, a major paper in *Science* in 1987 alleged that "duplication of a subsection of the critical segment of chromosome 21 that is duplicated in Down syndrome may be the genetic defect in Alzheimer's disease," a finding that newspapers celebrated for identifying the "genetic cause" of Alzheimer's.[184] Alongside the buildup to the Human Genome Project in the 1990s, a grand undertaking that raised the prospect of "reading" directly from the decoded "Book of Life," genetic explanations like this were appealing.[185] And even if reports of a singular genetic cause were ultimately overblown, studying how Down syndrome and Alzheimer's were related promised additional understanding of the latter's emergence.[186]

Diagnosing Alzheimer's in individuals with Down syndrome, however, had long challenged clinicians. Limitations to communication capacity meant that many traditional screening tests for Alzheimer's, which looked in part for impaired social abilities, could not be easily applied. In 2000, the Irvine team published findings with Irvine Down syndrome expert Ira Lott and University of Oslo anatomist Reidun Torp on efforts to molecularly date the appearance of senile plaques in the brain, drawing on preserved tissue from deceased individuals with Down syndrome as well as mixed-breed dogs and beagles.[187] Focused primarily on improving molecular dating techniques, the study avoided drawing broader comparisons between Alzheimer's in dogs and in humans. But if limited communication abilities were a key justification for developing the beagle tasks in the first place, there was reason to believe that bringing those learning tasks back across the species boundary might improve midlife diagnoses of Alzheimer's in people with Down syndrome.

This notion fit within an emergent paradigm known as "comparative neuropsychology," a combination of earlier work, such as Scott and Fuller's intelligence studies in dogs, with new understandings of brain structure and function. In one of the first major papers in the field, neurologist Marlene Oscar-Berman and psychiatrist Stuart Zola-Morgan had shown in 1980 that it was possible to test for Korsakoff's syndrome, a severe retrograde amnesia associated with alcoholism, by using learning tasks designed originally for nonhuman animals.[188] The majority of conventional comparative psychology, Oscar-Berman and Zola-Morgan had noticed, involved designing tests to study cognition in other animals and applying those results to homologous parts of the human brain. The reverse, they emphasized, seldom followed: human neuropathology was "rarely re-examined in the context of experimental paradigms which are known to be valid and reliable tests of nonhuman functional breakdown following brain damage."[189]

In turn, Oscar-Berman and Zola-Morgan modified a WGTA device, like the one used to test the Toronto beagles, in order to work with humans, and they redesigned reversal learning tests originally produced for nonhuman primates for their human patients. As we have seen, the use of animals as scientific proxies required the construction of specialty laboratory prosthetics. Yet here it was not a case of technologies enabling the use of nonhumans as proxies for humans, but rather the redeployment of tests and devices designed for that proxy relation back to humans.[190] This reversal demonstrated a second kind of "salvage" methodology: earlier tests designed for divergent aims were repurposed to support new experimental and clinical purposes.

Comparative neuropsychology proposed a partial dismantling of the presumed cognitive superiority of humans, one that had shaped views of nonhuman intellectual inferiority and framed much early comparative research. Tests that were produced initially to explore nonhuman animal intelligence, which the absence of language or higher cognitive abilities made it difficult to study, were recognized instead for their value in studying people who appeared to lack those same capacities. As Oscar-Berman summarized in 1994, "Experimental paradigms known to be reliable and valid tests of behavioral deficits in brain damaged nonhuman animals, can be adapted successfully for use with people."[191] This adaptation allowed for comparative studies of intelligence across the phylogenetic scale and in so doing suggested a broader continuum of ability on which living beings could be placed. Rather than on the thresholds of absolute intellectual difference that occupied earlier psychologists, the focus now was on appreciating points of similarity.

In 1989, Oscar-Berman and University of Toronto neurologist Morris Freedman had taken spatial and visual tasks designed for nonhuman animals and applied them to patients with Alzheimer's and Parkinson's, seeking to differentiate the anatomical and neuropsychological mechanisms behind the two conditions.[192] Freedman, in turn, joined the Irvine-Toronto collaboration in order to assist in designing a study of Down syndrome, led by neuropsychologist Linda Nelson, that would make use of methods previously tried on beagles and primates as well as the WGTA device from earlier experiments. Study participants received small penny rewards for successfully completing tasks, such as repeatedly identifying one of two household objects associated with a reward, and verbal praise, such as "That's great!"[193] For completing similar tasks, beagles would have received rewards of food.

Results across a range of tasks indicated that performance was highly correlated with a participant's score on the "Dementia Questionnaire for Learning Disabilities," a screening tool developed by Heleen Evenhuis and colleagues at the former Hooge Burch Institute to diagnose dementia in those with Down syndrome.[194] The Alzheimer's–Down syndrome work is thus legible as an attempt to approach, by novel methods, cognitive decline in individuals with different intellectual capacities. However, although Head and many others became prominent figures in the study of Alzheimer's and Down syndrome, further attempts to apply the beagle tests have been minimal. In their paper, Nelson and colleagues noted "important limitations" in translating animal tasks to humans. Many aspects of the study were challenging for individuals with Down syndrome and yielded a wide range of error scores that researchers believed would not exist in the performance of individuals without Down syndrome. This made it difficult to "generalize the findings" to individuals with "normal aging" or "other groups of adults with low IQ."[195]

The application of beagle tests to individuals with Down syndrome, despite optimistic and sincere intentions, raises uncomfortable questions about interspecies comparison. Were the tests humanizing animals by stressing interspecies parallels in cognition? Or did they risk animalizing human beings, members of a group that has historically been discriminated against on the basis of inferior performance on intellectual measurements?[196] Philosopher Jeff McMahan once proposed that dogs and some disabled humans have equal cognitive capacities, in part to criticize the presumption that people with disabilities lead an "unfortunate life."[197] But as theorist Michael Lundblad notes, drawing on a critical response to McMahan from philosopher Eva Feder Kittay, even a positively intended comparison may not be able to "stand up to the weight of history or the harm it can still produce."[198] Reflecting on a rhetorical comparison of dogs and children with Williams syndrome—that the latter might wag tails if they had them—Mariam Motamedi Fraser similarly cautions that the history of ableism could "make one feel not less but *more* uneasy" about the statement.[199] In this view, the application of tasks designed on a deficit model of beagle intelligence to those with Down syndrome seems only to join nonhuman and disabled populations together as inferior in contrast to an ostensibly "normal," or normative, cognition.

But Kittay's argument is additionally revealing: "I simply cannot know enough about what it is like to be a dog," she writes, "to think like a dog, to sense the world like a dog, to know how to compare my own, my daughter's, or any human being's intelligence to that of a dog."[200] What is at stake, in short, are both the epistemic and ethical limits of proxy relations. The application of beagle tests to those with Down syndrome confirms the profound tensions in using dogs as cognitive proxies. While the question of similarity in the context of pharmaceuticals was one of apparently neutral physiological differences—are the tumors of beagles *like* those of human beings—transferring cognitive tasks along the phylogenetic scale returned researchers once more to the phenomenological mire of distinguishing something like objective differences in cognitive ability. The metaphors that draw nonhuman animals and individuals with disabilities together threaten to overwhelm the quotidian work of scientific research.

The point is not that such studies were unique, but that they were coextensive with the longer, risky history of species projection. For over half of a century, scientists had imagined themselves and other humans, their pasts and futures, inside dogs. The animals functioned not only as scientific tools or prosthetics, but also as imaginative devices that could aid in understanding and making sense of what it means to be human. Much of this work was autobiographical, attempts by scientists to get a handle on the biological conditions of their mortality, but so too did research turn outward, with the animals serving as tools to study human others. The redeployment of beagle tests to study individuals with Down syndrome, which relied on a developed sense of the dogs as proxies for human cognition, represented species projection in the most literal sense. Now, however, ideas about dogs were projected backward onto human beings, along a ladder running from beagles to macaques to us. As the gap that separates the properly human from the properly canine closes, those individuals rhetorically and socially positioned in the interstices remain in fraught positions for future comparison.

Enrichments and Environments

If the connection between Alzheimer's and Down syndrome proved less fruitful than some originally hoped, another shift in the focus of the beagle studies offered a different, unexpected promise. Moving gradually away from the neuropathologically centered search for biomarkers, such as amyloid plaques and neurofibrillary tangles, researchers instead began to ask whether the *environmenticity* of experimental dogs—namely, the fact that dogs were defined through their relationship to human beings and their sharing of a general environment—might reveal insights into some of the alternative causes of Alzheimer's or even offer hints at possible nutritional interventions.[201] In this way, the experiments reflected a shifting sense about what made dogs "good" for science. Formerly pure, standardized experimental tools, dogs had now become valuable because they were usefully contaminated by life in a shared world.

In the early 2000s, the Toronto-Irvine researchers began supplementing the diet of their dogs with antioxidant-enriched foods, inspired by growing interest among biologists and gerontologists in the role of reactive oxygen species, so-called free radicals, in aging. Originating in the 1950s, the "free radical theory" asserted that a buildup of oxidative damage caused much of the incapacitation of aging. Although initially limited in popularity, it returned with a vengeance in the 1990s as a large cohort of researchers began to argue that mitigating the damage from free radicals could dramatically slow the aging process.[202] Experiments in fruit flies and mice appeared to bear this out, and some in the Alzheimer's field grasped onto the idea that vitamin E supplementation might delay or even reverse symptoms in human patients.[203] "In the last few years," explained *New York Times* science correspondent Jane Brody in 1994, "the word 'antioxidant' has moved rapidly from the domain of chemists and biochemists into common use, at least among health-conscious Americans."[204]

While early studies with rats had shown mild benefits from vitamin and other antioxidant food supplements, even late in life, the Toronto-Irvine researchers argued that results in beagles would offer more meaningful comparisons to "age-dependent cognitive dysfunction" in people.[205] In 2002, Brody reported on their early results, which "strongly suggest that some of the more debilitating effects of age on the brain" could be "averted or at least eased" by antioxidant-rich foods.[206] According to Milgram, who is quoted in the article, "Antioxidants are the best suggestion of a possibly useful intervention." It was an appealing argument to consumers as well as scientists, and studies such as these spurred a major advertising push toward "antioxidant-rich" foods and beverages, which flooded the United States in the 2000s, including Vitaminwater's "XXX" flavor.

An additional player had been important in pushing the beagle researchers toward antioxidants: Hill's Pet Nutrition, based in Topeka, Kansas. Hill's, a former rendering plant–turned–dog food company, was well known for producing both Prescription Diet, a special line of food for dogs with specific diseases, formulated by veterinarian Mark Morris in the 1940s, and the follow-up Science Diet product range, designed by Morris's son, Mark Morris Jr., in the 1970s.[207] Unlike the narrower market of its predecessor, Science Diet was aimed at the growing proportion of regular pet owners willing to pay higher prices for dog food if it came stamped with the imprimatur of science.[208] In January 2002, researchers associated with Hill's presented at the nineteenth annual North American Veterinary Conference in Orlando, Florida, on a one-year, in-home study that found substantial improvements in the behavior of older dogs receiving antioxidant-enriched food. The study relied on pet owners to report on their dogs' progress—because only an owner, after all, could perceive subtle signs of aging—and reported stunning recoveries in tested pets, including the cessation of home soiling.[209]

Hill's, which saw "brain" food as a significant new market area, had released what they called "Prescription Diet Canine b/d" dog chow the year prior. Early media reports tied the product explicitly to Anipryl, which "served many older dogs, but has not been of help in all cases," according to one newspaper.[210] Antioxidant-enriched food was therefore a plausible alternative, or supplement, for those whom Anipryl failed, and Hill's would soon add a special "antioxidant blend" to many of its products. Health-conscious pet owners could treat their dogs with food rather than drugs, just as they often preferred to do for themselves. A collaborative paper from the Toronto-Irvine beagle group, Lovelace's Bruce Muggenburg, and Steven Zicker of Hill's, which reported on studies of both elderly beagles from Lovelace and beagles from the internal Pet Nutrition Colony at Hill's, had argued that aging dogs kept on the special brain diet displayed improved performance on a number of learning tasks.[211] Hill's supported the research, as did the National Institute on Aging and the US Army.[212]

Parsing the data more closely, however, it was clear that food was insufficient on its own. A longitudinal, two-year study on the Lovelace beagles found that meaningful environmental enrichment, including housing the dogs in pairs for socialization, giving them extensive walks for exercise, and offering them cognitively challenging tasks, was key to realizing the full, CDS-fighting effects of antioxidant foods.[213] Environmental and nutritional enrichment also seemed to produce increased levels of brain-derived neurotrophic factor (BDNF), a protein that acts on neurons in the central and peripheral nervous systems and seemed to decrease with age.[214] Indeed, despite Hill's own emphasis on its groundbreaking scientific diet, work done by Head and others made clear that a beagle's best bet for resisting the encroaching threat of cognitive decline was a healthy, active lifestyle filled with social interaction and cognitive challenges. Zicker presented work at the tenth meeting on Canine Cognition in 2005, an annual event begun by Cotman, Milgram, and Muggenburg back in 1994, confirming that "the combination of dietary antioxidants and environmental enrichment was synergistic and resulted in the least amount of cognitive decline" over a thirty-month study period.[215] Diet could not, by itself, fight off CDS.

The longitudinal study of beagles, as a press release from the National Institutes of Health noted, was "among the first to examine the combined effects" of diet and exercise in fighting Alzheimer's.[216] It would hardly be the last, as recent decades have seen cascading attention to "lifestyle" prevention for comparable conditions. In this way, the shift from focusing on Anipryl to Hill's b/d mirrored the shift from a narrower focus on brain localization to a broader view of environmental causes in Alzheimer's research, from silver bullet pharmaceuticals to long-term, environmentally oriented prevention. As Margaret Lock shows, that shift was met with an accelerated pace of reporting from mainstream publications, particularly the *New York Times*, which offered updates on new research and an implicit promise to readers that they could delay the worst by keeping up-to-date with developments.[217]

But research on environmental and nutritional enrichment also demonstrates how the value of laboratory dogs came to emphasize their environmenticity. Beagles were still thought of as high-quality laboratory organisms, but they were especially useful because they lived, played, and breathed in the same kinds of ways that human beings did. Mutual conformations meant they were not just beagle-shaped bags of conditioned reflexes, but brainy individuals unavoidably entangled with the world, simulating cognitive decline on an accelerated timeline. Even if their intellectual challenges were doggy and reflected a canine umwelt, there remained a clear sense of analogy. Elderly dogs should live with others, play, and exercise; people should take walks and maybe try a brain-training game.

The studies additionally reveal how the pet industry became a meaningful player in research with human implications, a partial reversal of the original situation with Anipryl. As it turned out, the "entanglement" approach to Alzheimer's fit comfortably with the emergent double market in human and pet pharmaceuticals and specialized nutrition products. In 2003, Milgram, who had always been attracted to the possibility of commercializing his research, founded a company, CanCog, "a contract research organization providing non-invasive research and development services in the companion animal market," with Ontario veterinarian Gary Landsberg, who had aided Hill's in initial tests of their CDS-fighting food.[218] CanCog's first executive chairman was David Stevens, the former CEO of DAHI. Today, the market for canine pharmaceuticals is larger, by multiple magnitudes, than when Milgram first talked with Shulman.

The beagle collaboration survived CanCog's arrival, but Cotman and Head became the more visible faces of further studies of a canine model for Alzheimer's. By his own admission, Milgram's work became increasingly business oriented. At the tenth meeting on Canine Cognition, his coauthored summary of the conference argued that an "underlying raison d'être for studying cognition in aging dogs is to develop interventions for companion animals demonstrating cognitive dysfunction."[219] The pet market was a reason in itself for research. For some others, however, the emphasis was still on thinking about dogs as models for understanding human disease. In 2008, Cotman and Head summarized the previous decade of research by noting that dogs showed "age-dependent losses in cognitive function" and a "progressive accumulation of neuropathology" that made them "uniquely well-suited to studies of dietary, environmental and AD-pathology targeted interventions."[220]

In retrospect, the practical nature of living and working with dogs seemed to make this more expansive vision for the canine model almost inevitable. Experimental beagles need to be well exercised and fed, the finding of every dog colony since Utah. Their participation in experiments relied on feeding and exercise, because rewards were typically edible and tests of performance involved movement. In studies with beagles, the environment could not help but seep in, not least because researchers recognized that sad, lonely dogs made for bad research subjects. As John Paul Scott noted on his lab's studies of canine intelligence, "A dog that is timid of apparatus or other strange things, will do poorly on the apparatus, irrespective of how much intelligence it has, and an animal not motivated by that kind of reward that you want to give it—say, food—will also do poorly."[221] In a sense, researchers had been slow to realize what beagles and other dogs were telling them for decades: a healthy life was produced not just by micro-interventions targeted at individual bodies, but by the maintenance of a conducive environment. Vibrant, long lives were sustained by opportunities for socialization, clean air, excellent nutrition, and attentive health care. Here was the latent utopianism of laboratory animal research, one that the singular focus of experimental systems made difficult to grasp.

Companion Science and the Disappearing Laboratory Dog

Investment in studies of aging and the diseases of old age, including Alzheimer's, has grown continually over the last few decades, not least because of fears that dementia will be one of the twenty-first century's next great public health "epidemics."[222] Millions of individuals around the world, living longer than earlier generations, fear the dissolution of self promised by dementia. And while fruit flies, worms, and rodents have long been popular for basic studies of aging, their utility in projects that aim to increase "healthspan"—healthy longevity, rather than longevity as such—is increasingly questioned.[223] Instead, many now call to take up dogs for aging research.

In 2017, researchers from the Senior Family Dog Project at Eötvös Loránd University in Hungary established a Canine Brain and Tissue Bank in order to collect biological material for future studies. As the project leads explained, "Dogs, in particular, are gaining increasing attention in translational research on complex phenomena, like aging, cancer, and neurodegenerative diseases."[224] Citing work by the Irvine-Toronto collaboration and others, the Dog Project researchers argued that "two major limitations" of traditional animal models, notably the lack of similar neuropathology and differences in the lived environment, "could be addressed by involving companion dogs in dementia research."[225] Amid the environmentalization of aging research, dogs are positioned as ideal proxies because they are some of the only animals that live a lifestyle "like ours." Subsequently, their value and acquisition have changed in important ways.

By the 1990s, utilization of specialty-bred dogs in scientific research, especially in the United States, was declining across the board. Between 1978 and 1987, the number of dogs used in NIH-funded studies was 187,464; between 1998 and 2007, the number decreased to 69,223. Dogs remained more popular than cats, at a roughly three-to-one ratio that grew slightly from 1973 to 2007, but the great atomic beagle colonies were now closed. Those who still required beagles paid a premium for quality and secrecy from licensed, commercial "Class A" breeders like Marshall. A secondary set of dealers, designated "Class B" in American regulations, sold dogs acquired from various sources, such as individuals, hobby breeders, pounds, and shelters, but these "random source" dogs represented only 4 percent of the animals used in research during fiscal year 2007 in the United States, according to government figures.[226]

Studying aging in dogs obviously requires older dogs. Yet producing older dogs was not traditionally a prime focus for commercial breeders. Earlier scientists wanted young, "normal" animals: it was because Utah and Davis planners could not buy enough young beagles, for instance, that they had to breed their own back in 1951. Older dogs, on the other hand, who were often used in breeding, could occasionally be acquired for discounted prices. The early Toronto beagles were examples. Many smaller breeders also adopted out retired champions, sires, or dams—or simply kept them as pets. The control animals at Lovelace were, effectively, elaborately well-cared-for pets living quietly on until their natural ends. Their expected value lay in establishing baselines for comparison, but for those interested in cognitive deterioration, these dogs and their surviving compatriots at Davis were an ideal cohort of research subjects.

Their ancestry was known, and their lives had been similar. They were healthy and, importantly, *old*. Their experimental promise, in other words, was tied to traits that would have been considered defective in earlier views of what made a "good" laboratory dog. Value accrued to the old and "abnormal," rather than the young and "standard." Quite a few of the beagles that took part in Alzheimer's studies were thus not the Beagle™ commodities advertised by Marshall, but last year's models, sitting idly on the shelf.

This salvage turn to older beagles echoed the discursive architecture of the accumulation of pound dogs for research. Previous generations of scientists had argued that unclaimed pound dogs would be euthanized and should at least be used in research that might improve human lives first.[227] In parallel, some now argued that older animals should help scientists understand brain function or physiological changes rather than simply being killed. As their colonies closed in the 1980s, researchers at Utah, Davis, and Argonne had made the same point: their beagles were now most useful for work in gerontology, and studies of canine cognition might have drawn on hundreds of additional research subjects if they had gained steam earlier.

As two major changes restructured the modern life sciences landscape—tighter budgets for laboratory animals and pressure to reduce the number of animals in use—this renewed approach to salvage accumulation of dogs also encouraged novel forms of cooperation and sharing. Because old dogs were newly valuable but not always especially available, effective coordination could allow their use in multiple studies rather than a single project. Lesley Sharp, following Marilyn Strathern, calls such practices "collaborative creativity" and argues that they represent a now-common response to external demands such as the "3Rs": to "reduce," "replace," and "refine" the use of live animals in science.[228] Such collaborative creativity represented its own change to traditional styles of research, where animals were expected to largely stay in one facility and on one project.

The atomic beagle labs, for instance, were outwardly interdisciplinary: veterinarians supported physicists and biologists in conducting and maintaining the wide variety of experiments needed by the Atomic Energy Commission. Yet beyond the networks of conferences, workshops, and friendly travel among government-sponsored radiobiologists, beagles were rarely subject to the open forms of sharing and cooperation that Kohler has shown characterized the "moral economy" of fruit fly researchers.[229] While flies could be cheaply and easily shipped to collaborators, Andersen could not do the same with his dogs. However, some of the last beagles at Davis were used by anatomists, veterinary behaviorists, and Alzheimer's researchers in separate disciplines and on separate campuses. The dogs at Lovelace enabled a collaborative research program between academic and industry groups in New Mexico, Kansas, and Toronto, while also completing long-term studies on inhalation toxicology. The struggle to continue working with dogs produced a more extensively interconnected network of shared experimental materials. Political economy inspired a new moral economy.

If, as Simon Werrett has shown, modern science relied from its earliest days on "thrifty" efforts to "make the most" of available experimental materials, thrift became a condition of beagle research with the disappearance of large external funding sources and the imposition of stricter care regulations.[230] Today, when researchers in one laboratory euthanize a dog in order to study its brain, for instance, a second group conducting research on liver function and a third focused on cardiovascular health might each request access to those parts of the beagle. "The line often stretches down the hall," one researcher noted in an interview, only somewhat metaphorically. If a beagle must be killed, scientists look to use everything but the bark. New tissue banks and databases, such as the Senior Family Dog Project's, have also allowed researchers to reexamine or reuse samples from dogs in earlier studies, a secondary form of salvage.[231]

Recently, scientists have become interested not just in older laboratory dogs, but in the far larger population of elderly pet dogs. If older dogs are needed for studies, that same vast population targeted by Anipryl marketers has something to offer gerontologists as well. Avoiding both the significant expenses of purchasing animals and the skyrocketing costs of keeping dozens of dogs inside a lab for years, scientists have begun to frequently invite pet owners to volunteer their animals for research. I call this new relationship between lab animals and people *companion science*, a modified, multispecies variant of the older notion of participatory "citizen science."[232]

One of the most significant examples is the Dog Aging Project of the University of Washington, which began in 2014. The project, which has grown over the last decade, initially focused on rapamycin, a drug that appears to increase longevity in mice and delay some basic biological processes involved in aging.[233] Early results from the Washington work, which involved giving rapamycin or a placebo drug to enrolled pet dogs, suggested optimism about the medication's potential to modulate aging effects and extend the "healthspan" of both dogs and people. The Dog Aging Project is one of multiple projects to have emerged recently with the goal of studying aging in pet or companion dogs, alongside the Vaika study of sled dogs and the Golden Retriever Lifetime Study.

Collectively, they are heirs to the legacy of gerontological research at Davis and early experiments on Anipryl, which showed not only that a measurable form of cognitive decline exists in dogs, but that it might also be treated with pet-directed pharmaceuticals.

By specifically enrolling pet owners as test participants and educating them about the value of utilizing the study's product, research on Anipryl foreshadowed a shifting methodology in scientific projects involving dogs. Today, rather than purchasing commercially produced dogs or running internal colonies, more and more scientists rely on the companion science model of volunteered animals. This approach, in which dogs are both subjects and participants, extends far beyond the Dog Aging Project and is the flip side of beagle "salvage" science. It has grown in popularity, in part, because it solves two fundamental problems for researchers.

On the one hand, companion science allows laboratories to conduct experiments on large numbers of dogs without expending funding on housing, feeding, and maintaining animals in internal colonies. Purdue's David J. Waters thus noted in 2011 that the "pet dog paradigm" could obviate "the purchase and per diem costs typically associated with large animal research."[234] Those expenses are instead out-labbed to the pet owners themselves, who, given ever-escalating attention to the health of canine companions, now treat their dogs better than many owners did theirs at the time of the early Davis research. Within the microcosm of this experimental system, dog owners serve as a kind of volunteer veterinary technician. The marketing of Anipryl and other pet pharmaceuticals encouraged this through an insistence that pet owners needed to stay attentive to behavioral changes in aging animals, changes that they were most likely to notice. The financial savings of companion science, however, are no promise of a study's longevity: in January 2024, it was reported that the Dog Aging Project was at risk of losing nearly 90 percent of its funding in the coming years. The Vaika Project, on the other hand, which was initially funded principally by Russian philanthropies, was a casualty of the Russian invasion of Ukraine.[235]

Companion science also promises benefits for laboratory practice, however. Given increased attention among gerontologists to the role of the environment in aging processes, life outside of the laboratory can now be considered a more accurate mirror of the conditions of "normal" human aging. As Dog Aging Project codirector Matt Kaeberlein explains, "Companion animals, particularly pet dogs, represent an outstanding intermediate between laboratory animals and humans for understanding the basic mechanisms of aging."[236] Where dogs once represented a bridge in the species chain running from mice to humans, they are now seen to bridge laboratories and the outside world. This is a dramatic reversal from earlier visions in which laboratories represented the only places where pure aging in dogs could be studied. The absence of conventional laboratory colonies also offers a possible escape from animal rights controversy: clinical trials with companion animals "may be more acceptable," two researchers argued in 2003, than traditional drug testing.[237] The Dog Aging Project's website thus stresses that it does "not conduct" and is not "associated with, any research on laboratory or captive dogs."[238] Dog owners themselves are now the ones who must make decisions about sacrifice. While companion science participants continue to be "experimental dogs," they are no longer "laboratory dogs."

This broader revaluation of the environmenticity of dogs has also led pets to be highlighted, in recent years, as vital pieces in a revolution in "comparative oncology," the study of cancer in humans and nonhumans. Collaborations between veterinary and cancer researchers over the last decade have confirmed what Auerbach and others once argued: dogs get tumors just like people do, and often because of the same gene mutations and environmental influences. The shared environment makes humans and dogs not just companions but "companion model systems," one group noted.[239] Because dogs do not have a prescribed standard of care that they must fail before joining clinical trials of a new drug, companion animals can also test novel treatments without immune systems that have been devastated by chemotherapy, a long-standing problem for human trials.[240] Many researchers hope that studies in this area will yield a generation of new treatments for cancer. The Golden Retriever Lifetime Study of the Morris Animal Foundation, the research institution founded with profits from Mark Morris Sr.'s dog food sales, aspires to similar translational findings.[241]

At a minimum, the attention to older dogs has disseminated a reconfigured understanding of canine aging. According to the widely prevalent view, a dog's "human" age can be calculated by multiplying its calendar years by seven. While the exact origins of this notion are uncertain, many point to a popularization during the rise in dog ownership in the 1950s.[242] Albert M. Scheflen and Marcus Mason, compiler of the *Bibliography of the Dog*, noted in a 1953 article for the *Cornell Veterinarian*, for instance, that it was "commonly accepted that 1 year of a dog's life is equal to 7 human years."[243] Yet there are traces of the idea dating back to the 1920s, such as a paper by endocrinologist Knud Sand that asserts that "a dog's year is reckoned as equal to 7 human years."[244] By the 1930s, when life insurance for dogs became popular in England, companies such as Lloyds generally refused to write policies for dogs over seven based on the understanding that they were equivalent to human beings over sixty.[245]

That it is an old idea, however, does not mean it has gone unchallenged. The same year as Mason and Scheflen's article, French veterinary researcher Albert Lebeau influentially rejected the old maxim, arguing instead that dogs experience accelerated initial aging followed by a slower senescence. According to Lebeau's model, a two-year-old dog was around twenty-four human years, while a five-year-old dog was roughly thirty-six.[246] By 1958, the Gaines Dog Research Center had printed a "scale of equivalents" for use by veterinarians based on Lebeau's calculations, and the Davis beagle researchers confirmed the values in their own studies.[247] Scientists also recognized that smaller dogs age more slowly than larger dogs: a chihuahua born the same year as a cane corso, for example, is likely to live longer, which departs from a general trend that larger animals survive longer than smaller ones. The human-dog aging comparison has been further reformulated by studies of DNA methylation, a process that can signal the age of tissues and cells. In 2020, a group of researchers argued that a dog's general "human" age should be computed as the natural logarithm of its calendar years, multiplied by sixteen, plus thirty-one.[248] In this Rube Goldberg calculation, a two-year-old beagle is nearly forty-two. The broader result is a vision of canine life in which aging happens very quickly at the start but slows dramatically thereafter, with most of a dog's life an extended middle age.[249]

The older, surviving beagles at Davis and Lovelace presaged the companion science model, but the growth of that paradigm now threatens the long-term prospects of the beagle as a laboratory dog. Rather than Andersen's vision of future canine research, where purebred experimental beagles would generate consistent data across laboratories, it would appear that John Paul Scott and others were ultimately right about what was to come: attention to the messy diversity and genetic complexity of various breeds. Returning, in a roundabout manner, to Stockard's view that dogs exemplified the diversity of human forms, a variety of dogs aging in a number of ways now seem to better reflect the complicated nature of human aging. Perhaps there is no such thing as a standard dog, and no such thing as a standard human either.

There are likely beagles involved in studies of the Dog Aging Project: photographs from the lab's website in 2021 showed beagle members of the "Dog Aging Project Pack" (formerly the lab's "Citizen Scientists").[250] There will also almost certainly be beagle tissues in the Hungarian Brain and Tissue Bank and in trials of new cancer drugs. But it will be because many of those beagles were volunteered pets, not specially chosen laboratory animals. Research populations in companion science are constructed around the purchasing whims of individual human buyers or adopters, rather than support for a singular breed standard—even if, deep down, some scientists might wish otherwise. The shifting milieu of laboratory finances and ethical standards has dictated a different approach to the study of dogs, but one that is ultimately unsurprising: it was, after all, cost and complicated logistics that set the beagle on the path of scientific discovery in the first place, taking popular pet dogs and sending them out into the laboratory. Now many dog owners treat their animals as well as the early canine laboratories did, with scientific nutrition plans, daily exercise, socialization, and comfortable indoor housing. In the coming years, beagles may continue as experimental dogs, but their years could be close to over as the laboratory dog.

President Johnson playing with two of his beagles in the Oval Office. March 18, 1966. LBJ Library photo by Yoichi Okamoto. Public Domain.

Conclusion

THE ENDS OF BEAGLE SCIENCE

In October 2021, North Carolina representative Madison Cawthorn appeared before the US House of Representatives to deliver an impassioned and unwieldy critique of National Institute of Allergy and Infectious Diseases director Anthony Fauci's management of Covid-19. Among the varied claims in Cawthorn's speech, one was particularly provocative: Fauci, he argued, had greenlit gruesome experiments in which "sweet beagle puppies had their heads stuffed into crates so that sand flies could slowly strip away the skin from their bones."[1] It would emerge later that this was a conflation of studies, few of which had even tangential relations to Fauci, but the beagle attack gained traction.[2] Colorado representative Lauren Boebert tweeted a photo showing former president Donald Trump holding aloft a hefty beagle with the caption "Beagle Lives Matter," and other beagle criticisms peppered conservative media for weeks.[3]

"BeagleGate" was self-evidently opportunistic and died off gradually, but not before a flood of angry calls to Fauci's office.[4] Beagles then returned to the news in July 2022, when nearly four thousand were released from the facilities of contract research company Envigo following findings that the company had violated federal animal safety regulations. The plight of the beagles motivated an outpouring of support from dog lovers—Harry and Meghan, the Duke and Duchess of Sussex, adopted one—but the exodus also symbolized a victory for animal rights groups who have struggled to make the breed icons of animal experimentation.[5] For quite a few activists, freeing some beagles is the first step toward ending experimentation with dogs entirely. In June 2024, Envigo agreed to cease dog-breeding operations and pay $35 million, the "largest ever fine" in an animal welfare case.[6] Scientists, veterinary technicians, and industry groups who use laboratory beagles are likely to defend the continued necessity of their work, but growing attention suggests that the breed's time in laboratories might be coming to an end.

As this book has shown, such a transformation would be a significant one. Over the last century, beagles have been key participants in countless experimental systems. I have explored the most notable cases, but there are fascinating stories still to be told. Beagles never went to outer space, for instance—other than Snoopy, who became a symbol of the American space program—yet many lived as simulated cosmonauts in low atmospheric pressure for hundreds of days to test the environmental control systems of spacecraft.[7] We know how periodontitis develops and which oral hygiene interventions may prevent it, in part, because of the careful brushing of beagle teeth in places like Lexington, Kentucky, and Gothenburg, Sweden.[8] Although prominent, beagles were also just some of the countless canines to become centerpieces of modern research during a period in which the future of science appeared unavoidably entangled with the study of dogs.

In emphasizing the multivalent utility of dogs, some highly standardized and others much less so, the preceding chapters have resisted a story focused on the production and circulation of "model organisms." Such animals, plants, and fungi, from mice to Arabidopsis and neurospora, have been important for a range of projects, but "model organism" is often a misleading term: through a sleight of hand, it places many creatures under its umbrella and conceals much of the complex labor of interaction, manipulation, and representation that goes into making an organism an experimental or laboratory one. Although beagles have been proposed as "models" for solving a variety of problems, I have shown how their role extended far beyond that. Transforming from an amiable hunting companion to a distinctive laboratory tool, beagles served as multifaceted proxies for understanding human life and health and rearticulated the relationship between humans and dogs.

Proxies are a particular kind of tool: they are stand-ins, intermediaries between the controlled universe of experimentation and the messiness of the living world. This book has shown how a single kind of organism could have innumerable stand-in roles: eating irradiated food like Cold War children, smoking cigarettes like 1970s adults, and losing their memories like modern Alzheimer's patients. Using beagles in these ways required extensive imaginative and practical work to visualize human systems and behaviors in the dogs as well as organized efforts to alter their bodies, social worlds, and lived environments. All of this enabled beagles to become stand-ins for specific human forms of life, a process I call *species projection*. The term emphasizes how the human lives that laboratory organisms are called to represent remain intertwined with historically and contextually specific ideas about what and who people are.

Much writing about animal experimentation, whether supportive or critical, offers a strongly unidirectional story of human efforts to transform and use other creatures—with occasional caveats about animal resistance. It was "super-dominance" over canine ways of life that enabled researchers to put beagles to so many uses, a relation of power that is not distinct to the laboratory but part of the structural inequity between humans and domestic dogs.[9] Yet this book has also emphasized how beagles shaped the forms of life of their users through powerful encounters between one living being and another. Researchers are changed by the organisms they work with, regardless of their actual or perceived control over them. Laboratories had to conform to the rhythms of canine life or risk producing useless data, and laboratory cultures embraced dogs and dogginess, as beagle jokes and imagery became elements of scientific self-understanding. "The dog is always right," Marvin Goldman used to say; "our job is to find out why he's doing it."[10] These *conformations* between beagles and scientists produce a form of interspecies intimacy in which knowledge and care, sacrifice and violence have been densely entangled.

"To interrogate the use of proxies is to ask," writes Dylan Mulvin, "*to whom or to what do we delegate the power to represent the world?*"[11] Before proxies become common sense, they are the subject of contestation: where the scientific dog was once obviously multiple, support consolidated around beagles as *the* scientific dog only through experience, collaboration, and material exchange. The momentum of use in earlier projects was vital, but so were cultural forces, such as Snoopy and beagle derbies.[12] Especially important were politics and money, and the book has stressed the political economy of laboratory animal use: local regulations on pound dogs, harmonizing international regulatory regimes, and neoliberal budgets cuts all enabled or limited the modes of production and use of laboratory dogs. Historians and animal studies scholars need to forefront these systems of production and exchange in order to grasp the pasts and futures of laboratory organisms.

So, what is a laboratory dog? In an analysis of advertisements from lab suppliers, Arnold Arluke noted that commercial laboratory animals are frequently depicted as three semicontradictory entities. They are "classy chemicals," pure surfaces for scientific knowledge production, and "consumer goods," carefully produced scientific devices, but they are also "team players," docile and friendly scientific companions.[13] Describing the multiple roles of dogs in Ivan Pavlov's laboratory, Daniel Todes adds that the dogs were "simultaneously technologies, physiological objects of study, and products."[14] This book has shown how these varying facets of experimental organisms were brought together in beagles: laboratory beagles were standardized proxies produced at varying scales, imagined both as pure objects for knowledge production and as globally uniform technologies, while nevertheless remaining living, singular entities with vital encounter value. I have also shown how the *environmenticity* of dogs, the fact that they seem quintessentially defined by sharing an environment with people, became increasingly central to their perceived scientific utility.

What any single experimental system using dogs accomplished was not always certain. Such is the fate of much scientific research. But one undeniable result of over a century of thinking about beagles and other breeds was a newfound understanding of dogs and how humans should relate to them. This book charted how the initial use of beagles and dogs as proxies for humans has been partially displaced by an interest in understanding dogs themselves, on something like their own terms. How did dogs evolve? How do their brains work? What would dogs be like without people? While much of this research still defines *Canis familiaris* by a relationship to humans—dogs-as-companions, rather than dogs-in-themselves—contemporary ideas about the physical, emotional, and cognitive complexity of canine life have undermined the traditional uses of dogs as experimental organisms. Such similar and capable companions deserve more than a laboratory life, many feel. If the ends of beagle research were many, what might the end of beagles in research mean for the future of science?

Enigmas of Beagle Pain

One long-standing critique of animal experimentation argues that research with nonhuman animals is not simply unethical, but useless. We learn nothing from animal experiments, this line of argumentation goes, because so much is subject to varying interpretations and uncertain implications.[15] A number of possible "alternatives," from organ chips and cell cultures to computer simulations, are routinely presented as more promising options. Scientists often disagree with these points, and the viability of many alternatives remains tied to research with living organisms: to date, for instance, organ chips still presuppose continued research with living animal models.[16] Even those further afield reveal the coil of alternative and tradition: the much-lauded Virtual Heart, a computer program introduced in 1997 to allow students to study that vital organ in three dimensions ("dissection without death," per the *San Francisco Chronicle*), was predictably based on a dog and produced with assistance from Davis veterinarians, drawing on decades of knowledge from the radiobiology beagles.[17] As with Garrett's electronic dog, the virtual is never entirely severed from the living being. Nevertheless, a number of pharmacologists and toxicologists who found themselves debating the continued utility of beagle studies in the 1970s and 1980s were left with major questions about what *exactly* they prove.

In the late 1980s, a new class of antibiotics known as fluoroquinolones came onto the market. Produced by introducing fluorine into the chemical structure of quinolone, fluoroquinolones promised improved effectiveness and safety compared to early quinolone antibiotics, such as NegGram, which were potent against ailments like urinary tract infections but yielded to antimicrobial resistance. The first fluoroquinolone available to the public, Kyorin Pharmaceutical / Merck's norfloxacin, and those that followed, including Miles/Bayer's ciprofloxacin (or "Cipro"), received excited endorsements: they offered "freedom from hospitals," because patients could take drugs home rather than endure lengthy stays.[18] Fluoroquinolones were, microbiologist Richard Wise noted, "the proverbial magic bullets."[19] The arrival of Cipro, the *Times* of London added in March 1987, seemed to confirm that "every time bacteria evolve a resistance, it seems, the chemists devise something new to outflank them."[20] Starting in 1987, a deluge of quinolones fell on the world, "rivaled only by the number of introductions of all β-lactam antibacterials (penicillins, cephalosporins, and carbapenems)."[21]

But there were problems. In the 1970s, safety analyses in animals found a high risk of quinolone-generated lesions in the articular cartilage, the connective tissue that helps our joints move effectively. A report in 1977 from researchers at May and Baker suggested that multiple drugs in the class induced a lameness in beagles "so severe" that it forced researchers "to kill them on humane grounds."[22] Because articular cartilage occupies strategic locations in the body, and because of limited capacity for repair, it appeared uniquely prone to degenerative lesions.[23] Studies further found that juvenile beagles, analogized to human children, had the highest risk of arthropathic damage, compared to dogs of other ages. Caution followed about prescribing fluoroquinolones to pediatric populations.

Yet additional studies of "quinolone-induced arthropathy" (QAP) revealed the complications of interspecies scaling. The condition had appeared in an array of experimental animals, including mice, rats, marmosets, guinea pigs, and ferrets.[24] But in a 1992 review of earlier work, experimental toxicologist Alec Gough and his colleagues noted that "no unequivocal clinicopathologic data demonstrate that humans are susceptible to this enigmatic drug-induced lesion."[25] Much of the argument for danger in humans, in other words, rested on an assumption about interspecies translation: the more animals were susceptible to QAP, the more likely humans were, too.

Further complicating matters, it was beagle puppies who appeared particularly susceptible to QAP. The majority of persuasive studies on the dangers of fluoroquinolones involved beagles, because their "pharmacokinetic profiles" were "similar" to those of humans, and "the fact that dogs are highly susceptible to quinolone toxicities"; but whether beagle puppies were *really* similar to human children was difficult to ascertain on the basis of existing research.[26] According to Gough and colleagues, the mysterious prevalence of canine QAP might even present beagles as a model for better understanding the articular cartilage itself, with relevance for the treatment of osteoarthritis. In addition, a wide variety of quinolones on the market made it difficult to compare studies that used different species and different compounds. If one group found that marmosets taking norfloxacin did not experience QAP, whereas beagles taking ofloxacin did, was that proof of the safety of norfloxacin, evidence of species-specific sensitivities, or something else entirely?

In 1989, the same year that an international moratorium began on using fluoroquinolones in pediatric populations, representatives from Bayer approached prominent cystic fibrosis clinicians to suggest a public case to the FDA about further fluoroquinolone tests.[27] Late that year, the FDA invited Bayer and others to a Medical Advisory Committee meeting to discuss possible clinical trials in two pediatric populations: those with cystic fibrosis and those with cancer. In both populations, a powerful antibiotic might be essential to battling opportunistic infections. For Bayer, which owned Cipro, the trials represented a key opportunity to get the drug to a new market. Presenting data from compassionate use in 634 pediatric patients in Europe, the company argued that its product was safe. The bet would pay off handsomely: Cipro sales reached a peak of nearly two billion euros in 2001.[28]

Many studies endorsing the safety of fluoroquinolones were industry sponsored or industry funded. Given the centrality of fluoroquinolone antibiotics to global health, with many listed as World Health Organization "Essential Medicines," the possibility of finding alternative, marketable formulations remains a subject of intensive focus.[29] But whether fluoroquinolones are safe for children can seem to depend on whom one asks: a French study in 2003 found higher rates of arthralgias among those taking fluoroquinolones, concurring with continued caution about their use.[30] A separate summary in 2009 suggested that Cipro, the most well studied of the drugs, was relatively safe in pediatric populations but cautioned against expanded use because of separate concerns about antibiotic resistance.[31]

In 2008, the FDA added its first box warning about the risk of tendonitis and tendon rupture. Further warnings followed: in 2016, the FDA suggested that quinolones should be used as drugs of last resort because the risk of "serious side effects," including permanent damage to tendons, muscles, joints, and nerves, "generally outweighs the benefits."[32] Two years later, those precautions were strengthened by requiring labeling about the risks of mental health issues and hypoglycemic coma. In December 2018, the agency added a warning about aortic aneurysms.[33] Such warnings have, in turn, meaningfully decreased prescription fills of Cipro and other fluoroquinolones.[34] Once hailed as a revolutionary shift, fluoroquinolones have faltered.

The FDA warnings about permanent joint damage seemed to confirm the beagle findings, but questions about similarity between dogs and people remain. Bayer's push for human trials partially obviated difficult debates about interspecies similarity, and studies remain uncertain about why canine joints seem so acutely vulnerable to QAP.[35] Despite decades of research, dogs are surprisingly mysterious creatures. To take one further example: in 1983, a number of beagles purchased from Marshall Farms by pharmaceutical company Hoffmann–La Roche began to display intermittent pain, fevers, and other symptoms.[36] Initially attributed to the effects of an experimental compound, this "beagle pain syndrome" was also identified in dogs at other unrelated facilities. Now more commonly known as "steroid responsive meningitis-arteritis," the exact origins of the condition, which appears in other breeds as well, remain unclear, exacerbating lingering questions about the causes and meanings of beagle pain.[37]

Dogs respond to many drugs just like humans do: the compounds follow similar paths through their bodies and, when properly scaled, are eliminated in parallel fashion. Human medications can work in companion dogs, at appropriate doses, for these reasons. But in practice, things are complicated, and the ease of species projection runs into one of biology's fundamental tensions: unification versus organismic holism. If, as biologist Jacques Monod once claimed, what is true for *E. coli* is true for an elephant "except more so," it is clear that what is true for a beagle is not always even true for a human being.[38] The uniqueness of whole, living organisms remains a fascinating paradox.

Reassembling the Beagle Coalition

The beagle protests of the last few years were not the first time the breed's utilization inspired passionate activism and vitriol. The chemical weapons testing controversy of 1973–75 was the most significant public flashpoint, but many of the researchers introduced in this book dealt with criticism and opposition. Public concern for dogs motivated institutions such as the AEC to site laboratories in out-of-the-way places, like Lovelace: being far from major population centers had obvious benefits when it came to housing hundreds of howling beagles. But difficulties grew in maintaining large-scale beagle facilities as groups such as the Animal Liberation Front (ALF) and People for the Ethical Treatment of Animals (PETA) adopted increasingly confrontational tactics during the 1970s and 1980s. Oscar Auerbach's smoking research was picketed by protestors who likened his lab to Auschwitz, and half a century later his sons still remembered threatening nighttime calls to their home.[39] Battles over beagles even reached the dryly storied pages of the *Encyclopedia Britannica* when, in the early 1990s, Michael Fox's update to the entry for "Dogs" included reference to the use of beagles in research. An inundation of criticism caused editor Robert McHenry to delete references to testing.[40]

Lab break-ins placed further strain on scientists and institutions. One night in April 1980, a number of unidentified individuals broke into the Davis beagle colony and opened many of the dog cages. The ensuing liberation was brief and bittersweet: scientists quickly found the yelping hounds, who had limited interest in leaving the place they seemed to consider home.[41] A few years later, budget cuts to the Department of Energy during the Reagan administration freed the dogs once and for all. In 1988, the ALF took responsibility for a more prominent break-in at the lab of UC Irvine researcher Robert Phalen, having taken thirteen beagles used in studies of the toxicity of smog. Celebrity game show host Bob Barker applauded the activism, but the Irvine theft revealed persistently opposed visions of what laboratory life might mean to a dog.

Despite a significant reward for information leading to their return, the beagles never came back to Irvine. According to FBI memos, their remains were found buried in the California desert, close to the Salton Sea, and identified with help from the lab's lead animal technician, who entered counseling over the loss of thirteen of her friends, each with a unique personality and a name: Addy, Bonnie, Cindy, Daphne. "IRONICALLY," the FBI concluded, ALF members had killed the dogs.[42] The lab's research progress and that of a study of sleep apnea that shared animal housing were devastated. Like many before him, Phalen rejected criticisms of beagle research: the lab was certified by veterinary inspections, and his dogs were treated better than most pets. They were "like little astronauts when they go through a study, and I think they enjoy it," he had explained at the time.[43] Most or all would have been adopted out after the study's conclusion. But Barker called Phalen's suggestion ridiculous: "Would you believe for one moment that an animal would enjoy breathing smog and running on a treadmill?" Although research like Phalen's had convinced organizers of the 1984 Summer Olympics in Los Angeles that it was safe for people to do basically that, the obscure phenomenology of dogs and the riddle of what a dog experiences remained matters of passionate contest.

For many universities, lab break-ins forced dramatic changes in security protocols at biological facilities and millions of additional dollars in expenses. Quite a few eventually eliminated internal dog colonies entirely. Today, the surviving large beagle colonies are mostly relegated to pharmaceutical companies, contract research purveyors, and for-profit breeders. These facilities, such as Envigo's, are closely guarded and architecturally anonymous. Yet in 2001, the ALF took responsibility for scaling the fences at Marshall's North Rose facility, taking thirty beagles and ten ferrets.[44] "Camp Beagle," a protest begun in 2021 outside of a Marshall facility in Cambridgeshire, England, remains active at the time of writing in 2024. As part of these efforts, activists from Animal Rising, an offshoot of the broader Extinction Rebellion movement, broke into the Marshall facility in December 2022 and released twenty dogs from their cages.[45] Marshall has sought a permanent injunction against the protestors.[46]

Encounters with beagles, their jovial and responsive behavior even in the face of discomfort, made the dogs valuable to researchers. Many scientists also loved working with beagles not because they were sadists, but because the dogs were charming and fun to be around. Yet the mainstreaming of animal rights discourses, changing sentimentality about pets, and developing conceptions of dogs as cognitively complicated beings has made those once-valuable traits targets of intense criticism. All of this, in combination with declining funding and increasingly stringent animal care requirements, has caused the relative number of beagles used in science to shrink steadily since the mid-1970s. There were around two hundred thousand dogs in worldwide scientific research in 2015, but a survey of veterinary studies in 2022 found that perhaps as few as 2 percent of them were beagles.[47] Until quite recently, that decline went hand in hand with limited public interest, and beagles in science largely fell from the spotlight, just as they were eliminated from the pages of the *Encyclopedia Britannica*.

Over the last fifteen years, however, debates about the breed's future in research have gained renewed enthusiasm owing to advocacy from a loose coalition of nonprofit and activist groups. One of them, the Beagle Freedom Project (BFP), first reached the public spotlight in June 2011 when attorney Shannon Keith and other volunteers sheltered nine beagles formerly held in a California laboratory.[48] By 2015, the originally Los Angeles–based organization was finding success pushing "beagle bills," such as the one I saw debated in Massachusetts, in other states including Minnesota, Nevada, New York, and Connecticut.[49] The BFP's initial strategy focused on legislation and assembling adopters for former laboratory dogs—I tried to take one in myself during the early stages of research for this book—but over time, the organization also emphasized public records requests that revealed information about dogs used in government-funded research, training individuals in how to file Freedom of Information Act requests of their own.[50]

Larger debates erupted in 2017, after whistle-blowers reported irregularities in animal care measures within experiments funded and conducted by the VA. Public criticism grew after some of the more unnerving studies—including experiments that involved severing canine spinal columns to study rehabilitation measures for wounded veterans—were seized upon by the media. "Gruesome dog testing outrage," announced *USA Today* in September.[51] While many of the relevant projects were within the VA's mission to improve health and well-being for veterans, others had been funded with a more capacious view of their possible benefits, the same wide-ranging research vision that supported Auerbach's smoking studies. "Part of our mission is to push the envelope constantly in search of medical advancements that will help improve the lives of disabled veterans," explained David Shulkin, then head of the VA.[52] Where dogs were once celebrated for their sacrifice to ensure the safety of American GIs, in laboratories as much as battlefields, war dogs and lab dogs have become discursively separated, with combat sacrifice distinguished from its laboratory counterpart. In the Amazon Prime series *Reacher*, starring Alan Ritchson as a former Army investigator, there is no clearer sign of an inhumane personality, even amid senseless killing of people, than casual cruelty to dogs, who have given their lives to protect soldiers.

The debate over VA dog experiments echoed controversies from 1997 and 1998, when PETA used audio and video collected by undercover investigators to accuse contract research organization Huntingdon Life Sciences of torturing research beagles. Huntingdon's lawsuit, brought in response, placed a rare spotlight on PETA's investigational tactics.[53] By 2017, however, the beagle accusations grew from an isolated matter of laboratory animal treatment to a general, antiresearch talking point, particularly for conservative members of Congress. Florida representative Matt Gaetz drew on the plight of the VA dogs to call for eliminating public science funding, cosponsoring Virginia representative Dave Brat's Preventing Unkind and Painful Procedures and Experiments on

Respected Species (or PUPPERS) Act.[54] For Gaetz, who has defended a limited government view, dogs offered a new avenue of attack. The idea that pain should be prevented for "respected" species also highlights how far dogs have separated themselves from other laboratory creatures in the economy of sentiment.

Learning the lesson of the closure of the atomic beagle colonies—that funding shortages can be as powerful as even the most committed activists—the White Coat Waste Project (WCWP), an advocacy group that emerged in 2013, argues that animal research "wastes" American tax money.[55] That line of argumentation is not new, having reappeared throughout decades of criticism of animal experimentation in the United States.[56] New Jersey senator Harrison Williams, for instance, received a number of constituent letters in the late 1960s protesting Auerbach's beagle research, following an information bulletin from the National Catholic Society for Animal Welfare that criticized "vast sums of the taxpayers' money" being "poured into animal-using experimentation."[57] But Anthony Bellotti, the Republican strategist who developed PR campaigns against Obamacare and Planned Parenthood before founding WCWP and turning his sights to animal welfare, sought to knit together "a new, unlikely coalition of fiscal conservatives and liberal activists" in order to "end federal funding for research involving dogs and other animals by targeting people's pocketbooks in addition to their heartstrings."[58] Yet the group's focus remained on the conservative side of that coalition: White Coat Waste's first major piece of public exposure came in a Glenn Beck–executive produced documentary called *Socialized Science,* and the group was criticized for circulating conspiratorial talking points during the Covid-19 pandemic.[59]

Although mainstream conservatives often aligned themselves against animal-protection regulations in earlier decades, especially in the context of agriculture, more and more Republican members of Congress have taken up the defense of beagles. The Congressional Animal Protection Caucus is predominantly made up of Democrats, but it now claims sixteen Republican members as well.[60] Where the animal rights movement was once seen as so radical that it was associated with terrorism, it now appears increasingly compatible with small-government, antiscience ideologies.[61] A growing realignment of the modern animal rights movement, once strongly tied to feminist, ecological, and anticapitalist priorities, has witnessed a big-tent approach targeting the pocketbooks of conservatives and corporations. "Joining the animal-protection movement as a conservative can feel a little like being a defector," wrote Humane Society member John Connor Cleveland in the *National Review* in 2016. "But as I've grown more familiar with the issues over time, one thing has become abundantly clear: Animal welfare is a fundamentally conservative cause."[62] The celebration of Vanda's criticism of beagle-testing requirements, introduced in chapter 3, shows how this alliance might grow.

In many calls to limit funding for animal research, beagles have emerged as the key faces of laboratory victimization. "The Department of Veterans Affairs (VA) is still conducting painful experiments on beagles, hounds, and even 5-month-old puppies," proclaimed WCWP's "Dial for Dogs" campaign.[63] Cawthorn's concern about Fauci's beagles also emerged directly from WCWP advocacy. On the group's "Contact Congress" page in 2024, beagles appear as the representative animal for campaigns in support of the COST Act, the PUPPERS Act, and Violet's Law.[64] White Coat Waste and the Beagle Freedom Project have found synergy in the struggle to release dogs from laboratories.[65] The BFP, in turn, has won support from a number of prominent conservatives, notably Lara Trump, who shares a beagle with husband Eric Trump and made adoption advocacy a signature issue during Donald Trump's presidency. "Any voice she can give to elevate this issue or just grab the attention of new Republican allies, which we need, or even her father-in-law at the White House, we'll happily take," explained Kevin Chase, BFP's head of operations.[66] The absence of major beagle protection policy during Trump's first term in office likely disappointed such aspirations, but another chance now presents itself.

Contested Icons

The future of laboratory beagles remains open. As a cause célèbre for a major segment of the contemporary animal rights movement, beagles are "the most popular poster child" in debates over lab animal welfare, icons within the sentimental structure of laboratory life.[67] In a 2018 story about laboratory dogs, former *Intercept* editor Glenn Greenwald and reporter Leighton Akio Woodhouse argued that an imminent "collision is coming between the rapidly evolving scientific understanding of the capacity of animals to suffer, emote, and possess self-consciousness . . . and the legalized tolerance for mass animal abuse."[68] That collision would be paradoxical, in hindsight, because the atomic beagle studies were once lauded by early leaders of the animal welfare movement: Robert Bay, veterinarian for the Utah project, won the Animal Welfare Institute's first ever Albert Schweitzer Medal for his work.[69]

Yet whether laboratory beagles will disappear is a different question from the fate of experimental dogs generally. Despite criticisms, many believe that the benefits from working with dogs continue to outweigh the moral and emotional costs. The National Academies of Sciences, Engineering, and Medicine's ad hoc response committee released a substantial report on the future of VA research with dogs in early 2020, revealing substantial dissent about what that future should entail. "Initially we thought it would be fairly straightforward to answer the primary question, 'whether dogs are or will continue to be necessary for any type of biomedical research directly related to the VA's mission,'" wrote committee chair and vice chair Rhonda Cornum and W. Ron DeHaven. "What we all learned," they continued, "is that while facts are always facts, the emphasis that each individual places on each fact and the interpretation of a collection of facts leading to conclusions were widely disparate within this group." Such disagreement, which reads like an idealized definition of academic science studies, left the committee incapable of reaching a clear-cut conclusion about the necessity of dog studies: "Despite sincere efforts by all to reach consensus, it was not possible."[70]

On the one hand, the committee agreed that dogs were no longer "the preferred model" for work on diabetes, narcolepsy, imaging, or primary pharmacological research. Many of the programs sponsored by the VA would need to look for alternative animal models. On the other hand, the report acknowledged that dogs remained "scientifically necessary" in specific areas of active VA work such as "mechanistic insights of premature ventricular contraction-induced cardiomyopathy" and "development and testing of implantable devices to stimulate respiration and cough in spinal cord injury." Speculatively, the committee added that approaches and treatments might appear for "unknown, new, or reemerging diseases or disorders," ones that could be tested only in dogs.[71] Covid-19 took the public spotlight just a few months after the report's publication, and dogs' limited susceptibility may have saved beagles and other breeds from a larger role in studies of the disease.[72] Because the committee's purview was limited to topics that the VA itself was likely to investigate, it did not offer conclusions about the broader value of dogs in science. Yet the report's findings make clear that prominent investigators could not rule out dogs' utilization for a wide set of questions in the coming years. The vision of biomedical progress tied to experimentation with dogs suggested by the "*Really* Man's Best Friend" exhibit in 1946 retained some of its power.[73]

For example, the report urged additional VA use of "companion dogs," which remain "promising models" of conditions such as "obesity, diabetes, infectious disease, Alzheimer's disease, osteoarthritis, hereditary glaucoma, cardiomyopathy, thoracic spinal cord injury, and cancer."[74] The justification was not necessarily comparative—the report offered no suggestion that companion dogs were better than those produced for research purposes—but here, as elsewhere, companion science was framed explicitly as a more ethical and pragmatic alternative to traditional experimentation with dogs. Such a clear recommendation in an official report reveals how companion science has emerged as one of the most promising avenues for continued research with large animals. If the VA wants to work further with dogs, those animals may have to be pets volunteered by their owners. The trend is likely to accelerate, and with more former research beagles being "rehomed" or adopted, some might even find themselves returning to labs in a new guise as companion scientists. Beagles were the eighth most popular dog on the AKC's 2022 list, so the number of potential participants remains significant.[75]

Economic, political, and moral concerns threaten the end of beagles as laboratory dogs. In most cases, the once-grand colonies of beagles, heralded as fundamental breakthroughs in modern science, have now disappeared. Biobanked materials from those studies, such as tissues preserved from early atomic research, now offer a way of working with dogs by proxy for those who cannot house them, as the age of dogs has given way to the careful use of aging dogs.[76] But there are also signs that the old posture of scientists in relation to animal rights protest, which involved keeping a low profile and avoiding public debate, is changing. During interviews for this book, numerous veterinarians and veterinary technicians argued that members of their profession needed to become vocal advocates once more for the work that they were carrying out. In a field where practitioners struggle with compassion fatigue and a serious mental health burden, many see opening to dialogue as essential to their own futures.[77]

Even if beagles disappear from research entirely, they are likely to continue playing crucial roles in constructing and demarcating human identity and difference. That work is clear in the case of the US Beagle Brigade, a troop of detector dogs that began sniffing for contraband at Los Angeles International Airport in 1984 under the auspices of the Department of Agriculture, before being transferred in 2003 to the authority of Customs and Border Protection (switching their green jackets for blue ones).[78] "Don't be fooled by his fuzzy face," began Katy Tur's report on the Beagle Brigade for the *NBC Evening News* in October 2012. "Linus is a federal agent, and he will sniff you out."[79] The dogs remain key actors in America's border security regime by, in the words of Heather Paxson, deeming "some edible materials as appropriate for absorption by the national body politic while rejecting others as filthy, adulterated, infected, misbranded or other-wise unfit."[80] The remarkable olfactory capacities of beagles have also positioned them as possible "disease sniffers" now that the use of nonhuman animals as "sentinels" of illness has grown in attractiveness.[81] In this work, beagles serve as living, prosthetic counterparts to x-ray machines within the American security state, defining borders between a safe America within and a potentially dangerous world beyond.

Politically, beagles retain their symbolic function as well. Over the last two centuries, the once-foreign beagle became quietly synonymous with a popular notion of American identity—hardy, earnest, and loyal. In 1964, supporters of presidential candidate Barry Goldwater began to appear at events with elaborate "Beagles for Barry" signs, a not particularly subtle jab at President Lyndon Johnson, who was seen picking up one of his pet beagles by the ear. "The people of this land have had enough of being mauled and made to yelp like helpless beagles on the White House lawn," Goldwater intoned in a speech the same year.[82] While the attacks on Johnson never entirely stuck, in part because it was hard to question his deep commitment to the dogs, beagles were metaphorically aligned with good American citizenship. To mistreat beagles, per the Goldwater campaign, was to mistreat Americans themselves—although a suggestively *white* America for a campaign that, legal scholar Ian Haney López argues, helped to inaugurate the use of coded, racist "dog whistle" politics.[83]

More recently, beagles made a major reappearance in Raphael Warnock's campaign during Georgia's US Senate special election in 2020 and 2021. Multiple of Warnock's most viral advertisements featured Alvin, a small, light-colored beagle, "tugging a puffer-vest-clad Mr. Warnock for an idealized suburban stroll—bright sunshine, picket fencing, an American flag."[84] Alvin was not actually Warnock's dog, and campaign strategists admitted that he was introduced intentionally to depict Warnock as a "safe" choice for voters, many of whom probably learned belatedly about Alvin's fictional kinship ties. For divergent purposes, the commercials drew on the same lingering connection between beagles and (white) American identity that Goldwater had. Warnock's warm embrace of the placid dog, which fooled many into believing Alvin was his pet, reads as an embrace of "normal" America.

The very American status of the political beagle might appear to conflict with the more abstract role of beagles as a global scientific standard. But as the preceding pages have shown, the "cultural" beagle and its "scientific" counterpart were produced and shaped by similar forces during a period in which American research and institutions fundamentally shaped international science. Beagles were simultaneously local and global, which was part of what made them useful and available to scientists, and enmeshed with the history of capitalism in the twentieth century. The path of beagles paralleled that of the US dollar: one became the world's reserve currency following the Second World War and Bretton Woods Agreement of 1944, the other the world's reserve research dog a few years after. In both cases, the strength and expansion of postwar American economic and scientific power enabled local standards to become transnational ones, linking together countries and markets. Following their global distribution and the centralization of their production by large corporate actors, beagles emerged as scientific commodities that could be trusted to produce good results wherever they were used. But unlike dollars, beagles remained "lively" animal capital, biotechnologies that could perform for researchers and build complicated relationships. The capacity to affect and be affected was central to their value, and practices of species projection allowed the dogs to serve as vehicles, simultaneously, for basic experimental inquiry and for the imagination. The rich semiotic possibilities revealed by political beagles confirms why experimental beagles remain icons in contested debates about the proper form of scientific inquiry.

Whether beagles finally leave the lab or stay for decades more, the number of dogs that participated in research programs over the last century remains substantial. The global total, which is especially difficult to ascertain because of limited disclosure requirements, is certainly in the hundreds of thousands, if not the millions. Although that number is smaller than the comparable toll for laboratory rodents or the hundreds of millions of animals slaughtered each year for food, it represents an enormous loss of life. One scientist interviewed for this book expressed a hope that I would make clear the benefits from much of this research: "sacrifice," in its scientific usage, suggests that the death of a laboratory organism comes for the sake of a greater good. But that calculation is a very difficult one to make. I have shown that some dogs and humans indeed profited from this research, whether through safer drugs, for instance, or an awareness of cancer risks. But in other areas, the benefits remain far more ambiguous, not least because intense class stratification has kept benefits from reaching many in need. Quite a few dogs were also sacrificed without clearly advancing knowledge in a meaningful way. This seesawing balance between health and risk, care and violence, value and loss, defines the scientific relation with nonhuman organisms. Solutions to the problem of animal experimentation require solutions to the problem of social order.[85]

On September 17, 1956, a few years after work began at Davis, the *San Francisco Chronicle* published a short meditation on the new task beagles had been called on to do. It concluded with an apology: "We are sorry that it is necessary for the beagle to act as a guinea pig in an experiment that involves our safety and well-being. He is the victim of man's inability to put a leash on the atom and, even more to the point, on himself. It is a tough test of the friendship of man's best friend. But, perhaps, the beagle is the best one to undergo it. If any friendship could survive the estimated 15 to 20 years of this test, it would be the friendship of the beagle toward his irresponsible master."[86] As much as humanity's friendship with dogs has transformed in the decades since, we remain caught in the same logic of apologetic necessity and irresponsible mastery. Whether in determining dangers from radiation, cigarettes, or drugs, dogs were undeniably victims of our inability to "put a leash" on ourselves. The friendship between dogs and humans could have taken other paths—a beagle at every firehouse, perhaps—and it might still. Some now hope to return the favor of the dog's scientific sacrifice with the pharmaceutical gift of a few more years. But beagling and history share in the absence of predetermined destinations. The difficulties and the joys are in the search, so off we go, together.

ACKNOWLEDGMENTS

"The cardinal crime any beagle can commit is to quit," wrote George Whitney in *This Is the Beagle*. If you are holding this book on pages printed by the University of Chicago Press or, more likely, reading it electronically, then I have avoided the high crime of beagledom once and for all. Much to my chagrin, it has not been just a casual stroll to this point.

To answer the most common question first: I do not have a beagle. My dog, Laszlo, is many things, but "purebred" is not one of them. According to a genetic test from Darwin's Ark, he is first and foremost an American pit bull terrier, to a lesser extent a border collie. Nearly a quarter of his DNA is "unknown," and the rest of it carries traces of the German shepherd, labrador, chihuahua, chow chow, and more. Even with the known unreliability of consumer canine genetics tests, Laszlo is a walking reminder of the complicated human relationship and history with dogs. He loves Frisbee and a small plush we call "George," but there's not a bit of beagle in him.

I have lived with dogs almost all my life—none of them beagles—and my experiences with Van Gogh, Douglas, Darwin, and Laszlo undoubtedly set some of the foundations for this book. Dogs have made my life richer in innumerable ways, but I never expected to study or write a book about them, so a host of other people deserve credit for inspiring this work and supporting me over the years that went into writing it.

On a practical level, I am grateful to archivists and librarians, who are the backbone of academic history. Some who aided me include Nicole Milano (Medical Center Archives, New York–Presbyterian / Weill Cornell Medicine), Jocelyn Wilk (Columbia University Archives), Peter Corina (Rare and Manuscript Collections, Cornell), Fina Martinez-Myers (Nuclear Testing Archive), David Buhler (University of Utah Archives),

Ruth Chan and Sean Heyliger (National Archives at San Francisco), Gayle O'Hara (MASC, Washington State University), Robert Franklin (Hanford History Project), Kyle Hovious (University of Tennessee), Jessica Murphy and Stephanie Krauss (Center for the History of Medicine, Harvard University), Bob Vietrogoski (Rutgers Medical Library), Nicholas Webb (New York Medical College), Bethany Antos (Rockefeller Archive Center), Tim Pennycuff (University of Alabama at Birmingham Archives), Peggy Balch and Anna Kaetz (University of Alabama at Birmingham Special Collections), Andy Harrison (Johns Hopkins Chesney Medical Archives), and Mary Linnemann (University of Georgia Archives). Because a significant portion of my research was conducted during and after Covid-19, when visiting archives became challenging, I am especially grateful to those who shared materials electronically at little cost, including the faceless interlibrary loan workers at libraries around the world.

I spoke or emailed with a number of people who told me about their lives, family, or work during the research and writing of this book. They include Bruce and Dick Auerbach, Lorna Coppinger, Elizabeth Head, Roger McClellan, Bill Milgram, Robert Phalen, Daniel Promislow, Stanley Saxe, Martin Seligman, Sheldon Steinberg, Christa Studzinski, Thomas Uhde, and Charles Whitney. Veterinarians and veterinary technicians graciously offered their time and thoughts, and Allison Ostdiek was especially open in discussing her profession and work with animals. I consulted Scott Podolsky on the history of clinical trials and Robert Proctor on the history of smoking. Jeremy Beckham shared his MA thesis on beagles and a number of primary source documents. The staff of Darwin's Ark invited me to speak and introduced me to their work.

A few institutions offered space and funding to support my work. I spent five years in the History of Science program at Harvard, inauspiciously beginning months before the 2016 election and graduating during a world-historic pandemic. Peter Galison and David Jones steered my early thinking in vital ways. Janet Browne offered important support during a challenging period, and Linda Schneider was tireless in helping me to untangle varied bureaucratic quandaries. At MIT, Stefan Helmreich, David Kaiser, and Harriet Ritvo generously offered their time and eyes to my writing. From more or less the beginning of my academic career, Stefan has been an invaluable mentor. During this period, the Harvard Graduate Students Union and Boston DSA helped me to never forget the politics and inequality that structure life in the academy and beyond. Two years at the Institute on the Formation of Knowledge (IFK) at the University of Chicago allowed me both time away from beagles and the chance to return. I thank Shadi Bartsch-Zimmer for bringing me to Chicago and Stefanie White for ensuring my time there was comfortable. This book was finished during a postdoctoral membership at the Institute for Advanced Study (IAS), and I am deeply appreciative of consummate Vikings fan Myles Jackson, not least for always ensuring that I was having fun in the process.

At the University of Chicago Press, Karen Darling supported the project through to its completion, and Fabiola Enriquez Flores assisted with a number of vital tasks, including permissions for the images included in the book. I am especially grateful for the two anonymous reviewers of the manuscript, who offered a great deal of exceedingly useful feedback, especially about the structure of the final chapters.

A number of friends and colleagues read, heard, or assisted with early versions of this material. They include but are not limited to Leah Aronowsky, Erik Baker, Alyssa Botelho, Cameron Brinitzer, Danielle Carr, Angelica Clayton, Hannah Conway, Max Ehrenfreund, Jordan Howell, Gus Lester, Yvan Prkachin, Asa Seresin, Simon Torracinta, Claire Webb, and Jongsik Christian Yi. Sarah Pickman shared materials from the archives at Yale that I could not consult directly; my research assistant Sammy Zimmerman did the same at the University of Chicago. Eszter Zimányi helped with a few Hungarian translation questions. Postdocs at IFK, including Tal Arbel, Katherine Buse, Jordan Bimm, Iris Clever, Isabel Gabel, Mel Jeske, Hannah Moots, and Andre Uhl, offered an unforgettable academic community. Nicholas Robbins helped me think about environmenticity at the IAS. For over eight years, David Munns has balanced supportive advice with a steady stream of humor. Many other colleagues offered valuable comments and questions at academic conferences. To any left unnamed, I extend my gratitude nonetheless.

If family dogs were the initial nudge, the intellectual inspiration for this book was a conversation, many years ago, with Sophia Roosth, who suggested I look into the peculiar history of atomic pigs. That meandering path eventually brought me to beagles. As a mentor across multiple institutions, she has been responsible more than anyone else for putting me on the trail that culminated in this book. She is one of the most exemplary scholars I have met and manages the rare balance of being an engaged teacher and mentor as well. Someday she will have to excuse me for all the emails.

Three of my childhood friends made the years during which I worked on this project deeply enjoyable: Isaac always had a bed, couch, or floor to sleep on and a memorably bizarre film; Grant joined my improbable path to White Sox fandom while opening some supremely crushable bottles; and Peter was invariably available for grousing about university life. Sam was the best roommate I never expected to have, and Forrest and Michael offered unfailing comradery of their own.

My parents introduced me to the academic world and have always supported my pursuit of less than remunerative interests. My mother has been, unquestionably, my biggest fan, and my father an inspiration of a different sort in his earnest commitment to a life of teaching. I have always walked in a trail made easier by the steps of my brother, not least from the boxes of books he left from college, and I hold him responsible for opening doors for me that once appeared firmly closed. Last and certainly not least, I struggle to imagine a world without Katie, one of the most remarkable humans on this planet, whose contributions to my life push at the limits of language. She neither typed nor edited this book, but she has listened to me ramble about nearly everything in it and never tired, which is all the more impressive. I might have written a book without all these people, but it would hardly be worth thinking about.

NOTES

Preface

1. Rebecca Rubin, "'John Wick' Franchise Crosses $1 Billion Globally," *Variety* (blog), May 19, 2023, https://variety.com/2023/film/news/john-wick-franchise-box-office-billion-dollar-1235618570/. The epigraphs for this book are from James Frank Sullivan, "Light Infantry of the Hunting Field," ed. Alfred Edward Thomas Watson, *Badminton Magazine* 2, no. 10 (May 1896): 589; and J. Newell Stannard, "Interview with Robley D. Evans," 1978, 9, J. Newell Stannard Papers, Betsey B. Creekmore Special Collections and University Archives, University of Tennessee, Knoxville.

2. Michelle Megna and Ashlee Valentine, "Pet Ownership Statistics 2024," *Forbes Advisor*, January 3, 2024, https://www.forbes.com/advisor/pet-insurance/pet-ownership-statistics/.

3. hannah strong (@thethirdhan), "When I Say I Got That Dog in Me This Is What I Mean https://T.Co/5SESaZWxEH," Twitter (now X), November 2, 2023, https://twitter.com/thethirdhan/status/1720202805909102837.

4. Perhaps because beagles have served experimenters in smaller numbers than mice or rats, their role in science has gone comparatively unexplored. Exceptions include Eva Giraud and Gregory Hollin, "Care, Laboratory Beagles and Affective Utopia," *Theory, Culture and Society* 33, no. 4 (July 2016): 27–49, https://doi.org/10.1177/0263276415619685; Eva Giraud and Gregory Hollin, "Laboratory Beagles and Affective Co-productions of Knowledge," in *Participatory Research in More-Than-Human Worlds*, ed. M Bastian et al., Routledge Studies in Human Geography (Oxford: Routledge, 2017); Lesley A. Sharp, *Animal Ethos: The Morality of Human-Animal Encounters in Experimental Lab Science* (Berkeley: University of California Press, 2018); Jeremy Beckham, "Radioactive Beagles" (master's thesis, University of Utah, 2018). Most works that cover the history of dogs in science approach it as a small part of larger changes. Much of the best recent work has also focused on the history of dogs in Victorian Britain, rather than more recent developments. See, for instance, Philip Howell, *At Home and Astray: The Domestic Dog in Victorian Britain* (Charlottesville: University of Virginia Press, 2015), https://www.jstor.org/stable/j.ctt13x1rng; Michael Worboys, Julie-Marie Strange, and Neil Pemberton, *The Invention of the Modern Dog: Breed and Blood in Victorian Britain* (Baltimore: Johns Hopkins University Press, 2018); Michael Worboys, *Doggy People:*

The Victorians Who Made the Modern Dog (Manchester: Manchester University Press, 2023). Edmund Russell's recent history of the greyhound applies scientific concepts to human and canine history, rather than history to canine science. See Edmund Russell, *Greyhound Nation: A Coevolutionary History of England, 1200–1900* (Cambridge: Cambridge University Press, 2018).

5. Søren Kierkegaard, *Parables of Kierkegaard*, ed. Thomas C. Oden (Princeton: Princeton University Press, 1978), 21. Sophia Roosth brought the passage to my attention.

6. On Pavlov, see Daniel P. Todes, *Pavlov's Physiology Factory: Experiment, Interpretation, Laboratory Enterprise* (Baltimore: Johns Hopkins University Press, 2001); Daniel P. Todes, *Ivan Pavlov: A Russian Life in Science* (Oxford: Oxford University Press, 2014).

7. Donna Haraway proposed a "big book," *Birth of the Kennel*, as a canine complement to Michel Foucault's classic account of the emergence of modern medicine, *Birth of the Clinic*. It would show how the transformation of canine lives through selective breeding and pet keeping was vital to the history of experimental practice and laboratory culture. See Donna J. Haraway, *The Companion Species Manifesto: Dogs, People, and Significant Otherness*, Paradigm 8 (Chicago: Prickly Paradigm, 2003), 61. "The practices and actors in dog worlds, human and non-human alike, ought to be central concerns of technoscience studies," Haraway argues in *Companion Species Manifesto*, 3.

8. Andrew Pickering, *Science as Practice and Culture* (Chicago: University of Chicago Press, 1992), 3n1.

Introduction

1. John Paul Scott and John Langworthy Fuller, "Research on Genetics and Social Behavior: At the Roscoe B. Jackson Memorial Laboratory, 1946–1951—a Progress Report," *Journal of Heredity* 42, no. 4 (July 1951): 191–97.

2. Kerstin Lindblad-Toh et al., "Genome Sequence, Comparative Analysis and Haplotype Structure of the Domestic Dog," *Nature* 438, no. 7069 (December 2005): 803–19, https://doi.org/10.1038/nature04338. On Dog10K, see Elaine A Ostrander et al., "Dog10K: An International Sequencing Effort to Advance Studies of Canine Domestication, Phenotypes and Health," *National Science Review* 6, no. 4 (July 1, 2019): 810–24, https://doi.org/10.1093/nsr/nwz049.

3. This book follows *Merriam-Webster's Collegiate Dictionary* in using lowercase for dog breeds, which may be confusing to readers more familiar with the capitalization practices of the American Kennel Club and other groups.

4. "Visualization of the Beagle Standard," The National Beagle Club of America, n.d., https://www.nationalbeagleclub.org/page-18097#Coat, accessed July 10, 2024. The organization's logo, a sprinting lagomorph, reminds us of the breed's abiding quarry.

5. John Otho Paget, *Beagles and Beagling* (London: Hutchinson, 1923), 27.

6. Margaret Derry has traced broader concerns around the production of "beautiful" dogs and their limited ability to work, particularly in connection to the Collie. See Margaret E. Derry, *Bred for Perfection: Shorthorn Cattle, Collies, and Arabian Horses since 1800* (Baltimore: Johns Hopkins University Press, 2003), 63.

7. Douglas H. Appleton, *The Beagle Handbook*, Dog Lover's Library (London: Nicholson and Watson, 1959), 3–4.

8. Appleton, 2; *Oxford English Dictionary*, s.v. "beagle (n.), Etymology," September 2023, https://doi.org/10.1093/OED/2503540648. According to the *OED*, the derivation remains obscure.

9. The play was a possible inspiration for Shakespeare's *Comedy of Errors*. See Geoffrey Bullough, ed., *Narrative and Dramatic Sources of Shakespeare*, vol. 1, *Early Comedies, Poems, Romeo and Juliet* (London: Routledge and Kegan Paul, 1957), 27.

10. Another linguistic note: although the word "dog" now carries a relatively gender-neutral implication, many American writers understood a "dog" to be male and a "bitch" to be female until well into the twentieth century. This is clear from preserved stud books, where an animal's "sex" was often "dog." I use "dog" in its more general variant throughout.

11. Skepticism about such miniature beagles is common in early American histories of the breed. See, for instance, Bradford S. Turpin, *The Beagle and the Field Trials*, Popular Dogs of the Day 1 (Baltimore: F. J. Skinner, 1900), 9. These were likely examples of fanciful thinking or underfeeding. Until the twentieth century, creating "small" versions of animals often relied on stunted growth.

12. Michael Worboys, Julie-Marie Strange, and Neil Pemberton, *The Invention of the Modern Dog: Breed and Blood in Victorian Britain* (Baltimore: Johns Hopkins University Press, 2018), 2; Derry, *Bred for Perfection*, 49; Michael Worboys, *Doggy People: The Victorians Who Made the Modern Dog* (Manchester: Manchester University Press, 2023), 17–27.

13. Worboys, Strange, and Pemberton, *Invention of the Modern Dog*, 2. It is nevertheless worth noting that visual differentiation of mixed-breed dogs remains far more difficult than many imagine. See, for instance, Lisa M. Gunter, Rebecca T. Barber, and Clive D. L. Wynne, "A Canine Identity Crisis: Genetic Breed Heritage Testing of Shelter Dogs," *PLoS One* 13, no. 8, https://doi.org/10.1371/journal.pone.0202633.

14. Paget, *Beagles and Beagling*, 29.

15. George D. Whitney, *This Is the Beagle* (Jersey City, NJ: T.F.H., 1955), 11.

16. Paget, *Beagles and Beagling*, 29.

17. Paget's suggestion that there were multiple, unchanged beagle progenitors would be inaccurate from a modern standpoint, but the view was once popular. Charles Darwin and his "bulldog," Thomas Henry Huxley, argued that dogs likely had multiple evolutionary origins, implying that breeds were not all even the same species. Many thinkers of the time underestimated the dramatic changes produced in dogs during the Victorian era. See Worboys, Strange, and Pemberton, *Invention of the Modern Dog*, 163.

18. Paget, *Beagles and Beagling*, 29.

19. Georges Canguilhem, *Knowledge of Life*, trans. Stefanos Geroulanos and Daniela Ginsburg, Forms of Living (New York: Fordham University Press, 2008), 22.

20. Jacques Derrida notes that "the Animal" does not exist except in a plurality. Derrida's "Animot," a singular noun with a plural-sounding ending, akin to *animaux*, signals this multiplicity. See Jacques Derrida, *The Animal That Therefore I Am*, ed. Marie-Louise Mallet, trans. David Wills (New York: Fordham University Press, 2008). Russell also criticizes what he calls a "statue history of breeds," which sees dog breeds as uniform, isolated, and static. See Edmund Russell, *Greyhound Nation: A Coevolutionary History of England, 1200–1900* (Cambridge: Cambridge University Press, 2018), 6.

21. Beagle trials organized by the National Beagle Club near Aldie, Virginia, for example, were originally "stag" affairs, and log cabins were constructed for men. A "Squaw camp" about five hundred yards away was added when "the '20s of F. Scott Fitzgerald brought in the wives," and by the 1930s the cabins and camp were united. See "The Institute Corporation," n.d., F. Ambrose Clark Rare Book Room, MS RBR B 561 .I578, National Sporting Library and Museum, https://archive.org/details/National-Beagle-Club-History-The-Institute-Corporation. Women beaglers, on the other hand, were more prominent early on in activities at

Burlingame in California, where Mrs. C. Frederick Kohl held the title MFB or "Mistress Fair of Beagles." See E. G. B. Fitzhamon, "Golf Prize Waiting for W. H. Crocker," *San Francisco Examiner*, April 19, 1916, 11.

22. These typically followed a pattern of the owner's or kennel's name and then the dog's: Bolman's Laszlo, for instance. Larger trends are difficult to pinpoint, but there was a noticeable interest in racist and feminizing names across the early twentieth century.

23. James Frank Sullivan, "Light Infantry of the Hunting Field," ed. Alfred Edward Thomas Watson, *Badminton Magazine* 2, no. 10 (May 1896): 595. James Frank Sullivan was best known as a cartoonist and illustrator at *Fun*.

24. A dedication to Edward S. Herancourt at the front of Henry Wilson Prentice's collection of beagle knowledge uses the term. See Henry Wilson Prentice, *The Beagle in America and England* (DeKalb, IL: H. W. Prentice and W. A. Powel, 1920), https://www.biodiversitylibrary.org/item/78997#page/8/mode/1up.

25. Harry Collins overviews the notion of "form of life" as a sociological concept in Harry Collins, *Forms of Life: The Method and Meaning of Sociology* (Cambridge, MA: MIT Press, 2019). For the connections between "form of life" and "life form" see Stefan Helmreich and Sophia Roosth, "Life Forms: A Keyword Entry," *Representations* 112, no. 1 (November 1, 2010): 27–53, https://doi.org/10.1525/rep.2010.112.1.27.

26. Sullivan, "Light Infantry of the Hunting Field," 588.

27. Dogs such as Atomic Sue competed in contests during the 1950s. Sadly, the longtime headquarters of the club was destroyed by fire in September 2014. "Atomic Beagle Club in Solway Burns Overnight," *Knoxville News Sentinel*, September 22, 2014, http://archive.knoxnews.com/news/local/atomic-beagle-club-in-solway-burns-overnight-ep-628266793-354234391.html/. One of the club's only surviving members, Marvin Blair, donated much of its accumulated funding to St. Jude Hospital in 2019. See Robert Norris, "Rockford Man Delivers 'Transformational' Gift to St. Jude Hospital," *Daily Times*, July 30, 2019, https://www.thedailytimes.com/news/rockford-man-delivers-transformational-gift-to-st-jude-hospital/article_abad4137-0936-5447-928b-4c34e17f7298.html.

28. Stefan Helmreich suggested thinking about "conformation" along these lines. The term expresses part of what Russell designates by "coevolution" in his history of greyhounds: "Any change in frequency in any trait . . . is evolution." While humans undeniably coevolved with dogs and actively selected for a diversity of canine types, Russell's definition is extremely broad. Conformation, on the other hand, is meant to pinpoint the finer practical and cultural forces operating on shorter timescales and in particular contexts. See Russell, *Greyhound Nation*, 10. It connects with Haraway's concept of "becoming-with." As she writes, "If we appreciate the foolishness of human exceptionalism, then we know that becoming is always becoming with, in a contact zone where the outcome, where who is in the world, is at stake." See Donna J. Haraway, *When Species Meet* (Minneapolis: University of Minnesota Press, 2008), 244. Conformation also makes explicit the shaping of cultural and bodily forms.

29. Russell makes a parallel point regarding greyhounds as biotechnologies. See Russell, *Greyhound Nation*, 8.

30. Folke André, "Jakt Med Drivande Hund," in *Jakten Som Hobby*, ed. Gunvor Grenholm (Stockholm: Forum, 1954).

31. Richard Sandomir, "Remembering the One and Only Uno," *New York Times*, September 24, 2018, sec. Sports, https://www.nytimes.com/2018/09/24/sports/uno-westminster-dog.html.

32. Appleton, *Beagle Handbook*, 1.

33. L. E. Wolfe, "About Dogs," *Coshocton (Ohio) Tribune*, September 26, 1946, 5.

34. Walt Little, "Little Quotes," *Bakersfield Californian*, April 4, 1950, 29. Snoopy's first strip was October 4, 1950. On the history of *Peanuts*, see Blake Scott Ball, *Charlie Brown's America: The Popular Politics of Peanuts* (New York: Oxford University Press, 2021).

35. "Dog's Prestige at New High," *Peoria Heights Herald*, September 1, 1944, 7.

36. The "Project Hot Dog" moniker appears in a number of sources, for example, Ed Salzman, "UC Men Find Key to Death by Radiation," *Oakland Tribune*, August 9, 1959, 1.

37. Allen C. Andersen, "The Dog—a Facet of Human Society" (Noon Topics, San Francisco, CA, November 13, 1963), 2–3, carton 1, folder 6, UCSF Committee on Arts and Lectures Records, AR 2015–17, University of California, San Francisco, Archives and Special Collections.

38. Donna J. Haraway, *Primate Visions: Gender, Race, and Nature in the World of Modern Science* (New York: Routledge, 1989), 5.

39. Erika Lorraine Milam, *Creatures of Cain: The Hunt for Human Nature in Cold War America* (Princeton, NJ: Princeton University Press, 2019), 8.

40. Giorgio Agamben claims that human beings are fundamentally "anthropomorphous": *Homo sapiens* "must recognize himself in a non-man in order to be human." Giorgio Agamben, *The Open: Man and Animal* (Stanford, CA: Stanford University Press, 2004), 26. Emmanuel Levinas writes of Bobby, a dog encountered during his time as a prisoner of war in Germany during World War II: "For him—there was no doubt—we were men." Cited in Derrida, *Animal That Therefore I Am*, 114.

41. When studying scientific organisms, scholars have typically focused on practices of representation and standardization. Philosophers of science Rachel A. Ankeny and Sabina Leonelli orient their summary of the literature on model organisms around questions of what model organisms represent, how, and for whom they do so. See Rachel A. Ankeny and Sabina Leonelli, *Model Organisms*, Elements in the Philosophy of Biology (Cambridge: Cambridge University Press, 2020), 1. Historians of science Angela N. H. Creager and Karen A. Rader focus their respective histories of the tobacco mosaic virus and lab mouse around questions of standardization, production, and circulation. Angela N. H. Creager, *The Life of a Virus: Tobacco Mosaic Virus as an Experimental Model, 1930–1965* (Chicago: University of Chicago Press, 2002); Karen A. Rader, *Making Mice: Standardizing Animals for American Biomedical Research, 1900–1955* (Princeton, NJ: Princeton University Press, 2004).

42. As Derrida argues, human self-recognition and autobiography permeate the sciences as much as they do ontology, ethics, and law. See Derrida, *Animal That Therefore I Am*, 89. Sophia Roosth has shown how contemporary research is actively informed by a bidirectional relationship between science and science fiction. See Sophia Roosth, *Synthetic: How Life Got Made* (Chicago: University of Chicago Press, 2017), 109.

43. On modeling and anthropomorphism, see Sandra D. Mitchell, "Anthropomorphism and Cross-Species Modeling," in *Thinking with Animals: New Perspectives on Anthropomorphism*, ed. Lorraine J. Daston and Gregg Mitman (New York: Columbia University Press, 2005), 102. James R. Swearengen identifies six basic types of animal model: "induced" (or experimental), "spontaneous" (or natural), genetically modified, negative, orphan, and surrogate. See James R. Swearengen, "Choosing the Right Animal Model for Infectious Disease Research," *Animal Models and Experimental Medicine* 1, no. 2 (June 2018): 101.

44. As anthropologist Lesley Sharp has shown in studies of xenotransplantation and laboratory animal management, moral norms and forms of behavior are interconnected with cultural preconceptions and scientific work. See Lesley A. Sharp, *Animal Ethos: The Morality of Human-Animal Encounters in Experimental Lab Science* (Berkeley: University

of California Press, 2018); Lesley A. Sharp, *The Transplant Imaginary: Mechanical Hearts, Animal Parts, and Moral Thinking in Highly Experimental Science* (Berkeley: University of California Press, 2014).

45. On primates, see Megan Glick's exploration of "infrahumanism" in science. Megan H. Glick, *Infrahumanisms: Science, Culture, and the Making of Modern Non/Personhood* (Durham, NC: Duke University Press, 2018), 3–4.

46. Harriet Ritvo, *The Animal Estate: The English and Other Creatures in the Victorian Age* (Cambridge, MA: Harvard University Press, 1987), 84.

47. Thomas W. Laqueur, "What Are Dogs Doing in Eighteenth-Century British Art?" 25th Lewis Walpole Library Lecture, Lewis Walpole Library, Yale University, October 2022, https://library.yale.edu/news/video-what-are-dogs-doing-eighteenth-century-art. See also Thomas W. Laqueur and Alexander Nehamas, "Can a Dog Really Be a Man's Best Friend? An Exchange between Humans," *Qui Parle* 27, no. 2 (December 2018): 435–55.

48. Mikhail Bulgakov, *Heart of a Dog* (New York: Grove, 1982). Franz Kafka's short story "Investigations of a Dog" uses the quasi-scientific musings of a lonely dog to explore the phenomenological constraints on knowledge, available in Franz Kafka, *The Complete Stories* (New York: Schocken Books, 1988). On dogs and human projects of social formation, see Chris Pearson, *Dogopolis: How Dogs and Humans Made Modern New York, London, and Paris* (Chicago: University of Chicago Press, 2021); Andrew A. Robichaud, *Animal City: The Domestication of America* (Cambridge, MA: Harvard University Press, 2019).

49. Jacob F. Rivers III and Jeffrey Makala, eds., *In Dogs We Trust: An Anthology of American Dog Literature* (Columbia: University of South Carolina Press, 2019), xviii–xix.

50. Justyna Włodarczyk, *Genealogy of Obedience: Reading North American Dog Training Literature, 1850s–2000s* (Leiden: Brill, 2018), esp. 15–18 on intelligence and trainability. On pitties, see Erin C. Tarver, "The Dangerous Individual('s) Dog: Race, Criminality and the 'Pit Bull,'" *Culture, Theory and Critique* 55, no. 3 (September 2, 2014): 273–85, https://doi.org/10.1080/14735784.2013.847379. Experience with my own dog, who looks both like a pitt and a border collie, has reinforced this.

51. Literary theorist Ivan Kreillkamp calls literary animals "anthroprostheses," tools used to "define the non-animality of the human." See Ivan Kreilkamp, "Anthroprosthesis, or Prosthetic Dogs," *Victorian Review* 35, no. 2 (2009): 37, https://doi.org/10/gg7h4f; Ivan Kreilkamp, *Minor Creatures: Persons, Animals, and the Victorian Novel* (Chicago: University of Chicago Press, 2018); Keridiana W. Chez, *Victorian Dogs, Victorian Men: Affect and Animals in Nineteenth-Century Literature and Culture* (Columbus: Ohio State University Press, 2017).

52. Hans-Jörg Rheinberger, *An Epistemology of the Concrete: Twentieth-Century Histories of Life* (Durham, NC: Duke University Press, 2010), 7. The extremely close relationship between dogs and humans makes beagles a conspicuous example, but the concept has wider relevance for scholarly efforts to understand animal experimentation. As Tiago Saraiva argues in the case of Germany, "organisms that breeders of plants and animals produced through new practices of the sciences of heredity" were "as important as human bodies" in producing the material and ideological infrastructures of fascism. See Tiago Saraiva, *Fascist Pigs: Technoscientific Organisms and the History of Fascism* (Cambridge, MA: MIT Press, 2016), 2.

53. Heidi G. Parker, Abigail L. Shearin, and Elaine A. Ostrander, "Man's Best Friend Becomes Biology's Best in Show: Genome Analyses in the Domestic Dog," *Annual Review of Genetics* 44, no. 1 (2010): 309–36, https://doi.org/10.1146/annurev-genet-102808-115200. The "best in show" joke and other dog puns are common among contemporary researchers.

54. Eduardo Kohn, *How Forests Think: Toward an Anthropology beyond the Human* (Berkeley: University of California Press, 2013), 136.

55. Dylan Mulvin, *Proxies: The Cultural Work of Standing In* (Cambridge, MA: MIT Press, 2021), 4. My thinking on proxies builds on this work.

56. L. V. Bertalanffy, O. Hoffman-Ostenhof, and O. Schreier, "A Quantitative Study of the Toxic Action of Quinones on Planaria Gonocephala," *Nature* 158, no. 4026 (December 1946): 948–49, https://doi.org/10.1038/158948a0. Cheryl Logan has attributed the popularity of "model organism" to American biomedical work in the 1970s: Cheryl A. Logan, "Commercial Rodents in America: Standard Animals, Model Animals, and Biological Diversity," *Brain, Behavior and Evolution* 93, nos. 2–3 (2019): 79, https://doi.org/10.1159/000500073. But the term seems to have been popularized long before, initially in German. Hoffman-Ostenhof used "Modellorganismus," for instance, in subsequent papers without his coauthors. See O. Hoffmann-Ostenhof and Else Kriz, "Untersuchungen über bakteriostatische Chinone und andere Antibiotica," *Monatshefte für Chemie und verwandte Teile anderer Wissenschaften* 79, no. 5 (September 1, 1948): 410–20, https://doi.org/10.1007/BF00918555. Before the 1940s, "model organism" appeared in English with a different, moral connotation: a perfectly functioning body as *model* organism.

57. It also avoids the problem of reifying an active actor's category. Meunier has suggested thinking about some model organisms instead as "platforms." While beagles might be profitably considered from this angle, the contemporary buzz around platform-everything brings additional baggage to the term. See Robert Meunier, "Stages in the Development of a Model Organism as a Platform for Mechanistic Models in Developmental Biology: Zebrafish, 1970–2000," *Studies in History and Philosophy of Science Part C: Studies in History and Philosophy of Biological and Biomedical Sciences* 43, no. 2 (June 1, 2012): 522–31, https://doi.org/10.1016/j.shpsc.2011.11.013.

58. The procedure is common, both in scientific contexts and as a cosmetic choice by dog owners, but numerous groups have pushed to eliminate the practice. See, for instance, Matthew Hamity, "Cosmetic and Convenience Surgeries on Companion Animals: The Case for Laws with Bite to Protect a Dog's Bark," *Contemporary Justice Review* 19, no. 2 (April 2, 2016): 210–20, https://doi.org/10.1080/10282580.2016.1171574.

59. Robert Musil, *The Man without Qualities*, vol. 1, trans. Sophie Wilkins and Burton Pike (New York: Vintage, 1996), 624.

60. Nicole Nelson calls these "epistemic by-products": findings that served ends other than the ones originally unanticipated. See Nicole C. Nelson, *Model Behavior: Animal Experiments, Complexity, and the Genetics of Psychiatric Disorders* (Chicago: University of Chicago Press, 2018), 120–21.

61. Mariam Motamedi Fraser critically analyzes the dog's "species story": Mariam Motamedi Fraser, *Dog Politics: Species Stories and the Animal Sciences* (Manchester: Manchester University Press, 2024), 70–80.

62. Marilyn Strathern, *Relations: An Anthropological Account* (Durham, NC: Duke University Press, 2020), 168.

63. Emily E. Bray et al., "Early-Emerging and Highly Heritable Sensitivity to Human Communication in Dogs," *Current Biology*, June 3, 2021, https://doi.org/10.1016/j.cub.2021.04.055.

64. Donna J. Haraway, "Value-Added Dogs and Lively Capital," in *Lively Capital: Biotechnologies, Ethics, and Governance in Global Markets* (Durham, NC: Duke University Press, 2012), 93–120; Haraway, *When Species Meet*, 46–47.

65. Lévi-Strauss writes in the context of totemism, "We can understand, too, that natural species are chosen not because they are 'good to eat' but because they are 'good to think with.'" See Claude Lévi-Strauss, *Totemism*, trans. Rodney Needham (Boston: Beacon, 1963), 89.

66. Sharp notes that the term emerged in English-language usage in the first decade of the twentieth century but offers no theory for its origin. Nor does Michael Lynch, "Sacrifice and the Transformation of the Animal Body into a Scientific Object: Laboratory Culture and Ritual Practice in the Neurosciences," *Social Studies of Science* 18 (1988): 265–89, https://doi.org/10.1177/030631288018002004. There are traces of "sacrifice" dating back to the late eighteenth century in Britain: "If the substance of nerve be reproduced, certainly a period longer than the above must be necessary for this process; but to mark the precise point of time when the line is to be drawn, would *require the sacrifice of more animals than a question of mere curiosity could justify*," John Haighton wrote to the Royal Society in 1795. John Haighton, "VII. An Experimental Inquiry concerning the Reproduction of Nerves," *Philosophical Transactions of the Royal Society of London* 85 (December 1795): 196, https://doi.org/10.1098/rstl.1795.0009, emphasis mine). Haighton, known even in his time as a ruthless experimenter, was one of the first I could locate to use the term regularly. But "sacrificed" gained special popularity following widespread usage of "sacrifié" in the French physiological literature, in which the animal was thought to give its life to and for science. Biology undergraduates in one of my courses used "sac" conversationally in relation to their work with zebrafish.

67. Science studies scholars argue "sacrifice" fails to account for the complexity of human-animal relationships. Sharp suggests that the implications of ritualized behavior and a "sacred" relationship between researcher and organism exclude "other lexical and analytical possibilities" for understanding the morality of animal experimentation. Sharp, *Animal Ethos*, 130. Haraway, too, rejects the "idiom of sacrifice" for implying an animal victimhood. Haraway, *When Species Meet*, 75–77.

68. Radhika Govindrajan, *Animal Intimacies: Interspecies Relatedness in India's Central Himalayas* (Chicago: University of Chicago Press, 2018), 48. For Govindrajan, "sacrifice" implies not a static, dyadic ritual, but imbricated practices of relatedness and mutual obligation. Emma Roe and Beth Greenhough have sought to reconcile the presence of "care" and "harm" in laboratory animal research by showing how they are "situated, co-constitutive practices." See Emma Roe and Beth Greenhough, "A Good Life? A Good Death? Reconciling Care and Harm in Animal Research," *Social and Cultural Geography* 24, no. 1 (2021): 62, https://doi.org/10.1080/14649365.2021.1901977.

69. Charles Darwin, *The Descent of Man, and Selection in Relation to Sex*, 2nd ed., 1874, Project Gutenberg, http://www.gutenberg.org/files/2300/2300-h/2300-h.htm.

70. For an extended analysis of Darwin's role in vivisection debates, see Rob Boddice, *Humane Professions: The Defence of Experimental Medicine, 1876–1914* (Cambridge: Cambridge University Press, 2021), 20–41.

71. R. L. Chaput, R. T. Kovacic, and E. L. Barron, *Performance of Trained Beagles after Supralethal Doses of Radiation* (Bethesda, MD: Armed Forces Radiobiology Research Institute, Defense Nuclear Agency, February 1972), 1.

72. Haraway, *When Species Meet*, 80.

73. Clive D. L. Wynne, "Dogs' (*Canis lupus familiaris*) Behavioral Adaptations to a Human-Dominated Niche: A Review and Novel Hypothesis," in *Advances in the Study of Behavior*, vol. 53 (Cambridge, MA: Academic, 2021), 139.

74. Marc Bekoff and Jessica Pierce, *Unleashing Your Dog: A Field Guide to Giving Your Canine Companion the Best Life Possible* (Novato, CA: New World Library, 2019), 5. For an extended critical evaluation of this notion, see Fraser, *Dog Politics*, 172–73.

75. The phrase was later taken up in advertising copy for scientific dog breeder Marshall Farms. See Arnold Arluke, "'We Build a Better Beagle': Fantastic Creatures in Lab Animal Ads," *Qualitative Sociology* 17, no. 2 (1994): 143–58.

76. Jean-Pierre Vaissaire, *Le chien, animal de laboratoire* (Belgium: Editions Vigot Frères, 1972), 176. Unless otherwise noted, translations here and elsewhere are mine.

77. This book joins contemporary historians in focusing on the circulation of knowledge and artifacts. Throughout, I have drawn on methodologies from new histories of global capitalism, histories of science and capitalism, and studies of biocapital and animal capital. On circulations and flows of knowledge in the history of science, see Warwick Anderson, "From Subjugated Knowledge to Conjugated Subjects: Science and Globalisation, or Postcolonial Studies of Science?," *Postcolonial Studies* 12, no. 4 (December 1, 2009): 389–400, https://doi.org/10/cxkbpj; Warwick Anderson, "Remembering the Spread of Western Science," *Historical Records of Australian Science* 29 (2018): 73–81. On the history of capitalism, see for instance, Sven Beckert, *Empire of Cotton: A Global History* (New York: Vintage Books, 2015). Scholars of science have only recently turned directly to how markets and commodities intersect with and shape the practice of science. That broader turn is charted in Lukas Rieppel, Eugenia Lean, and William Deringer, "The Entangled Histories of Science and Capitalism," *Osiris* 33 (2018): 1–24. Bruno Latour was skeptical of the notion of a history of "capitalism," a perspective that still influences STS research. See Bruno Latour, *Reassembling the Social: An Introduction to Actor-Network-Theory* (Oxford: Oxford University Press, 2007). However, the centrality of capital and markets, particularly after the 2008 recession, have made Latour's earlier dismissals of capitalism as an object of study difficult to maintain. As the introduction to a conversation including Latour noted, "the ecologies of both humans and nonhumans are being radically made and unmade according to the logics of capitalism." Bruno Latour et al., "Anthropologists Are Talking—about Capitalism, Ecology, and Apocalypse," *Ethnos* 83, no. 3 (May 27, 2018): 587, https://doi.org/10.1080/00141844.2018.1457703.

78. Some might see my account as "diffusionist": of practices originating in the West and spreading elsewhere. If this book does center American science and scientists, it does so with good reason: the beagle story reinforces how the thing we call "science" was influentially shaped by the hegemonic force of twentieth-century American research, politics, and power. My aim is to show, nonetheless, how these very American research tools gradually shed their national contexts to become part of globalized science and capitalism. It is not an account of inevitable diffusion, but of fractious movement, propelled by scientific imperatives and powerful political and economic actors.

79. Robert E. Kohler, *Lords of the Fly: Drosophila Genetics and the Experimental Life* (Chicago: University of Chicago Press, 1994), 11. This account appears, for instance, in, Donna J. Haraway, *Modest_Witness@Second_Millennium.FemaleMan_Meets_OncoMouse: Feminism and Technoscience* (New York: Routledge, 1997), 98.

80. Robert Kohler, "Lords of the Fly Revisited," *Journal of the History of Biology* 55, no. 1 (March 1, 2022): 15–19, https://doi.org/10.1007/s10739-022-09671-y. I developed this suggestion in an earlier article: Brad Bolman, "Dogs for Life: Beagles, Drugs, and Capital in the Twentieth Century," *Journal of the History of Biology* 55, no. 1 (2022), 147–79, https://doi.org/10.1007/s10739-021-09649-2.

81. Anthropologists have used the term "biocapital" to describe the contemporary transformation of living substances into sources and producers of capitalist value, exemplified by the biotechnological commodification of DNA and other substances. An excellent overview is Stefan Helmreich and Nicole Labruto, "Species of Biocapital, 2008, and Speciating Biocapital, 2017," in *The Palgrave Handbook of Biology and Society*, ed. Maurizio Meloni et al. (London: Palgrave Macmillan, 2018). Capitalism, some note, has arguably always relied on the production of value from nonhuman animals, whether in agriculture, industry, or beyond. The question is how to name the specific effects that occur when animals are drawn into circuits of contemporary commodification. Nicole Shukin offers "animal capital" to describe the semiotic and material ways that nonhuman animals become forms of wealth, while Sarah Franklin showed how Dolly the Sheep's cloning tied her to an older lineage of agricultural capital, rendering Dolly "breed wealth" in multiple senses. See Nicole Shukin, *Animal Capital: Rendering Life in Biopolitical Times*, Posthumanities 6 (Minneapolis: University of Minnesota Press, 2009); Sarah Franklin, *Dolly Mixtures: The Remaking of Genealogy* (Durham, NC: Duke University Press, 2007). In a study of the exotic animal market, Rosemary-Claire Collard adopts the term "lively capital" (taken from Haraway) to describe "a 'stock' of living objects that generate, or have the potential to generate, value of some kind." Rosemary-Claire Collard, *Animal Traffic: Lively Capital in the Global Exotic Pet Trade* (Durham, NC: Duke University Press, 2020), 143.

82. Amy Nelson, "What the Dogs Did: Animal Agency in the Soviet Manned Space Flight Programme," *British Journal for the History of Science: Themes* 2 (2017): 86; David P. D. Munns and Kärin Nickelsen, *Far beyond the Moon: A History of Life Support Systems in the Space Age* (Pittsburgh: University of Pittsburgh Press, 2021), 21–22.

83. The parallel between standard animals and other standards was intentional from early boosters, such as Milton Greenman of the Wistar Institute. Greenman saw value in bringing Taylorism and consistency to science. See Logan, "Commercial Rodents in America," 74.

84. Dogs needed to be "immutable mobiles," as Latour once called those objects of scientific practice that remain (arguably) constant and true as they move. Because of surgical manipulation and local differences, they were just as often mut(e)able mobiles. See Bruno Latour, "Visualisation and Cognition: Drawing Things Together," in *Knowledge and Society: Studies in the Sociology of Culture Past and Present*, ed. Elizabeth Long and Henrika Kuklick (Greenwich, CT: JAI, 1986).

85. For "salvage accumulation," see Anna Lowenhaupt Tsing, "Salvage Accumulation, or the Structural Effects of Capitalist Generativity," *Cultural Anthropology Generating Capitalism*, March 30, 2015, https://culanth.org/fieldsights/salvage-accumulation-or-the-structural-effects-of-capitalist-generativity. I apply the term to the pound debates in Brad Bolman, "In the Animal House: Rabies, Labor, and Salvage Dogs in Jim Crow Birmingham," *Historical Studies in the Natural Sciences* 52, no. 5 (November 2022): 589–628. This usage differs from two other recent invocations of the term. See Joanna Radin, *Life on Ice: A History of New Uses for Cold Blood* (Chicago: University of Chicago Press, 2017); Nicolas Langlitz, *Chimpanzee Culture Wars: Rethinking Human Nature alongside Japanese, European, and American Cultural Primatologists* (Princeton, NJ: Princeton University Press, 2020), https://www.degruyter.com/document/doi/10.1515/9780691204260-010/html.

86. On the capitalist shift to mass production, see, for instance, David A. Hounshell, *From the American System to Mass Production, 1800–1932: The Development of Manufacturing Technology in the United States* (Baltimore: Johns Hopkins University Press, 1984); Joshua

B. Freeman, *Behemoth: A History of the Factory and the Making of the Modern World* (New York: W. W. Norton, 2018).

87. This differs markedly from the case of Jackson Labs, where Rader shows that a shift occurred from internal research to marketing animals to others. See Rader, *Making Mice*, 133–34.

88. The first trademarked lab animal in America was the WISTARAT, in 1942. Beagles got their trademark ahead of Haraway's OncoMouse®. See Haraway, *Modest_Witness@Second_Millennium.FemaleMan_Meets_OncoMouse.*

89. Rader, *Making Mice*, 99. On the growth of monopoly capital, see Harry Braverman, *Labor and Monopoly Capital: The Degradation of Work in the Twentieth Century* (New York: Monthly Review Press, 1998).

90. I have avoided extensively covering this because an excellent overview exists in Ryan Noah Shapiro, "Bodies at War: National Security in American Controversies over Animal and Human Experimentation from WWI to the War on Terror" (PhD diss., Massachusetts Institute of Technology, 2018).

91. Haraway, "Value-Added Dogs and Lively Capital," 95. I concur with Collard's suggestion that "encounter value" is less a distinct third "mode" of value than a crucial component of use value—one that was central to the work of scientific research. See Collard, *Animal Traffic*, 143.

92. Haraway, *When Species Meet*, 71.

93. Peter Lovenheim, "The Puppy Farm," *Plain Dealer Magazine*, June 10, 1979, 21. This book is limited in its engagement with the thoughts and concerns of animal caretakers. Most left a far more limited paper trail than their publishing laboratory counterparts, which makes reconstructing their actions and beliefs challenging. Sharp engages some in *Animal Ethos*, as do the contributors to *Researching Animal Research: What the Humanities and Social Sciences Can Contribute to Laboratory Animal Science and Welfare*, ed. Gail Davies et al. (Manchester: Manchester University Press, 2024).

94. Eva Giraud insightfully notes that a focus on relationality and entanglement may undermine avenues for political action. See Eva Haifa Giraud, *What Comes after Entanglement? Activism, Anthropocentrism, and an Ethics of Exclusion* (Durham, NC: Duke University Press, 2019). Fraser adds that "entanglement" may be less fruitful in the context of dogs, because human-canine relations are powerfully shaped by notions of a transhistoric "bond." Fraser, *Dog Politics*, 220.

95. August Krogh, "The Progress of Physiology," *American Journal of Physiology* 90, no. 2 (October 1, 1929): 247. Krebs brought renewed attention to Krogh's publication in his own piece from 1975, "The August Krogh Principle," where he sought "to analyze the reasons why certain species lend themselves especially well to the study of certain problems." See Hans A. Krebs, "The August Krogh Principle: 'For Many Problems There Is an Animal on Which It Can Be Most Conveniently Studied,'" *Journal of Experimental Zoology* 194, no. 1 (1975): 221–22, https://doi.org/10.1002/jez.1401940115.

96. Krebs, "August Krogh Principle." "Le choix heureux d'un animal" also suggests a "successful" or "fortunate" choice. See Claude Bernard, *Introduction à l'étude de la médecine expérimentale* (Paris: J. B. Baillière et Fils, 1865).

97. Robert E. Kohler, *Lords of the Fly: Drosophila Genetics and the Experimental Life* (Chicago: University of Chicago Press, 1994).

98. Soraya de Chadarevian, "Of Worms and Programmes: Caenorhabditis Elegans and the Study of Development," *Studies in History and Philosophy of Science Part C* 29, no. 1 (1998): 81–105.

99. The apparent indifference of frogs to pain supported their use in experimental physiology, while the full embryonic development of zebrafish can occur in a petri dish. See Frederic L. Holmes, "The Old Martyr of Science: The Frog in Experimental Physiology," *Journal of the History of Biology* 26, no. 2 (1993): 311–28, https://doi.org/10.1007/BF01061972; David Jonah Grunwald and Judith S. Eisen, "Headwaters of the Zebrafish—Emergence of a New Model Vertebrate," *Nature Reviews Genetics* 3, no. 9 (September 1, 2002): nrg892, https://doi.org/10.1038/nrg892. The ease of breeding pigs and their ostensibly similar skin suggested a niche in studies of skin and dermatology. See Brad Bolman, "Pig Mentations: Race and Face in Radiobiology," *Isis* 112, no. 4 (December 2021): 694–716. The writing on these histories is extensive, but see especially Anita Guerrini, *Experimenting with Humans and Animals: From Galen to Animal Rights* (Baltimore: Johns Hopkins University Press, 2003); Bonnie Tocher Clause, "The Wistar Rat as a Right Choice: Establishing Mammalian Standards and the Ideal of a Standardized Mammal," *Journal of the History of Biology* 26, no. 2 (1993): 329–49, https://doi.org/10.1007/BF01061973; Meunier, "Stages in the Development of a Model Organism"; Rachel A. Ankeny, "The Natural History of *Caenorhabditis elegans* Research," *Nature Reviews Genetics* 2, no. 6 (June 1, 2001): 35076538, https://doi.org/10.1038/35076538.

100. Quoted in Paget, *Beagles and Beagling*, 34.

101. I draw the terms "momentum" and "emergence" from Manuel Delanda's theory of "assemblage history." See Manuel DeLanda, *Assemblage Theory* (Edinburgh: Edinburgh University Press, 2016).

102. Models "create anastomoses through their spreading," Rheinberger notes, establishing cross connections between adjacent channels of experimental inquiry. See Hans-Jörg Rheinberger, *Toward a History of Epistemic Things: Synthesizing Proteins in the Test Tube*, Writing Science (Stanford, CA: Stanford University Press, 1997), 109.

103. This is not the first use of "beagling" in science studies. To emphasize her attention to smell rather than visual metaphors in science, Sara Ann Wylie described a "beagling" approach. See Sara Ann Wylie, *Fractivism: Corporate Bodies and Chemical Bonds* (Durham, NC: Duke University Press, 2018), 48. But "beagling" here serves a function similar to Creager's invocation of radioisotopes as historical "tracers"; see Angela N. H. Creager, *Life Atomic: A History of Radioisotopes in Science and Medicine* (Chicago: University of Chicago Press, 2013), 4–5.

104. As Creager argues in her history of the tobacco mosaic virus, tracking the virus between fields of inquiry can "reveal the otherwise inconspicuous connections between biological experimentation and activities usually relegated to the domains of technology, politics, medicine, and agriculture." See Creager, *Life of a Virus*, 3. Where early histories of laboratory organisms were framed as explorations of "practice" and "material culture" in biology, Creager's work and other work that followed signaled a transition from studying "practice" to studying the "infrastructures" of scientific experimentation. See Angela N. H. Creager and Hannah Landecker, "Technical Matters: Method, Knowledge and Infrastructure in Twentieth-Century Life Science," *Nature Methods* 6, no. 10 (October 2009): 701–5, https://doi.org/10/b3z4gf; Hannah Landecker, "The Matter of Practice in the Historiography of the Experimental Life Sciences," in *Handbook of the Historiography of Biology*, Historiography of Science 1 (Cham, Switzerland: Springer International, 2018). Laboratory animal production and provision "form part of the little studied auxiliary infrastructure that has sustained the phenomenal growth of the biomedical sciences in the latter half of the twentieth century," argue historians Tone Druglitrø and Robert G. W. Kirk in "Building

Transnational Bodies: Norway and the International Development of Laboratory Animal Science, ca. 1956–1980," *Science in Context* 27, no. 2 (2014): 334.

105. Etienne Benson, "Animal Writes: Historiography, Disciplinarity, and the Animal Trace," in *Making Animal Meaning*, ed. Linda Kalof and Georgina M. Montgomery (East Lansing: Michigan State University Press, 2012), 3.

106. Gilles Deleuze and Félix Guattari, *What Is Philosophy?*, trans. Hugh Tomlinson and Graham Burchill (New York: Columbia University Press, 1994), 55.

107. Galison uses "mesoscopic" to distinguish his approach in *Image and Logic* from macroscopic and microscopic approaches. See Peter Galison, *Image and Logic: A Material Culture of Microphysics* (Chicago: University of Chicago Press, 1997), 61.

Chapter 1

1. Jane Stafford, "Mongrels Get Highest Prize of Dogdom: Whipple Award Conferred by General Kirk," *Syracuse Post Standard*, February 17, 1946, NewspaperArchive.

2. Despite promises at the time that the award would be conferred annually, I find no newspaper evidence that it was ever given again.

3. Jean Walrath, "2 UR Lab Dogs Go Snooty after Honors in New York," *Democrat and Chronicle*, February 17, 1946, B3.

4. Quoted in Madeline Ryttenberg, "2 Dogs Win Honor Collars for Aid in Medical Research," *Newsday*, February 8, 1946.

5. "Pair of Mongrel Canines Wins Military Awards," Salt Lake Tribune, February 7, 1946, 2.

6. Dave Terrill, "Just Dogs," *Daily Herald*, January 21, 1954, NewspaperArchive.

7. Leon Fradley Whitney, *The Truth about Dogs* (New York: Thomas Nelson and Sons, 1959), 100.

8. H. W. "Frog" Hayes, "Beagling Catches On Fast among Piedmont Sportsmen," *High Point Enterprise*, February 22, 1952, NewspaperArchive.

9. The joke appears in numerous issues of *Forest and Stream* from the early 1870s. See, for instance, *Forest and Stream*, September 25, 1873, 109.

10. "About Dogs," *Sonoma County Journal*, January 4, 1861. An announcement of Barnum's show appears in "Letter from New York," *Sacramento Daily Union*, June 11, 1862. The show was criticized both for woefully underperforming its promise of two thousand dogs and for being little more than an advertisement for New York's dog dealers.

11. Rusticus, "Letter to the Editor: Beagle Gossip," *Forest and Stream*, July 26, 1883.

12. Justyna Włodarczyk, *Genealogy of Obedience: Reading North American Dog Training Literature, 1850s–2000s* (Leiden: Brill, 2018), 54.

13. Razor, "Letter to the Editor: The Beagle Club," *Forest and Stream*, September 27, 1883.

14. A. C. Krueger, "Letter to the Editor: American English Beagle Club," *Forest and Stream*, November 15, 1883. Plans for incorporation were finalized in December, with an election carried out via *Forest and Stream*.

15. Rowett was born in Cornwall, England, and came to the United States in 1851. He served with distinction in the Illinois Infantry and turned to breeding thoroughbred horses and beagles after the Civil War. See Tom Emery, *Richard Rowett: Thoroughbreds, Beagles, and the Civil War* (Carlinville, IL: History in Print, 1997). Although many refer to Rowett as the first beagle importer, the exact date of his doing so remains uncertain. Multiple references to "good strains" of beagles in Morristown, New Jersey; Milford

Pike County, Pennsylvania; and Guernsey County, Ohio, appear in *Forest and Stream* by September 1873, which would precede the suggested date of Rowett's first import in 1876 and newspaper reference to Rowett's "imported beagles" in 1878. See "The Kennel," *Forest and Stream*, September 4, 1873.

16. O. W. Rogers, "Letter to the Editor: The Beagle Club," *Forest and Stream*, December 27, 1883.

17. "Answers to Correspondents" section, October 16, 1884, *Forest and Stream.*

18. "The American English Beagle Club," *Forest and Stream*, January 10, 1884.

19. "The Beagle Club's Name Altered," *Forest and Stream*, April 17, 1890, 255.

20. "English Kennel Notes," *Forest and Stream*, January 8, 1885.

21. There is inconsistency in the records about whether the AEBC was founded in 1883 or 1884, but newspapers from 1883 suggest the former date. See "Roundabout," *Fall River Daily Herald*, November 16, 1883; "Canines In Session," *Boston Globe*, April 3, 1888.

22. Bradford S. Turpin, *The Beagle and the Field Trials*, Popular Dogs of the Day 1 (Baltimore: F. J. Skinner, 1900), 12.

23. "The American Hare: Hunting with Hounds," *(Philadelphia) Times*, November 15, 1885, Sunday Morning edition, NewspaperArchive.

24. Herman F. Schellhass, "The Beagle Hound," in *The American Book of the Dog*, ed. G. O. Shields (Chicago: Rand, McNally, 1891), 270.

25. "The New National Beagle Club—Massachusetts Kennel Club Matters, Etc.," *Boston Herald*, May 11, 1890, Sunday edition, 18.

26. "New National Beagle Club," 18. The location is discussed in "National Beagle Club of America Minutes from April 3, 1890 to June 17, 1898," 1890, F. Ambrose Clark Rare Book Room, MS RBR B 561 .285 1890, National Sporting Library and Museum. Mechanics Hall was razed for the construction of the Prudential Center.

27. "Field Trial of Beagles," *Boston Herald*, October 26, 1890, Sunday edition, 18.

28. The image was based on Don, a hound owned by W. F. Rutter Jr. of Lawrence.

29. "Some Champion Dogs," *Philadelphia Times*, April 14, 1889, 16, NewspaperArchive.

30. "Field Trial of Beagles," 18.

31. James C. O'Connell, "How Metropolitan Parks Shaped Greater Boston, 1893–1945," in *Remaking Boston: An Environmental History of the City and Its Surroundings* (Pittsburgh: University of Pittsburgh Press, 2009), 174.

32. O'Connell, 177.

33. "Following the Beagles," *Boston Herald*, November 8, 1890, NewsBank.

34. "The National Beagle Club," *Boston Herald*, December 21, 1890, Sunday edition, NewsBank.

35. An initial newspaper report has "wrong," while a republication of the speech in 1921 has "mistaken." I have left the original report. "The National Beagle Club," *Boston Herald*, January 3, 1891, NewsBank; Eugène Lentilhon, *Forty Years Beagling in the United States* (New York: E. P. Dutton and Company, 1921), 3. The comment was likely a response to an unsigned note in *Forest and Stream* from September 1890 that "men who run beagles down in Jersey and about Philadelphia are quietly smiling" about the upcoming field trials. "They don't say that the field trials will be a failure, but they want to know how in the world they are going to be a success." See "[National Beagle Club Field Trials]," *Forest and Stream*, September 18, 1890.

36. Schellhass, "Beagle Hound," 273.

37. Quoted in Lentilhon, *Forty Years Beagling in the United States*, 3. Not long after, in April, at the first annual banquet in Boston, the issue of a Western subdivision was broached.

38. Joe M. Bassford Jr., untitled report, *Breeder and Sportsman*, January 19, 1889, 43, Biodiversity Heritage Library.

39. "Kennel Notes," *Boston Herald*, January 12, 1891, NewsBank.

40. "Field Trial of Beagles," *Boston Herald*, November 7, 1890, NewsBank.

41. Not everyone was sold: a group of "prominent dog men" met in September 1891 in Boston to discuss hosting separate field trials, since the activities of the National Beagle Club "do not appeal to so large a portion of the hunting world as would trials between pointers or setters." See "Field Trials for New England," *Boston Herald*, September 30, 1891, NewsBank.

42. "A Thousand Dollars for a Beagle," *New York Herald*, February 6, 1892, 8.

43. "Wide Interest in Trials of National Beagle Club of America," *(Knoxville) Journal and Tribune*, November 17, 1901, 2.

44. "These traditional field sports provided the basis for the earliest periodicals devoted to promoting outdoor activities; they were the source of the initial markets for specialized sporting equipment; and it was frequently the journalists and merchants of field sports who became product champions for the 'new' varieties of games that emerged during the middle third of the century," writes Stephen Hardy in "'Adopted by All the Leading Clubs': Sporting Goods and the Shaping of Leisure, 1800–1900," in *For Fun and Profit: The Transformation of Leisure into Consumption*, ed. Richard Butsch (Philadelphia: Temple University Press, 1990), 74.

45. Despite its historical significance, there is a paucity of scholarship on beagling, particularly after 1940. For beagling as a marker of wealth, see Christine Elise Garneau, "Establishing an Elite Sport: The Men and Hounds of the National Beagle Club of America, 1890–1940" (MA thesis, Simon Fraser University, 2012).

46. "Imported Beagle Hounds," *Evening Dispatch*, June 10, 1881, 1. As Kristin Hoganson argues, many Americans of the period leaned heavily on internationally imported goods to construct their domestic worlds. See Kristin L. Hoganson, *Consumers' Imperium: The Global Production of American Domesticity, 1865–1920* (Chapel Hill: University of North Carolina Press, 2010), 11, http://muse.jhu.edu/book/44128/.

47. "Pets at Thornfield," *(Baltimore) Sun*, September 9, 1901, ProQuest Historical Newspapers.

48. "Beagles Give Peninsula Set Exercise and Sport," *San Francisco Examiner*, December 17, 1916, ProQuest Historical Newspapers.

49. Andrew A. Robichaud, *Animal City: The Domestication of America* (Cambridge, MA: Harvard University Press, 2019), 160.

50. "Beagles Raising a Fad," *Newark Daily Advocate*, March 31, 1900.

51. "Millionaires' Fads," *(Washinton, DC) Sunday Star*, April 30, 1905, sec. part II, 1.

52. "Rockefeller's Beagles," *Brooklyn Daily Eagle*, April 6, 1902; "Rockefeller's Beagles Will Be Exhibited Here," *Pittsburgh Press*, March 7, 1909. In 1905, Rockefeller beagles were valued at $175 (roughly $6,000 today) each. See "Rockefeller's Beagles in South," *Sun*, February 8, 1905, 9, ProQuest Historical Newspapers.

53. Dexter Marshall, "William G.: The Rockefeller of Tomorrow," *Muncie Morning Star*, July 1, 1906, sec. Magazine, 3, ProQuest Historical Newspapers.

54. "Pet Dogs Whose Values Mount into Five Figures," *Pittsburgh Post*, February 11, 1906, 32, ProQuest Historical Newspapers.

55. "Latest Society Fad Is Beagle Hunting," *Philadelphia Inquirer*, October 26, 1913. For Burlingame, see Cholly Francisco, "Game Scarce Sport Fine at Initial Beagle Hunt," *San Francisco Examiner*, February 28, 1916, ProQuest Historical Newspapers.

56. Francisco, "Game Scarce Sport Fine."

57. Maximilian Foster, *Shoestrings* (New York: D. Appleton, 1917), 176.

58. For reference to a good kennel man, see Alexander Henry Higginson, "An All-Round Sporting Dog: The Beagle Growing Popular," *Outing Magazine* 43, no. 1 (October 1903): 40.

59. E. S. McClure, "Absorbing Fads of Some Millionaires," *Democrat and Chronicle*, April 30, 1905, 15.

60. The exact date of the German shepherd's arrival to the United States is debated, but press coverage of the breed after 1913 is consistent.

61. "Sustained Popularity of Beagle without Equal in United States," *Bluefield Daily Telegraph*, June 7, 1936, 3, NewspaperArchive.

62. The "melting eyes" descriptor is frequent in coverage of cockers from the era. See, for instance, Bob Becker, "Mostly about Dogs," *Chicago Tribune*, March 13, 1949, A5.

63. "Beagles Most Popular in Dogdom," *Brooklyn Daily Eagle*, January 29, 1954, 15.

64. George D. Whitney, *This Is the Beagle* (Jersey City, NJ: T.F.H., 1955), 12. Already in 1906, many felt that "an American dog means one born in the United States" and resented the sense that prizes typically went to the "latest dog from the ship." See "Pet Dogs Whose Values Mount into Five Figures," 32.

65. Glenn G. Black, *American Beagling* (New York: G. P. Putnam's Sons, 1949), 17.

66. Journalistic attention to pet hair and allergies grew noticeably in American newspapers during the late 1940s and early 1950s. See, for example, Beulah France, "Child Plagued by Colds Needs Good Checkup," *(Newport News) Daily Press*, November 20, 1951, 7.

67. Thomas L. Altherr, "'Mallards and Messerschmitts': American Hunting Magazines and the Image of American Hunting during World War II," *Journal of Sport History* 14, no. 2 (1987): 151–52; Andrea L. Smalley, "'Our Lady Sportsmen': Gender[,] Class, and Conservation in Sport Hunting Magazines, 1873–1920," *Journal of the Gilded Age and Progressive Era* 4, no. 4 (October 2005): 355–80, https://doi.org/10/fb322s. See also, for the significance of hunting in American history, Thomas L. Altherr and John F. Reiger, "Academic Historians and Hunting: A Call for More and Better Scholarship," *Environmental History Review* 19, no. 3 (1995): 39–56, https://doi.org/10/bc3v9q.

68. Altherr, "Mallards and Messerschmitts," 151.

69. "Patience, Please," *Field and Stream* 47, no. 6 (October 1942): 6.

70. *Recreation for Industrial Workers: A Guide for Workers in Charge of Recreation in Industrial Plants* (New York: National Recreation Association, 1945), 23. For the ad, see "Meet the Champs!," *Albuquerque Journal*, March 19, 1945.

71. J.H., "Dogs and Rationing," *Field and Stream*, May 1943, 99.

72. "Dogs Play Part in Reconditioning of Disabled GI's," *Brewster Standard*, May 31, 1945. For the history of seeing-eye dogs in postwar rehabilitation, see Julie Anderson and Neil Pemberton, "Walking Alone: Aiding the War and Civilian Blind in the Interwar Period," *European Review of History: Revue européenne d'histoire* 14, no. 4 (December 1, 2007): 459–79, https://doi.org/10.1080/13507480701752128; Monika Baár, "Prosthesis for the Body and for the Soul: The Origins of Guide Dog Provision for Blind Veterans in Interwar Germany," *First World War Studies* 6, no. 1 (January 2, 2015): 81–98, https://doi.org/10/ggrqwn; Neil Pemberton, "Cocreating Guide Dog Partnerships: Dog Training and Interdependence in 1930s America," *Medical Humanities* 45, no. 1 (March 1, 2019): 92–101, https://doi.org/10/ggrqwp.

73. "Wounded Vet's Beagle Wins 1st in Field Trials," *Alton Evening Telegraph*, April 23, 1945, 9.

74. Susan D. Jones, *Valuing Animals: Veterinarians and Their Patients in Modern America* (Baltimore: Johns Hopkins University Press, 2002), 117. For Lyons, see John Lyons, "Sports Lines . . . ," *Victoria Advocate*, May 25, 1955, 13.

75. Anna M. Waller, *Dogs and National Defense* (Washington, DC: Department of the Army Office of the Quartermaster General, 1958).

76. J.H., "Post-war Plans," *Field and Stream* 49, no. 10 (February 1945): 102.

77. Black, *American Beagling*, 30.

78. Black, 34.

79. Lyons, "Sports Lines . . . ," 13.

80. Philip J. Pauly, "The Beauty and Menace of the Japanese Cherry Trees: Conflicting Visions of American Ecological Independence," *Isis* 87 (1996): 51–73. Lukas Rieppel has also shown how "intellectuals from the United States routinely turned to the natural world as a source of national pride." See Lukas Rieppel, *Assembling the Dinosaur: Fossil Hunters, Tycoons, and the Making of a Spectacle* (Cambridge, MA: Harvard University Press, 2019), 53.

81. "Main Line Man Has Done Much to Popularize Beagles," *Philadelphia Inquirer*, March 11, 1917, 23.

82. "'Doc' Dopey," *Lawton Constitution*, August 19, 1948, 6.

83. Jack Melder, "Beagles Lead the Way," *New Castle News*, March 21, 1955, 8.

84. Remarks from February 1955, published as Nelson N. Foote, "A Neglected Member of the Family," *Marriage and Family Living* 18, no. 3 (1956): 216.

85. Richard Kleiner, "22 Million Canines Doggone Big Business," *Indiana Evening Gazette*, May 22, 1952, 19.

86. Foote, "Neglected Member of the Family," 214.

87. Jones, *Valuing Animals*, 120. As an example, see K. C. Jerome, "What's So Bad about a Dog's Life," *Daily Herald*, November 7, 1954, 45. In Victorian Britain, this process began perhaps slightly earlier than in the United States, with department stores dedicating space to an assortment of canine accessories by the first decades of the twentieth century. See Sarah Amato, *Beastly Possessions: Animals in Victorian Consumer Culture* (Toronto: University of Toronto Press, 2015), 43.

88. Lizabeth Cohen, "A Consumers' Republic: The Politics of Mass Consumption in Postwar America," *Journal of Consumer Research* 31, no. 1 (2004): 236.

89. Bob Becker, "The Beagle—a Happy Hunter," *St. Louis Globe-Democrat*, October 20, 1940, ProQuest Historical Newspapers; Bob Brister, "Beagling Comes to Front as New Dog Sport over South," *Marshall News Messenger*, May 9, 1954, 29.

90. McIntyre's "New York by Day" column appeared in papers across the country. He used "beagling" in one form or another at least four times over the years. See, for example, O. O. McIntyre, "New York Day by Day," *Appeal-Democrat*, February 11, 1937, 10, Newspapers.com World Collection.

91. Becker, "Beagle—a Happy Hunter."

92. Leon Fradley Whitney to Charles Benedict Davenport, December 16, 1922, Charles Benedict Davenport Papers, box 95, folder 1, "Whitney, Leon F.," American Philosophical Society, Philadelphia.

93. Leon Fradley Whitney to Charles Benedict Davenport, December 21, 1922, Charles Benedict Davenport Papers, box 95, folder 1, "Whitney, Leon F.," American Philosophical Society.

94. Leon Fradley Whitney to Robert M. Yerkes, October 19, 1925, Robert Mearns Yerkes Papers (MS 569), box 52, folder 1007, Manuscripts and Archives, Yale University Library, New Haven, CT.

95. Cynthia Huff, "Victorian Exhibitionism and Eugenics: The Case of Francis Galton and the 1899 Crystal Palace Dog Show," *Victorian Review* 28, no. 2 (2002): 1–20, https://doi.org/10.1353/vcr.2002.0014. Chris Pearson notes that human ethnographical displays frequently shared space with dog shows at venues like the Crystal Palace in South London: "Scientific racism and dog breeding fed off each other to admonish the mixing of races in humans and breeds in dogs." Chris Pearson, *Dogopolis: How Dogs and Humans Made Modern New York, London, and Paris* (Chicago: University of Chicago Press, 2021), 33.

96. J. B. S. Haldane, "The Argument from Animals to Men: An Examination of Its Validity for Anthropology," *Journal of the Royal Anthropological Institute of Great Britain and Ireland* 86, no. 2 (1956): 1–14, https://doi.org/10.2307/2843990. On eugenics in Plato, see also D. J. Galton, "Greek Theories on Eugenics.," *Journal of Medical Ethics* 24, no. 4 (August 1, 1998): 263–67, https://doi.org/10.1136/jme.24.4.263.

97. See Włodarczyk, *Genealogy of Obedience*, 56.

98. For a recent reevaluation and rejection of this analogy, see Heather L. Norton et al., "Human Races Are Not Like Dog Breeds: Refuting a Racist Analogy," *Evolution: Education and Outreach* 12, no. 1 (July 9, 2019): 17, https://doi.org/10.1186/s12052-019-0109-y.

99. Charles Benedict Davenport to Leon Fradley Whitney, January 8, 1923, Charles Benedict Davenport Papers, box 95, folder 1, "Whitney, Leon F.," American Philosophical Society.

100. Leon Fradley Whitney to Charles Benedict Davenport, December 28, 1923, Charles Benedict Davenport Papers, box 95, folder 1, "Whitney, Leon F.," American Philosophical Society.

101. Leon Fradley Whitney to Charles Benedict Davenport, February 14, 1924, Charles Benedict Davenport Papers, box 95, folder 2, "Whitney, Leon F.," American Philosophical Society.

102. Leon Fradley Whitney to Charles Benedict Davenport, May 25, 1925, Charles Benedict Davenport Papers, box 95, folder 2, "Whitney, Leon F.," American Philosophical Society.

103. L. Whitney to Yerkes, October 19, 1925.

104. Wesley R. Coe to Robert M. Yerkes, November 28, 1925, box 52, folder 1007, Robert Mearns Yerkes Papers (MS 569), Manuscripts and Archives, Yale University Library.

105. On the GEB and research in this period, see also Tracy Sarah, "An Evolving Science of Man: The Transformation and Demise of American Constitutional Medicine, 1920–1950," in *Greater Than the Parts: Holism and Biomedicine, 1920–1950* (Oxford: Oxford University Press, 1998), 161–88; Adele E. Clarke, *Disciplining Reproduction: Modernity, American Life Sciences, and "the Problems of Sex"* (Berkeley: University of California Press, 1998); Adele E. Clarke, "Research Materials and Reproductive Science in the United States, 1910–1940," in *Physiology in the American Context: 1850–1940*, ed. Gerald L. Geison (Bethesda, MD: American Physiological Society, 1987).

106. "Stockard, Dr. Charles R. (Extract from Interviews with Doctor Rose)," March 18, 1925, General Education Board Records (FA058), box 703, folder 7235, Rockefeller Archive Center, Sleepy Hollow, NY.

107. George W. Corner, "Stockard, Charles Rupert," in *Dictionary of American Biography*, vol. 22, supp. 2 (New York: Charles Scribner's Sons, 1958), 633–35.

108. Letters between Whitney and Stockard do not appear to have been preserved; neither left a full collection of their papers to any institution. Whitney clearly recognized a rivalry with Stockard, however, and almost certainly felt bitter about the latter's usurpation of his experiment. Sending results of a canine study to Leslie C. Dunn in 1928, Whitney wrote, "May I beg you to see that these do not fall into the hands of Dr. Stockard

before publication. He is a competitor so to speak + you'll agree that would not be fair." See Leon Fradley Whitney to Leslie Clarence Dunn, April 1928, L. C. Dunn Papers, box 22, "Whitney, Leon F.," American Philosophical Society.

109. Stockard studied at MAMC both as an undergraduate and for an MA degree. A little later, he transferred to a better-salaried position at the nearby Jefferson Military College. John Crumpton Hardy, January 29, 1903, box 1, folder 16, Charles R. Stockard Papers, Medical Center Archives of New York–Presbyterian / Weill Cornell Medicine, New York.

110. Glenn W. Herrick, January 27, 1903, box 1, folder 16, Charles R. Stockard Papers, Medical Center Archives of New York–Presbyterian / Weill Cornell Medicine.

111. John Crumpton Hardy, January 29, 1903, box 1, folder 16, Charles R. Stockard Papers, Medical Center Archives of New York-Presbyterian / Weill Cornell Medicine.

112. He was one of the first to describe "critical periods" in organismic growth, moments of major transformation that shape later development. See Brad Bolman, "Critical Periods in Science and the Science of Critical Periods: Canine Behavior in America," *Berichte zur Wissenschaftsgeschichte* 45, nos. 1–2 (2022): 112–34.

113. It "would give me pleasure to serve in this capacity for you if I felt that I could properly function. . . . I really could not do justice to such a Committee as its chairman," Stockard replied. Charles Rupert Stockard to Clarence Cook (C. C.) Little, December 10, 1928, American Eugenics Society Records, box 20, "Stockard, Charles R.," American Philosophical Society.

114. On chicken eggs, see, for example, Charles Benedict Davenport to Charles Rupert Stockard, April 13, 1912, Charles Benedict Davenport Papers, box 87, "Stockard, Charles R.," American Philosophical Society. Stockard informed Davenport of a new study on "human forms" in February 1921, asking for statistics on bodily averages. In December 1923, Stockard told Davenport that his studies on development in guinea pigs were overstretching capacities and wondered whether Cold Spring Harbor might have room for a larger colony of animals. Davenport responded that he would recommend the Board of Managers of the Biological Laboratory "do what they can to assist." In 1925, Davenport was finally able to offer land he had been using for sheep-breeding experiments, but Stockard had located a site on "a farm up the Hudson River." See Charles Rupert Stockard to Charles Benedict Davenport, February 4, 1921, Charles Benedict Davenport Papers, box 87, "Stockard, Charles R.," American Philosophical Society; Charles Rupert Stockard to Charles Benedict Davenport, December 19, 1923, Charles Benedict Davenport Papers, box 87, "Stockard, Charles R.," American Philosophical Society; Charles Benedict Davenport to Charles Rupert Stockard, December 24, 1923, Charles Benedict Davenport Papers, box 87, "Stockard, Charles R.," American Philosophical Society; Charles Benedict Davenport to Charles Rupert Stockard, November 3, 1925, Charles Benedict Davenport Papers, box 87, "Stockard, Charles R.," American Philosophical Society; Charles Rupert Stockard to Charles Benedict Davenport, November 5, 1925, Charles Benedict Davenport Papers, box 87, "Stockard, Charles R.," American Philosophical Society.

115. Charles Rupert Stockard, "An Experimental Study of Racial Degeneration in Mammals Treated with Alcohol," *Archives of Internal Medicine* 10, no. 4 (October 1, 1912): 369–98, https://doi.org/10/bf7wmn; Charles Rupert Stockard, "Alcohol as a Selective Agent in the Improvement of Racial Stock," *British Medical Journal* 2, no. 3215 (1922): 255–60.

116. W. W. Brierley to Livingston Farrand, June 3, 1925, General Education Board Records (FA058), box 703, folder 7235, Rockefeller Archive Center.

117. Charles Rupert Stockard to Abraham Flexner, June 18, 1925, General Education Board Records (FA058), box 703, folder 7235, Rockefeller Archive Center.

118. In later years, the site was known as "Shrub Oak."

119. Michael Pettit, "Becoming Glandular: Endocrinology, Mass Culture, and Experimental Lives in the Interwar Age," *American Historical Review* 118, no. 4 (October 1, 2013): 1052–76, https://doi.org/10/ggkd7x.

120. "Cornell University—Dr. Stockard's Studies," February 13, 1930, 2, Rockefeller Foundation Records, projects, SG 1.1, series 100, box 81, folder 975, Rockefeller Archive Center.

121. Frank Blair Hanson, "Diary, May–December," 1934, 185, Rockefeller Foundation Records, Officers' Diaries, RG12, F-L, Hanson, Frank B., Rockefeller Archive Center.

122. Charles Rupert Stockard, "An Experimental Dog Farm for the Study of Form and Type," *Collecting Net*, August 29, 1931, 263.

123. Charles Rupert Stockard to Livingston Farrand, October 19, 1926, General Education Board Records (FA058), box 703, folder 7235, Rockefeller Archive Center.

124. Charles Rupert Stockard to Wickliffe Rose, March 30, 1925, 3, General Education Board Records (FA058), box 703, folder 7235, Rockefeller Archive Center.

125. Charles Rupert Stockard to Abraham Flexner, February 23, 1928, General Education Board Records (FA058), box 703, folder 7235, Rockefeller Archive Center.

126. Charles Rupert Stockard to Richard M. Pearce, April 29, 1929, Rockefeller Foundation Records, projects, SG 1.1, series 100, box 81, folder 975, Rockefeller Archive Center. To lay new concrete floors and build a drainage system to fight off further infection, Stockard requested another $8,000 in 1928. Stockard to Flexner, February 23, 1928.

127. Stockard, "Experimental Dog Farm," 262.

128. Charles Rupert Stockard to Richard M. Pearce, February 4, 1929, Rockefeller Foundation Records, projects, SG 1.1, series 100, box 81, folder 975, Rockefeller Archive Center.

129. Alan Gregg, "Visit to Dr. Stockard's Dog Farm," March 12, 1932, General Education Board Records (FA058), box 703, folder 7235, Rockefeller Archive Center.

130. Alan Gregg, "Diary," 1934, 15, Rockefeller Foundation Records, Officers' Diaries, RG12, F-L, Gregg, Alan, Rockefeller Archive Center. Stockard knew "of a man he thinks would be competent but suggests that he would need money to start with" but did not clarify his name.

131. Charles Rupert Stockard, *The Physical Basis of Personality* (New York: W. W. Norton, 1931), vii.

132. Charles Rupert Stockard, *Hormones and Structural Development*, Beaumont Foundation Lectures 6 (Detroit: Wayne County Medical Society, 1927), 39.

133. Stockard, 34.

134. Stockard, *Physical Basis of Personality*, 223.

135. See, for instance, "Scientist Compares Men and Dog Freaks," *New York Times*, April 20, 1928, 25; "'Personality' Depends on Your Size, Shape and Bodily Make-Up," *San Francisco Examiner*, May 24, 1931, AW5.

136. "Dr. C. R. Stockard, Cornell Biologist," *New York Times*, April 8, 1939, Rockefeller Foundation Records, projects, SG 1.1, series 100, box 81, folder 977, Rockefeller Archive Center. News of Stockard's death was carried widely, and obituaries bemoaned the loss of a man who once promised indefinite life extension. See "Dr. Stockard, of Rockefeller Institute, Dead," *New York Herald Tribune*, April 8, 1939, 12.

137. Paaske would receive minimal credit for the final monograph despite heavy involvement, and her name was routinely misspelled in correspondence. She is initially referred to as "Miss Trask." See Alan Gregg, "Dean Ladd—Telephone," March 28, 1939, Rockefeller Archive Center.

138. Charles Rupert Stockard, *The Genetic and Endocrinic Basis for Differences in Form and Behavior, as Elucidated by Studies of Contrasted Pure-Line Dog Breeds and Their Hybrids*, American Anatomical Memoirs 19 (Philadelphia: Wistar Institute of Anatomy and Biology, 1941). Gregg insisted the Rockefeller name not appear in the text out of a preference to "remain anonymous in records of this kind." See Alan Gregg to Joseph C. Hinsey, October 23, 1941, Rockefeller Foundation Records, projects, SG 1.1, series 100, box 81, folder 978, Rockefeller Archive Center.

139. Hans Grüneberg, "Dogs and Eugenics," *Eugenics Review* 34, no. 2 (July 1942): 62.

140. Grüneberg, 63. J. B. S. Haldane played a vital part in bringing Grüneberg to England.

141. Williams S. Ladd to Alan Gregg, October 16, 1937, Rockefeller Foundation Records, projects, SG 1.1, series 100, box 81, folder 976, Rockefeller Archive Center.

142. Grüneberg, "Dogs and Eugenics," 62.

143. Howard S. Liddell to William T. James, March 24, 1938, box 2, folder 4, William T. James Papers, University of Georgia Archives; William T. James to Howard S. Liddell, March 21, 1938, box 2, folder 4, William T. James Papers, University of Georgia Archives, Athens.

144. Alan Gregg, "Sat. May 28, 1938," May 28, 1938, Rockefeller Foundation Records, projects, SG 1.1, series 100, box 81, folder 977, Rockefeller Archive Center.

145. William T. James, "[Dog Farm Memorandum]," 1939, Rockefeller Foundation Records, projects, SG 1.1, series 100, box 81, folder 977, Rockefeller Archive Center.

146. James, 4. The latter suggestion seems to have been influenced by Gregg.

147. W. Horsley Gantt to Alan Gregg, May 9, 1939, Rockefeller Foundation Records, projects, SG 1.1, series 100, box 81, folder 977, Rockefeller Archive Center. Gregg had helped recruit him to Johns Hopkins years before from Pavlov's lab in Russia, and the two remained in contact.

148. Alan Gregg to Williams S. Ladd, June 14, 1939, Rockefeller Foundation Records, projects, SG 1.1, series 100, box 81, folder 977, Rockefeller Archive Center.

149. Howard S. Liddell to Williams S. Ladd, September 25, 1939, Rockefeller Foundation Records, projects, SG 1.1, series 100, box 81, folder 977, Rockefeller Archive Center. Liddell's work is covered more extensively in Robert G. W. Kirk and Edmund Ramsden, "Working across Species down on the Farm: Howard S. Liddell and the Development of Comparative Psychopathology, c. 1923–1962," *History and Philosophy of the Life Sciences* 40, no. 24 (2018): 23–24.

150. A typical version of the story is Associated Press, "Cornell Research on Dogs Reveals Nerve Quality That Explains Surprising Ability of Humans to Withstand Bombing Attacks," *Ithaca Journal*, April 4, 1941, 1. "Why Bombed Public Can Take It" comes from a copy kept by James and preserved at the University of Georgia Archives, with no bibliographic information.

151. Nellie Mathews Meyer to William T. James, April 5, 1941, box 2, folder 3, William T. James Papers, University of Georgia Archives.

152. Misspellings and ellipses are in the original. See Harold Faller to William T. James, "The Dogs Go to War," April 10, 1941, box 2, folder 3, William T. James Papers, University of Georgia Archives.

153. For the distribution of dogs, see William T. James, "Morphological and Constitutional Factors in Conditioning," *Annals of the New York Academy of Sciences* 56, no. 2 (1953): 171, https://doi.org/10/b45zpb. The early years of the Jackson Memorial project are recounted in John Paul Scott and John Langworthy Fuller, "Research on Genetics and Social Behavior: At the Roscoe B. Jackson Memorial Laboratory, 1946–1951—a Progress Report," *Journal of Heredity* 42, no. 4 (July 1951): 191.

154. "Cornell University—Stockard Dog Farm" (Rockefeller Foundation Filing Department, June 1939), Rockefeller Foundation Records, projects, SG 1.1, series 100, box 81, folder 978, Rockefeller Archive Center. In one of the more peculiar examples, farm research director Malcolm Gilman claimed that they had discovered penicillin before Alexander Fleming but discarded the results because of a vogue for maggot therapy. "Undoubtedly we were dealing with Penicellin [*sic*]. We also applied it to the purification of Sewerage." See C. Malcolm B. Gilman, June 7, 1943, General Education Board Records (FA058), box 703, folder 7235, Rockefeller Archive Center. A later RF officer noted of the claim: "Gilman's paper in so far as it has a relation to present work in penicillin recalls the remark about Columbus—namely that his performance was characterized by the fact that he did not know where he was going when he started out nor where he was when he got there, nor where he had been when he got back." See Alan Gregg, "Dr. Gilman," June 17, 1943, General Education Board Records (FA058), box 703, folder 7235, Rockefeller Archive Center.

155. Clive M. McCay, "Operating Problems," in *Transactions of the Tenth and Eleventh Conferences* (New York: Conference on Problems of Aging / Josiah Macy Jr. Foundation, 1949), 186.

156. On the emergence of laboratory animal practice, see Robert G. W. Kirk, "Subjugated Love: Aligning Care with Science in the History of Laboratory Animal Research," in Davies et al., *Researching Animal Research: What the Humanities and Social Sciences Can Contribute to Laboratory Animal Science and Welfare*, ed. Gail Davies et al. (Manchester: Manchester University Press, 2024).

157. Ladd to Gregg, October 16, 1937.

158. Robert G. W. Kirk and Edmund Ramsden, "'Havens of Mercy': Health, Medical Research, and the Governance of the Movement of Dogs in Twentieth-Century America," *History and Philosophy of the Life Sciences* 43, no. 4 (December 2, 2021): 126, https://doi.org/10.1007/s40656-021-00478-4; Bolman, "In the Animal House"; Rob Boddice, *Humane Professions: The Defence of Experimental Medicine, 1876–1914* (Cambridge: Cambridge University Press, 2021), 20–41.

159. "Prescription for Anti-vivisection: A Program of Action on State and Local Levels to Build Public Understanding of Medical Research and to Insure an Adequate Supply of Dogs for Experimental Use," circa 1949, box 7, folder 1, ISMR Papers, Special Collections Research Center, University of Chicago. For estimates, see Philip Bard, "Report of Committee to Advise on the Breeding of a Pure-Line Strain of Dogs for Scientific Purposes" (National Research Council, June 1939), 2, Rockefeller Foundation Records, projects, SG 1.1, series 200, box 151, folder 1855, Rockefeller Archive Center.

160. Quoted in Walrath, "2 UR Lab Dogs Go Snooty," B3.

161. Surgery Study Section, *Care of the Dog Used in Medical Research* (Washington, DC: National Institutes of Health, July 1949), 2. On the history of care in animal research more broadly, see Robert G. W. Kirk and Dmitriy Myelnikov, "Governance, Expertise, and the 'Culture of Care': The Changing Constitutions of Laboratory Animal Research in Britain, 1876–2000," *Studies in History and Philosophy of Science* 93 (June 1, 2022): 107–22, https://doi.org/10.1016/j.shpsa.2022.03.004.

162. C. C. Little joked in 1941 that their methods "if investigated would probably cause some very amusing fireworks." Clarence Cook Little to Frank Blair Hanson, February 13, 1941, Rockefeller Foundation Records, SG 1.1, series 200D, box 143, folder 1775, Rockefeller Archive Center.

163. On the history of maize, specifically, see Helen Anne Curry, "Taxonomy, Race Science, and Mexican Maize," *Isis* 112, no. 1 (March 1, 2021): 1–21. See also Robert E. Kohler, *Lords of the Fly: Drosophila Genetics and the Experimental Life* (Chicago: University of Chicago Press, 1994); Bonnie Tocher Clause, "The Wistar Rat as a Right Choice: Establishing Mammalian Standards and the Ideal of a Standardized Mammal," *Journal of the History of Biology* 26, no. 2 (1993): 329–49, https://doi.org/10.1007/BF01061973.

164. Frank Blair Hanson, "Diary," 1939, 128, Rockefeller Foundation Records, Officers' Diaries, RG12, F-L, Hanson, Frank B., Rockefeller Archive Center.

165. Frank Blair Hanson, "Diary," 1937, 203, Rockefeller Foundation Records, Officers' Diaries, RG12, F-L, Hanson, Frank B., Rockefeller Archive Center. The story of animal standards more broadly has been well documented by historian Robert G. W. Kirk. See, for instance, Robert G. W. Kirk, "'Wanted—Standard Guinea Pigs': Standardisation and the Experimental Animal Market in Britain ca. 1919–1947," *Studies in History and Philosophy of the Biological and Biomedical Sciences* 39, no. 3 (September 2008): 280–91; Robert G. W. Kirk, "A Brave New Animal for a Brave New World: The British Laboratory Animals Bureau and the Constitution of International Standards of Laboratory Animal Production and Use, circa 1947–1968," *Isis* 101, no. 1 (March 2010): 62–94; Robert G. W. Kirk, "'Standardization through Mechanization': Germ-Free Life and the Engineering of the Ideal Laboratory Animal," *Technology and Culture* 53, no. 1 (January 2012): 61–93.

166. Erwin Brand to Ludvig Hektoen, December 8, 1938, "Breeding Pure-Line Strain of Dogs for Scientific Purposes," Central Policy Files: 1932–39, National Academies Archives, Washington, DC.

167. Du Vigneaud's suggestion reported in Erwin Brand to McKeen Cattell, January 12, 1937, 2, box 16, folder, 16, McKeen Cattell Papers, Center for the History of Medicine in the Francis A. Countway Library, Harvard University, Boston. Du Vigneaud would go on to win the Nobel Prize in Chemistry in 1955 for his work on oxytocin.

168. Albert L. Barrows, "Office Memorandum No. 507," July 12, 1938, "Breeding Pure-Line Strain of Dogs for Scientific Purposes," Central Policy Files: 1932–39, National Academies Archives.

169. Leslie Clarence Dunn to H. P. Barss, December 27, 1938, "Breeding Pure-Line Strain of Dogs for Scientific Purposes," Central Policy Files: 1932–39, National Academies Archives.

170. Edmond R. Long to Albert L. Barrows, December 19, 1938, "Breeding Pure-Line Strain of Dogs for Scientific Purposes," Central Policy Files: 1932-1939, National Academies Archives.

171. Frank B. Hanson notes in his diary that Brand had sold dogs to Calvery, Howard Taylor at New York University, and George Whipple at Rochester. See Frank Blair Hanson, "Diary," 1941, 168, Rockefeller Foundation Records, Officers' Diaries, RG 12, F-L, Hanson, Frank B., Rockefeller Archive Center.

172. Erwin Brand to Ludvig Hektoen, December 8, 1938, 2, "Breeding Pure-Line Strain of Dogs for Scientific Purposes," Central Policy Files: 1932–39, National Academies Archives.

173. Bard, "Report of Committee to Advise on the Breeding of a Pure-Line Strain of Dogs for Scientific Purposes," 2.

174. Philip Bard to Ross G. Harrison, December 5, 1939, "Breeding Pure-Line Strain of Dogs for Scientific Purposes," Central Policy Files: 1932–39, National Academies Archives.

175. For the foxhound attempts, see Hanson, "Diary," 1937, 176. For the "standard beagle," see "Interview: FBH Oct. 3, 1939," October 31, 1941, Rockefeller Foundation Records, projects, SG 1.1, series 200, box 151, folder 1855, Rockefeller Archive Center.

176. Magnus Gregersen to Walter B. Cannon, January 12, 1937, box 113, folder, 1568, Walter B. Cannon Papers, Center for the History of Medicine in the Francis A. Countway Library, Harvard University.

177. Willard C. Rappleye to Frank D. Fackenthal, March 22, 1941, box 364, folder 12, Willard C. Rappleye Files, University Archives, Rare Book and Manuscript Library, Columbia University Libraries, New York.

178. Erwin Brand and Leslie C. Dunn, "Establishment and Maintenance of a Source of Standard Dogs for Scientific Purposes in a 'Laboratory for Animal Research' at the Columbia University Property, Nevis, Irvington-on-Hudson," 1940, series 2, box 76, folder 1448, Robert Mearns Yerkes Papers (MS 569), Manuscripts and Archives, Yale University Library.

179. Robert M. Yerkes to Erwin Brand, October 19, 1940, Robert Mearns Yerkes Papers (MS 569), series 2, box 76, folder 1448, Manuscripts and Archives, Yale University Library.

180. Hanson, "Diary," 1941, 94.

181. Bard to Harrison, December 5, 1939, 2.

182. Clarence Cook Little to Leslie T. Webster, January 25, 1941, Rockefeller Foundation Records, SG 1.1, series 200D, box 143, folder 1775, Rockefeller Archive Center. On Jackson, see, again, Karen A. Rader, *Making Mice: Standardizing Animals for American Biomedical Research, 1900–1955* (Princeton, NJ: Princeton University Press, 2004). As Allan Brandt shows in *Cigarette Century*, Little would go on to become an almost parodically pro-tobacco voice, citing his studies showing that cancer was caused by genetic factors. See Allan Brandt, *The Cigarette Century: The Rise, Fall, and Deadly Persistence of the Product That Defined America* (New York: Basic Books, 2009), 184. Little's early papers are held at the University of Michigan, but a valuable collection of later papers was removed from public collections at the University of Maine at the request of Little's family. A few of those documents are preserved online because of tobacco litigation.

183. Alan Gregg to Clarence Cook Little, January 3, 1944, Rockefeller Foundation Records, SG 1.2, series 200A, box 133, folder 1189, Rockefeller Archive Center. Gregg was encouraged to try Little, in 1943, by Harvard statistician Edwin Bidwell Wilson. See Edwin Bidwell Wilson to Alan Gregg, December 27, 1943, Rockefeller Foundation Records, SG 1.2, series 200A, box 133, folder 1189, Rockefeller Archive Center; Alan Gregg to Edwin Bidwell Wilson, December 30, 1943, Rockefeller Foundation Records, SG 1.2, series 200A, box 133, folder 1189, Rockefeller Archive Center.

184. Alan Gregg to Clarence Cook Little, January 3, 1944; Gregg to Wilson, December 30, 1943. See also Gregg Mitman, "Dominance, Leadership, and Aggression: Animal Behavior Studies during the Second World War," *Journal of the History of the Behavioral Sciences* 26 (January 1990): 3–16.

185. For additional background on these studies, see Diane B. Paul, "The Rockefeller Foundation and the Origins of Behavior Genetics," in *The Expansion of American Biology*, ed. Keith R. Benson, Jane Maienschein, and Ronald Rainger (New Brunswick, NJ: Rutgers University Press, 1991); Donald A. Dewsbury, "Origins of Behavior Genetics: The Role of the Jackson Laboratory," *Behavior Genetics* 39, no. 1 (January 1, 2009):

1–5, https://doi.org/10.1007/s10519-008-9240-1; Donald A. Dewsbury, "A History of the Behavior Program at the Jackson Laboratory: An Overview," *Journal of Comparative Psychology* 126, no. 1 (2012): 31–44, https://doi.org/10.1037/a0021376.

186. Robert M. Yerkes to Alan Gregg, March 31, 1944, 2, Rockefeller Foundation Records, SG 1.2, series 200A, box 133, folder 1189, Rockefeller Archive Center.

187. Alan Gregg, "Diary," 1943, 120, Rockefeller Foundation Records, Officers' Diaries, RG 12, F-L, Gregg, Alan, Rockefeller Archive Center.

188. Ellsworth Huntington and Leon Fradley Whitney, *The Builders of America* (New York: William Morrow, 1927), 58.

189. Robert M. Yerkes to Leon Fradley Whitney, October 14, 1925, Robert Mearns Yerkes Papers (MS 569), box 52, folder 1007, Manuscripts and Archives, Yale University Library. Whitney later wrote widely on animals, particularly dogs, for instance in Leon Fradley Whitney, "Heredity of the Trail Barking Propensity in Dogs," *Journal of Heredity* 20, no. 12 (December 1929): 561–62. He produced most of his explicitly eugenic work in the 1930s. See, for instance, Leon Fradley Whitney, *The Case for Sterilisation*, ed. Norman Haire (London: Bodley Head, 1935). There was, nevertheless, a persistent connection in his thinking between dog breeding and eugenics. A later text, *The Truth about Dogs* (Edinburgh: Thomas Nelson and Sons, 1959), includes a chapter titled "How Dogs Have Degenerated." Whitney's eugenics work is chronicled carefully in Paul A. Lombardo, *Three Generations, No Imbeciles: Eugenics, the Supreme Court, and* Buck v. Bell (Baltimore: Johns Hopkins University Press, 2008), 231. Whitney's son, George, was a dedicated beagler and wrote an extended treatise on the dogs. See G. Whitney, *This Is the Beagle.*

190. Gregg writes in his diary, "Carrel would like to see a long term experiment on dogs to determine whether by excellence of nutrition and care in breeding a race of dogs could be developed conspicuously above the ordinary in points of intelligence, nervous stability, and minimum susceptibility to disease." See Alan Gregg, "Diary," 1935, 22, Rockefeller Foundation Records, Officers' Diaries, RG 12, F-L, Gregg, Alan, Rockefeller Archive Center.

191. Clarence Cook Little to Alan Gregg, January 6, 1944, Rockefeller Foundation Records, SG 1.2, series 200A, box 133, folder 1189, Rockefeller Archive Center.

192. Clarence Cook Little to Alan Gregg, May 18, 1944, Rockefeller Foundation Records, SG 1.2, series 200A, box 133, folder 1189, Rockefeller Archive Center.

193. C. Sidney Burwell to Clarence Cook Little, March 15, 1944, Rockefeller Foundation Records, SG 1.2, series 200A, box 133, folder 1189, Rockefeller Archive Center; Clarence Cook Little to C. Sidney Burwell, March 24, 1944, Rockefeller Foundation Records, SG 1.2, series 200A, box 133, folder 1189, Rockefeller Archive Center.

194. Clarence Cook Little, "Proposed Plan for Investigation of the Genetics of Intelligence and Emotional Types in Mammals," 1944, 3, Rockefeller Foundation Records, SG 1.2, series 200A, box 133, folder 1189, Rockefeller Archive Center.

195. "Rockefeller Grant to Study Genetics of Intelligence and Emotion," *Journal of Heredity* 36, no. 5 (May 1945): 194.

196. Clarence Cook Little to Hans Grüneberg, October 16, 1944, Correspondence: Lidicker-Little, Grüneberg, Professor Hans, PP/GRU/46, Wellcome Collection, London.

197. Hans Grüneberg to Clarence Cook Little, December 10, 1944, 1, Correspondence: Lidicker-Little, Grüneberg, Professor Hans, PP/GRU/46, Wellcome Collection.

198. Clarence Cook Little to Alan Gregg, March 21, 1945, Rockefeller Foundation Records, SG 1.2, series 200A, box 133, folder 1190, Rockefeller Archive Center.

199. John Paul Scott, "A Challenge to the Eugenist," *Journal of Heredity* 27, no. 7 (July 1936): 261–64.

200. Vicari's role is noted in Stockard, *Physical Basis of Personality*, vii. Vicari's publications include, for instance, Emelia M. Vicari, "Observations on the Nature of the Para-Follicular Cells in the Thyroid Gland of the Dog: A Preliminary Note," *Anatomical Record* 68, no. 3 (June 1937): 281–85. Vicari's first name appears in some publications as "Emilia," in others as "Emelia." Both spellings are common in newspaper articles. Stockard used "Emilia," as does a photo from the Woods Hole Marine Biological Laboratory Archives, so I have followed that spelling. See "Emilia Vicari," n.d., Black Photo Album 2, Marine Biological Laboratory Archives, https://history.archives.mbl.edu/digital-collection/emilia-vicari.

201. "[Dr. Emilia Vicari]," *Bar Harbor Times*, October 23, 1941. Her award is noted in "Awarded Scholarship," *Norwalk Reflector-Herald*, June 17, 1940.

202. Elsbeth died in Harrow, a town in the northwest of Greater London. See Dan Lewis and D. M. Hunt, "Hans Grüneberg, 26 May 1907–23 October 1982," *Biographical Memoirs of Fellows of the Royal Society* 30 (November 1, 1984): 247, https://doi.org/10.1098/rsbm.1984.0008.

203. John Paul Scott and John Langworthy Fuller, *Genetics and the Social Behavior of the Dog* (Chicago: University of Chicago Press, 1965).

204. Staff of the Division of Behavior Studies, *Manual of Dog Testing Techniques* (Bar Harbor, ME: Roscoe B. Jackson Memorial Laboratory, 1950).

205. Bolman, "Critical Periods in Science and the Science of Critical Periods," 120.

206. Many of the documents cited concerning tobacco are available in the University of California, San Francisco, tobacco database; here, see Clarence Cook Little to Earl Leroy Green, October 25, 1956, Ness Motley Law Firm Documents, https://www.industrydocuments.ucsf.edu/docs/qqwc0040. The note on Little's dachshund-pug hybrids also appears in this document.

207. Today the site is home to the Jones Center, an ecological research facility.

208. "Illness of Dogs Leads to Lab," *Ithaca Journal*, January 6, 1951, 3.

209. John Merrill Olin to Robert W. Woodruff, July 14, 1977, James A. Baker Institute for Animal Health Records, no. 24-9-3598, folder 6, Division of Rare and Manuscript Collections, Cornell University Library, Ithaca, NY. For more on Teagle's commitment to Cornell, see Bennett H. Wall, *Teagle of Jersey Standard*, History of Business and Business Leaders (New Orleans: Tulane University Press, 1974), 323.

210. Alan Gregg, "Diary," 1949, 91, Rockefeller Foundation Records, Officers' Diaries, RG 12, F-L, Gregg, Alan, Rockefeller Archive Center.

211. John Merrill Olin to James A. Baker, November 26, 1962, James A. Baker Institute for Animal Health Records, no. 24-9-3598, folder 6, Division of Rare and Manuscript Collections, Cornell University Library.

212. Barbara J. Mitnick, *Geraldine Rockefeller Dodge* (Morristown, NJ: Geraldine R. Dodge Foundation, 2000), 66.

213. "World Champion Dog Owned Here," *North Adams Transcript*, April 1, 1929, 3.

214. Joan Harrigan, "The Legacy of Geraldine Rockefeller Dodge," *Morris and Essex History* (blog), n.d., http://www.morrisandessexkennelclub.org/m-e-history-mrs-dodge.html, accessed March 2024.

215. *News Letter* (Ithaca, NY: Research Laboratory for Diseases of Dogs), July 1950, 3, folder 5, James A. Baker Institute for Animal Health Records, no. 24-9-3598, Division of Rare and Manuscript Collections, Cornell University Library.

216. "Cornell Opens New Lab for Dog Diseases," *Springville Journal*, January 11, 1951, 7.

217. "Will Penetrate Bolivian Wilds," *Philadelphia Inquirer*, December 26, 1914, 7; "Long Journey: American Explorers Going to the South American Jungles," *Kingston Daily Gleaner*, January 4, 1915, 10.

218. Not to be confused with General Mills, General Foods was eventually purchased and combined with Kraft by Philip Morris Companies.

219. "Weds Yankee after Jilting Foreign Lover," *Kokomo Daily Tribune*, September 23, 1925, 16. The wedding came as a shock even to close friends because Nancy had been betrothed to Ernest Lardinelli Becci, a notable in Italy's Department of Interior. Their engagement was called off without explanation, leaving Becci briefly stranded in America with $37,500 in jewels. See "Miss Nancy Sayles Weds New Yorker," *North Adams Transcript*, September 22, 1925, 1.

220. "Cornell Opens New Lab for Dog Diseases."

221. *News Letter*, July 1950, 1.

222. "Cornell Research Laboratory for Diseases of Dogs," 1950, 1, folder 5, James A. Baker Institute for Animal Health Records, no. 24-9-3598 Division of Rare and Manuscript Collections, Cornell University Library.

223. *News Letter*, July 1950, 4.

224. *Institute Report* (Ithaca, NY: Cornell Research Laboratory for Diseases of Dogs, Veterinary Virus Research Institute), October 1960, 7.

225. Black, *American Beagling*, 4. The profusion of concepts of blood—"blue blood," in particular—can be seen as a protogenetic vocabulary before genetic concepts infiltrated the world of beaglers. But the terminology also revealed widespread and uneasy connections between early American dog breeders and the popular discourse of eugenics. Owners obsessed over "good breeding" and the necessity of eliminating unfit dogs, and they frequently extolled particularly potent breeding males or bitches. The record-keeping practices of many hunters directly paralleled the work of eugenicists such as Francis Galton in tracking the inheritance of valuable traits, all of which propelled projects such as Stockard's.

226. "Disease-Free Kennel Building Presented to Cornell U. by Gaines Dog Research Center," *Veterinary News* 13, no. 6 (December 1950): 12.

227. "Veterinarian Recommends Gaines Food," *Elyria Chronicle*, February 25, 1942, 8.

228. "[Note on Gaines Food Company]," *Brookfield Courier*, September 23, 1931, 1.

229. "How 'Going to the Dogs' Made Jobs," *Veterinary News* 9, no. 1, January 1945, n.p.

230. "Sherburne Gets New Industries," *Syracuse Herald-Journal*, October 23, 1947, 3. Clarence briefly managed the subdivision before turning his attention to thoroughbred horses. "Clarence F. Gaines, 88; Introduced a Dog Meal," *New York Times*, January 4, 1986, 28.

231. "Grocers See Gain in Dry Dog Food," *New York Times*, July 20, 1950, 36.

232. Herbert Murphy, "Cats and Dogs: Tabby and Towser Are Important Customers for Big Business," *Barron's*, April 7, 1952, 13.

233. "Nepken Bill Winner over Bradford Dog," *Era*, September 20, 1933, 6.

234. See Tom Quick, "Puppy Love: Domestic Science, 'Women's Work,' and Canine Care," *Journal of British Studies* 58, no. 2 (April 2019): 289–314, https://doi.org/10.1017/jbr.2018.181. Quick has shown that the domestic science of dog food became a prominent concern in Britain during the early twentieth century, with specialty milk foods sold to canine caretakers under the auspices of research-based nutrition. In America, Milk-Bone foods did become popular around the same time, but the midcentury commercialization of explicitly scientific dog chow was of a different character. Quick stresses that canine caretaking in Britain became a vital component of women's domestic work, but popular guides to dog ownership published in the United States also tended to frame their advice for men as much as women. In his widely used guide, George Watson Little, for instance, advised presumptively male "masters" in 1925

to provide a balanced diet of meat, vegetables, and biscuits to their dogs or to try new "balanced" rations that had "been lately on the market." See George Watson Little, *Dr. Little's Dog Book* (New York: Robert M. McBride, 1925), 50.

235. Marge Chase, "W. Armstrong Retires, Former Gaines Dog Food President," *Norwich Evening Sun*, November 1, 1961, 1.

236. "Brighter World For US Dogs, Too, Planned for End of War," *New Holland Clarion*, September 22, 1944, 7.

237. For more on National Dog Week and Judy, see Lisa Begin-Kruysman, *Dog's Best Friend: Will Judy, Founder of National Dog Week and Dog World Publisher* (Jefferson, NC: McFarland, 2014).

238. "Former Maywood Resident Heads Research Center," *Bellwood Proviso Township Herald*, July 27, 1944, 14.

239. John Held, "Gaines Kennels Names Vail as Guiding Spirit," *New York Herald Tribune*, March 18, 1945, B2.

240. "Former Maywood Resident Heads Research Center."

241. "Millions Spent for Dog Food Research," *Agitator*, July 16, 1947, 6.

242. "Wise Owner Teaches His Dog to Eat Only from Own Dish," *Kyle News*, May 8, 1942, 2.

243. "Title is Awarded to Cataldo Dog," *Spokane Daily Chronicle*, October 21, 1947, 2.

244. "To Pass on War Dog Memorial," *Farms Weekly Review*, June 27, 1945, 1.

245. Gaines Dog Research Center, *Basic Guide to Canine Nutrition* (White Plains, NY: General Foods Corporation, 1965).

246. Quoted in "Millions Spent for Dog Food Research."

247. Gaines Dog Research Center, *Touring with Towser: A Director of Hotels and Motor Courts That Accommodate Guests with Dogs* (New York: Gaines Dog Research Center, 1957), n.p.

248. Kathleen Szasz, *Petishism: Pets and Their People in the Western World* (New York: Holt, Rinehart, and Winston, 1969).

249. Angela N. H. Creager, *The Life of a Virus: Tobacco Mosaic Virus as an Experimental Model, 1930–1965* (Chicago: University of Chicago Press, 2002), 174.

250. *Institute Report* (Ithaca, NY: Cornell Research Laboratory for Diseases of Dogs, Veterinary Virus Research Institute), September 1952, 14–16.

251. *Institute Report* (Ithaca, NY: Cornell Research Laboratory for Diseases of Dogs, Veterinary Virus Research Institute), September 1951, 11.

252. Norman H. Topping, "The Development of Our Knowledge and the Importance of Virus Diseases in Man and Animals," *Veterinarian* 41 (1951): 154.

253. For background on the "manpower" problem, see David Kaiser, "Cold War Requisitions, Scientific Manpower, and the Production of American Physicists after World War II," *Historical Studies in the Physical and Biological Sciences* 33, no. 1 (2002): 131–59, https://doi.org/10.1525/hsps.2002.33.1.131.

254. Michael Bresalier and Michael Worboys, "'Saving the Lives of Our Dogs': The Development of Canine Distemper Vaccine in Interwar Britain," *British Journal for the History of Science* 47, no. 2 (June 2014): 308.

255. Bresalier and Worboys, 310–11.

256. See, for the British reception of Phisalix's vaccine, Alison Skipper, "The 'Dog Doctors' of Edwardian London: Elite Canine Veterinary Care in the Early Twentieth Century," *Social History of Medicine* 33, no. 4 (2020): 1233–58.

257. Bresalier and Worboys, 318–21.

258. On Stockard's use of the vaccine, see "Cheering News from Cornell to Dog Lovers," *Los Angeles Times*, August 2, 1932, 10.

259. R. B. McClelland, "The Need for Research on Virus Diseases of Pet Animals," *Cornell Veterinarian* 41 (1951): 209–10.

260. Donald H. Avery, *Pathogens for War: Biological Weapons, Canadian Life Scientists, and North American Biodefence* (Toronto: University of Toronto Press, 2013), 25.

261. Amanda Kay McVety, *The Rinderpest Campaigns: A Virus, Its Vaccines, and Global Development in the Twentieth Century* (Cambridge: Cambridge University Press, 2018), 70.

262. "2-Way Dog Vaccine Is Developed," *Syracuse Herald Journal*, June 26, 1952, 2.

263. Myles Newcomb, "Scientifically Prepared Vaccine for Distemper, Hepatitis Available," *Waterloo Sunday Courier*, December 8, 1957, 45.

264. "2 to Be Feted for Work in Dog Disease," *Post-Standard*, May 13, 1961, 26.

265. Walter R. Fletcher, "Dog's Best Friend Is No Man but a Cornell Research Group," *New York Times*, August 11, 1963, 151.

266. For more on disease specificity, see Charles E. Rosenberg, "The Tyranny of Diagnosis: Specific Entities and Individual Experience," *Milbank Quarterly* 80, no. 2 (2002): 237–60, https://doi.org/10.1111/1468-0009.t01-1-00003.

267. Leon Fradley Whitney and George D. Whitney, *The Distemper Complex* (Orange, CT: Practical Science, 1953), 8–9.

268. For the therapeutic revolution in the history of medicine, see Charles E. Rosenberg, "The Therapeutic Revolution: Medicine, Meaning, and Social Change in Nineteenth-Century America," *Perspectives in Biology and Medicine* 20, no. 4 (1977): 485–506, https://doi.org/10.1353/pbm.1977.0040.

269. Edmond J. Farris, "Introduction to the Conference on Animal Colony Maintenance," *Annals of the New York Academy of Sciences* 46, no. 1 (1945): 4, https://doi.org/10.1111/j.1749-6632.1945.tb36157.x.

270. See an early sketch of the history of American laboratory animal care: C. N. Wentworth Cumming and F. G. Carnochan, "The Laboratory Animal Business in the USA," in *The UFAW Handbook on the Care and Management of Laboratory Animals* (London: Universities Federation for Animal Welfare, 1957), 201.

271. "Glenn G. Black," *Plain Dealer*, November 2, 1963. The name sometimes appears as "Casco." Black, who opened the Kickapoo kennels, was influential in designing the nation's beagle field trial championship system. A 1926 article called him America's "king beagler." See Charles F. Grosse, "Doggy Doings," *Plain Dealer*, August 1, 1926, 25.

272. This may be one reason that my attempts to consult records concerning beagles and other dogs in connection with this book have been met with tepid enthusiasm from institutions principally concerned with sporting history.

273. "2 Wills Name University," *Ithaca Journal*, March 22, 1962, 9.

274. Howard E. Evans, "A Dog Comes into Being," *Gaines Dog Research Progress*, Fall 1956, 1.

Chapter 2

1. William Patrick, "The Hounds of Beagleville: They May Save Your Life," *Salt Lake City Tribune*, May 31, 1952, M2. For more on the Soviet bomb project and reactions to it, see David Holloway, *Stalin and the Bomb: The Soviet Union and Atomic Energy, 1939–1956* (New Haven, CT: Yale University Press, 1994); Michael D. Gordin, *Red Cloud at Dawn: Truman, Stalin, and the End of the Atomic Monopoly* (New York: Farrar, Straus and Giroux, 2009).

2. Patrick, "Hounds of Beagleville," M2.

3. James J. Brown, "Dogs Get Atom Doses to Save Lives of Men," *Sacramento Bee*, December 10, 1953, 29.

4. Allen C. Andersen, "The Dog—a Facet of Human Society" (Noon Topics, San Francisco, November 13, 1963), 3, carton 1, folder 6, UCSF Committee on Arts and Lectures Records, AR 2015–17, University of California, San Francisco, Archives and Special Collections.

5. Angela Creager is excellent on the expansive biological programs of the AEC during this period. See Angela N. H. Creager, *Life Atomic: A History of Radioisotopes in Science and Medicine* (Chicago: University of Chicago Press, 2013).

6. "Man's Best Friend Faces Strontium 90," *San Francisco Chronicle*, September 17, 1956.

7. Brown, "Dogs Get Atom Doses to Save Lives of Men," 29.

8. Betsy J. Stover and Clarence N. Stover Jr., "The Laboratory for Radiobiology at the University of Utah," in *Radiobiology of Plutonium* (Salt Lake City: J.W., 1972), 30.

9. Matthew Lavine, *The First Atomic Age: Scientists, Radiations, and the American Public, 1895–1945* (New York: Palgrave Macmillan, 2013), 137; Luis A. Campos, *Radium and the Secret of Life* (Chicago: University of Chicago Press, 2015), 242. See also Claudia Clark, *Radium Girls, Women and Industrial Health Reform: 1910–1935* (Chapel Hill: University of North Carolina Press, 1997).

10. Austin M. Brues, Hermann Lisco, and Miriam Finkel, *Carcinogenic Action of Some Substances Which May Be a Problem in Certain Future Industries* (Chicago: Argonne National Laboratory, 1946), 1.

11. Miriam Finkel, interview by James Newell Stannard, September 6, 1978, J. Newell Stannard Papers, MS.2020, University of Tennessee Libraries, Knoxville, Special Collections.

12. Wright H. Langham et al., "The Los Alamos Scientific Laboratory's Experience with Plutonium in Man," *Health Physics* 8 (1962): 753–60. The most comprehensive retrospective analysis of the work is Advisory Committee on Human Radiation Experiments, *Final Report of the Advisory Committee on Human Radiation Experiments* (Washington, DC: US Government Printing Office, 1995), https://bioethicsarchive.georgetown.edu/achre/final/. The studies are put in broader context by David S. Jones, Christine Grady, and Susan E. Lederer, "'Ethics and Clinical Research': The 50th Anniversary of Beecher's Bombshell," in *The Social Medicine Reader*, ed. Jonathan Oberlander et al., 3rd ed., vol. 1 (Durham, NC: Duke University Press, 2019).

13. The studies were revealed to the general public in investigative reports by Eileen Welsome in 1993 and later collected in a 1999 book. See Eileen Welsome, *The Plutonium Files: America's Secret Medical Experiments in the Cold War* (New York: Dial, 1999).

14. Clifford Ladd Prosser et al., *Clinical Sequence of Physiology Effects of Ionizing Radiation in Animals* (Oak Ridge, TN: Atomic Energy Commission, January 7, 1947), 1.

15. Clifford Ladd Prosser to Wright H. Langham, May 7, 1971, Nuclear Testing Archive, US Department of Energy OpenNet. The study results appear in E. Painter et al., *Clinical Physiology of Dogs Injected with Plutonium* (Chicago: US Atomic Energy Commission / University of Chicago, June 1946), 8. Prosser would later become one of the celebrated giants of the field of comparative physiology. See Michel Anctil, *Animal as Machine: The Quest to Understand How Animals Work and Adapt* (Montreal: McGill-Queen's University Press, 2022), 115.

16. Andrew H. Dowdy, Robert D. Boche, and F. W. Bishop, *Observations on Animals Exposed to Whole Body Roentgen Radiation in Divided Doses over Long Periods: Introduction*

and Techniques (Oak Ridge, TN: US Atomic Energy Commission, 1946), 3. This is one of the first declassified reports from Rochester.

17. The projects are overviewed in Robert D. Boche, "Effects of Chronic Exposure to X Radiation on Growth and Survival," in *Biological Effects of External Radiation*, ed. Henry A. Blair, National Nuclear Energy Series, div. 6, vol. 2 (New York: McGraw-Hill, 1954), 222–52. On Casarett's work, see George W. Casarett, "Effects of X-Rays on Spermatogenesis in Dogs," *Quarterly Technical Report, University of Rochester Atomic Energy Project*, August 1, 1950, 14–22.

18. See Austin M. Brues, *Comparative Chronic Toxicities of Radium and Plutonium* (Chicago: Argonne National Laboratory, April 1951), 10, HathiTrust. The Chicago/Argonne work is also discussed in Brues, Lisco, and Finkel, *Carcinogenic Action of Some Substances*; Painter et al., *Clinical Physiology of Dogs Injected with Plutonium*, 9.

19. George W. Corner, *George Hoyt Whipple and His Friends: The Life-Story of a Nobel Prize Pathologist* (Philadelphia: Lippincott, 1963), 181, HathiTrust.

20. Corner, 101.

21. Miriam Finkel and Asher Finkel, interview by James Newell Stannard, October 9, 1981, J. Newell Stannard Papers, MS. 2020, University of Tennessee Libraries, Knoxville, Special Collections.

22. James Newell Stannard, *Radioactivity and Health: A History*, ed. Raymond W Balman Jr. (Springfield, VA: National Technical Information Service, 1988), 410; Stover and Stover, "Laboratory for Radiobiology at the University of Utah," 31.

23. Both men were involved in criticality accidents with the same subcritical mass of plutonium, subsequently known as the "demon core." For more on the concept of permissible dose, see J. Samuel Walker, *Permissible Dose: A History of Radiation Protection in the Twentieth Century* (Berkeley: University of California Press, 2000).

24. Wright H. Langham to Thomas L. Shipman, March 2, 1950, 2, Nuclear Testing Archive.

25. Thomas L. Shipman to John Z. Bowers, May 19, 1950, Nuclear Testing Archive. For the note to Evans, see Wright H. Langham to Robley D. Evans, February 3, 1950, 1, Nuclear Testing Archive.

26. Wright H. Langham to Thomas L. Shipman, March 2, 1950, 2. For the note to Evans, see Wright H. Langham to Robley D. Evans, February 3, 1950, 1, Nuclear Testing Archive.

27. Langham to Shipman, March 2, 1950, 3.

28. I cover the history of pigs as skin models during this period, as well as the broader context of animal experimentation, in Brad Bolman, "Pig Mentations: Race and Face in Radiobiology," *Isis* 112, no. 4 (December 2021): 694–716

29. "Transcript: 16th Meeting, Advisory Committee for Biology and Medicine," June 11, 1949, 3, AEC Division of Biology and Medicine, Collection 1130, box 3218 (box 49), folder 8, US National Archives at College Park, MD.

30. Although typically spelled "Eniwetok" in planning documents from the period, I have deferred to the more common contemporary rendering of "Enewetak." George V. LeRoy, "Biological Test Program for J-Division, LANL," July 7, 1949, Nuclear Testing Archive, https://www.osti.gov/opennet/detail?osti-id=16365206. On the NMRI, see Naval Medical Research Institute (US), *The Naval Medical Research Institute, 1942–1962* (US Government Printing Office, 1966), 13. See also A. H. Hempelman, "Review of the Bio-medical Program in Operation Greenhouse," n.d., Nuclear Testing Archive.

31. Langham to Shipman, March 2, 1950, 3.

32. Thomas L. Shipman to Shields Warren, March 6, 1950, Nuclear Testing Archive. The facility, located at Hunter's Point, California, would become briefly important for subsequent dog studies.

33. Shipman to Shields Warren, March 6, 1950. Because a great deal of plutonium work at Los Alamos was directly connected with efforts to construct nuclear weapons, those perceptions were perhaps not entirely misguided.

34. Paul B. Pearson to Thomas L. Shipman, May 31, 1950, Nuclear Testing Archive.

35. Albert Ray Olpin to John Z. Bowers, October 13, 1949, box 60, folder 1, Albert Ray Olpin Presidential Records, University Archives and Records Management, University of Utah Libraries, Salt Lake City.

36. Wolfgang Saxon, "Dr. John Z. Bowers, 80, Is Dead; Medical Educator and Historian," *New York Times*, October 19, 1993, B10.

37. Albert Ray Olpin to John Z. Bowers, December 27, 1949, box 60, folder 2, Albert Ray Olpin Presidential Records, University Archives and Records Management, University of Utah Libraries. See also, for background on the Arco project, Jack M. Holl, "The National Reactor Testing Station: The Atomic Energy Commission in Idaho, 1949–1962," *Pacific Northwest Quarterly* 85, no. 1 (1994): 15–24.

38. Shields Warren to Maxwell Wintrobe, January 4, 1950, box 60, folder 2, Albert Ray Olpin Presidential Records, University Archives and Records Management, University of Utah Libraries.

39. John Z. Bowers to Albert Ray Olpin, April 27, 1950, box 60, folder 3, Albert Ray Olpin Presidential Records, University Archives and Records Management, University of Utah Libraries.

40. "The discussion which I would like to undertake with you is to find out whether or not the medical school in Salt Lake City might not be interested in obtaining an AEC grant to embark on a long term project of this sort," Shipman wrote to Bowers in May 1950, although it appears Bowers knew about Shipman's earlier proposal to Warren. See Shipman to Bowers, May 19, 1950.

41. On the challenges of managing AEC towns, see, for instance, John M. Findlay and Bruce Hevly, *Atomic Frontier Days: Hanford and the American West* (Seattle: University of Washington Press, 2011).

42. As Pearson explained to Shipman, "With John Bowers going to the University of Utah it seemed appropriate to explore the possibility of having the plutonium work done there." Pearson to Shipman, May 31, 1950.

43. Stannard, *Radioactivity and Health*, 484.

44. "Contract between the University of Utah and the US Atomic Energy Commission for Atomic Energy Work," August 14, 1950, box 33, folder 11, University of Utah Vice President for Health Science Records, University Archives and Records Management, University of Utah Libraries.

45. "Excerpt from Discussion of Basis for the Radium Standard Mostly by R. D. Evans at Meeting of Task Group 6 of NCRP SC-57 at Tempe, Arizona, April 27, 1979," April 27, 1979, J. Newell Stannard Papers, MS.2020, University of Tennessee Libraries, Knoxville, Special Collections. The latter quote is from Thomas L. Shipman to Shields Warren, September 16, 1950, Nuclear Testing Archive.

46. Nonetheless, a great deal of early documentation for the project was thrown out when records were moved decades ago at the University of Utah. The historical record is impoverished as a result.

47. Carl E. Rehfeld, Jeanne A. Blomquist, and Glenn N. Taylor, "The Beagles," in *The Radiobiology of Plutonium* (Salt Lake City: J.W., 1972), 48.

48. Chester A. Gleiser, "The Determination of the Lethal Dose 50/30 of Total Body X-Radiation for Dogs," *American Journal of Veterinary Research* 14, no. 51 (April 1953): 284–86; Chester A. Gleiser, "The Sequence of Pathological Events in Dogs Exposed to a Lethal Dose 100/30 of Total Body X-Radiation," *American Journal of Veterinary Research* 14, no. 51 (April 1953): 287–97.

49. Wright H. Langham to Austin M. Brues, September 6, 1950, Nuclear Testing Archive.

50. J. Newell Stannard, "Interview with Robley D. Evans," 1978, 6, J. Newell Stannard Papers, Betsey B. Creekmore Special Collections and University Archives, University of Tennessee, Knoxville.

51. Stover and Stover, "Laboratory for Radiobiology at the University of Utah," 35.

52. Wright H. Langham, "Notes of Utah Meeting on Chronic Dog Experiment," November 16, 1950, Nuclear Testing Archive.

53. Shipman to Warren, March 6, 1950, 1.

54. Brandt, born in Larchwood, Iowa, taught at his alma mater, the Iowa State College in Ames, before moving to government work, including with the Naval Ordnance laboratory and the AEC. An influential statistician, he was also passionate about dogs. He published at least once in *American Field*, on canine heredity, in 1927, and was a frequent dog show judge. See, for instance, "Central States Dog Trials Will Open Here Today," *Des Moines Register*, November 2, 1935, 8. Correspondence between Ronald Aylmer Fisher and T. C. Hawley notes Brandt's experience with gun dog trials. See Ronald Aylmer Fisher to T. C. Hawley, November 19, 1947, R. A. Fisher Digital Archive, Ronald Aylmer Fisher Papers (MSS 0113), Archives and Special Collections, University of Adelaide, https://drmc.library.adelaide.edu.au/dspace/handle/2440/67737?mode=full. Brandt's centrality in locating a beagle provider for the laboratory is evidenced by a second note—after the December 14, 1950, planning meeting for Utah—that "Dr. A. E. Brandt again promised to look into the possibility of procuring beagles in the New York area. He agreed to pass this information on to Dr. Bowers." See "Minutes of December 14, 1950 Meeting, Subject: Chronic Toxicity of Plutonium, Radium and Mesothorium in Dogs," circa 1951, Nuclear Testing Archive.

55. Rehfeld, Blomquist, and Taylor, "Beagles," 48.

56. Sidney Farber, "Animal Colony Maintenance—Financing and Budgeting—Viewpoint of the University," *Annals of the New York Academy of Sciences* 46, no. 1 (1945): 111.

57. Warren Moscow, "US Pays $35 Each for Stray Dogs for Use in Atomic Research Work," *New York Times*, December 28, 1951, 1; Christine Stevens, "Animals in Science Research," *New York Times*, January 12, 1952, 12; "Bootlegged Beagles Bring Highest Prices from AEC," *Newsday*, January 19, 1952, 2.

58. As stated by Stover and Stover, "Laboratory for Radiobiology at the University of Utah," 36.

59. Stannard, *Radioactivity and Health*, 484.

60. Acquired on April 3, 1951, from A. Clyde Clark; see Rehfeld, Blomquist, and Taylor, "Beagles," 50.

61. The first quote appears in Rehfeld, Blomquist, and Taylor, 49. See also James Lyon, "Utah Housing Hounds for AEC Research," *Utah Daily Chronicle*, February 23, 1953, 4.

62. Patrick, "Hounds of Beagleville."

63. Bernard J. Snyder, *Aircraft Nuclear Propulsion: An Annotated Bibliography* (US Air Force History and Museums Program, May 3, 1996), xi, https://media.defense.gov/2014/Oct/14/2001329848/-1/-1/0/AFD-141014-032.pdf.

64. Robert LeBaron, "Review of the Aircraft Nuclear Propulsion Program" (Washington, DC: Committee on Atomic Energy, June 14, 1950), Nuclear Testing Archive.

65. NEPA Medical Advisory Committee, *Radiation Biology Relative to Nuclear Energy Powered Aircraft* (Oak Ridge, TN: NEPA Project, January 5, 1950), 5–7.

66. NEPA Medical Advisory Committee, 4.

67. Robert S. Stone, "Irradiation of Human Subjects as a Medical Experiment," January 31, 1950, 1, Nuclear Testing Archive.

68. Stone, 7.

69. "Transcript: 16th Meeting, Advisory Committee for Biology and Medicine," 8–9. For background on the malaria studies, see Nathaniel Comfort, "The Prisoner as Model Organism: Malaria Research at Stateville Penitentiary," *Studies in History and Philosophy of Science Part C: Studies in History and Philosophy of Biological and Biomedical Sciences* 40, no. 3 (September 1, 2009): 190–203, https://doi.org/10.1016/j.shpsc.2009.06.007.

70. Cited in Advisory Committee on Human Radiation Experiments, *Final Report of the Advisory Committee on Human Radiation Experiments.*

71. Shields Warren to Robert S. Stone, July 11, 1949, Nuclear Testing Archive. Warren's abstention is noted in James E. McCormack to Joseph C. Aub, December 7, 1950, Nuclear Testing Archive.

72. James E. McCormack to Astrid Kraus, "Problem of Experimentation with Human Volunteers," July 25, 1950, Nuclear Testing Archive.

73. M. C. Leverett, from the Fairchild NEPA Division, noted that the push for human experimentation was considered "a waste of energy" by February 1951. See M. C. Leverett to Shields Warren, February 15, 1951, Nuclear Testing Archive.

74. George Sacher, "Comments on Langham's 'Notes on Utah Meeting on Chronic Dog Experiment,'" November 21, 1950, Nuclear Testing Archive.

75. Austin M. Brues to Wright H. Langham, November 20, 1950, Nuclear Testing Archive.

76. W. A. Selle, "Chronological Review of Important Events in the History of NEPA's Effort to Secure Support for Its Recommendation on Human Experimentation," n.d., 5, Nuclear Testing Archive.

77. Comfort notes how prisoners during World War II served as both model organisms and experimental technicians, underscoring the parallelism between multiple experimental species. See Comfort, "Prisoner as Model Organism."

78. A. P. Gagge to Armstrong, "Medical Research Activities of NEPA Division, Fairchild Aircraft and Engine Corporation," March 7, 1951, 3, Nuclear Testing Archive.

79. Atomic Energy Commission, "January 1951," Monthly Status and Progress Reports (US Atomic Energy Commission, February 26, 1951), 28, Nuclear Testing Archive.

80. Project No. 2 was the radiology laboratory constructed at the University of California, San Francisco, under the direction of Robert Stone. Robert S. Stone to John B. deC. M. Saunders, January 22, 1960, 1, Nuclear Testing Archive.

81. Robert G. Sproul to Atomic Energy Commission, April 24, 1951, Nuclear Testing Archive. See also Stannard, *Radioactivity and Health*, 430.

82. "Atomic Visitor: AEC Scientist Visits the Aggie Campus," *California Aggie*, September 28, 1950, 6, California Digital Newspaper Collection, Center for Bibliographic Studies and Research, University of California, Riverside.

83. "Progress Report to the Joint Committee on Atomic Energy," December 21, 1951, 75, Nuclear Testing Archive; *Eleventh Semiannual Report of the Atomic Energy Commission* (Washington, DC: US Atomic Energy Commission, January 1952), 36, Nuclear Testing

Archive. See also Roy C. Thompson, *Life-Span Effects of Ionizing Radiation in the Beagle Dog: A Summary Account of Four Decades of Research Funded by the US Department of Energy and Its Predecessor Agencies* (Richland, WA: Pacific Northwest Laboratory, Department of Energy, 1989), 6. It is not clear whether Thompson located the first four annual reports from Davis, which do not appear in his list of facility reports. See Thompson, *Life-Span Effects of Ionizing Radiation in the Beagle Dog*, 318.

84. *Progress Report: Unclassified Project No. 4 Contract No. A.T-11-1-Gen-10: The Effect of X Radiation on the Work Capacity and Longevity of the Dog, 1951* (Davis: University of California, Davis, 1952), 1, in author's collection; *Progress Report: Unclassified Project No. 4 Contract No. A.T-11-1-Gen-10: The Effect of X Radiation on the Work Capacity and Longevity of the Dog, 1952* (Davis: University of California, Davis, 1952), 1, in author's collection.

85. See also Helen E. Longino, *Science as Social Knowledge: Values and Objectivity in Scientific Inquiry* (Princeton, NJ: Princeton University Press, 1990), 14.

86. For one among many examples, see Steven A. Book, "The Canine as a Model in Radiobiologic Research," in *The Canine as a Biomedical Research Model: Immunological, Hematological, and Oncological Aspects*, ed. Moshe Shifrine and Floyd D. Wilson (Springfield, VA: Technical Information Center, US Department of Energy, 1980), 316.

87. *Progress Report . . . 1951*, 1.

88. *Progress Report . . . 1952*, 1.

89. *Progress Report . . . 1951*, 1.

90. *Progress Report . . . 1951*, 2.

91. David Bruner, interview by James Newell Stannard, November 14, 1979, 3, University of Tennessee Libraries, Knoxville, Special Collections.

92. *Progress Report: Unclassified Project No. 4 Contract No. A.T-11-1-Gen-10: The Effect of X Radiation on the Work Capacity and Longevity of the Dog, 1953* (Davis: University of California, Davis, 1953), 1, in author's collection.

93. Allen C. Andersen, *A.E.C. Contract AT 11-1 Gen 10: Project No. 4; Fourth Annual Progress Report for the Period of July 1, 1954 to June 30, 1955* (Davis: University of California, Davis, 1955), 9–10. For more on the Fatigue Lab, see Robin Wolfe Scheffler, "The Power of Exercise and the Exercise of Power: The Harvard Fatigue Laboratory, Distance Running, and the Disappearance of Work, 1919–1947," *Journal of the History of Biology* 48, no. 3 (August 1, 2015): 391–423, https://doi.org/10.1007/s10739-014-9392-1.

94. Andersen, *A.E.C. Contract AT 11-1 Gen 10: Project No. 4; Fourth Annual Progress Report for the Period of July 1, 1954 to June 30, 1955*, 10.

95. *Progress Report . . . 1952*, 5.

96. *Progress Report . . . 1953*, 4.

97. Andersen, *A.E.C. Contract . . . Fourth Annual Progress Report*, 2.

98. Stannard, *Radioactivity and Health*, 414.

99. Brown, "Dogs Get Atom Doses to Save Lives of Men," 29.

100. Associated Press, "Dog Cancer Victim Gets Radioactive Iodine Treatment," *Washington Post*, August 3, 1951, 19.

101. *Report on Project 4 (1953–1954)* (Davis: University of California, Davis, 1954), 6, in author's collection. This could have been Owen M. Payne, president of the International Beagle Federation, or Glenn Black, whose *American Beagling* (New York: G. P. Putnam's Sons) came out to widespread interest in 1949, but no name is given in the report. The group is identified as the "American Beagle Breeders Association," but no organization existed under that name.

102. *Progress Report . . . 1953*, 6.

103. Vicki Feack, "UCD Scientists Enlist Beagles' Aid in Radiation Studies," *Davis Enterprise*, March 4, 1968, 8.

104. "AEC to Present Beagle Colony," *California Aggie*, April 5, 1962, California Digital Newspaper Collection, Center for Bibliographic Studies and Research, University of California, Riverside.

105. *Report on Project 4 (1953–1954)* (Davis: University of California, Davis, 1954), 4. For the later installations, see Andersen, *A.E.C. Contract . . . Fourth Annual Progress Report*, 2.

106. I discuss some of the prison-like features of beagle enclosures in Brad Bolman, "How Experiments Age: Gerontology, Beagles, and Species Projection at Davis," *Social Studies of Science* 48, no. 2 (2018): 232–58, https://doi.org/10.1177/0306312718759822.

107. Bruner, interview by James Newell Stannard.

108. Allen C. Andersen and George H. Hart, "Kennel Construction and Management in Relation to Longevity Studies in the Dog," *Journal of the American Veterinary Medical Association* 126, no. 938 (1955): 366. *Cowdry's Problems of Aging* was "the most authoritative treatise on the biology of aging," according to a 1952 reviewer. See "Cowdry's Problems of Aging—Biological and Medical Aspects—Third Edition," *California Medicine* 77, no. 5 (November 1952): 356. For broader context, see W. Andrew Achenbaum, "Handbooks as Gerontologic Maps," *Gerontologist* 36, no. 6 (1996): 825–26, https://doi.org/10.1093/geront/36.6.825.

109. Andersen and Hart, "Kennel Construction and Management," 366.

110. Kavita Sivaramakrishnan, *As the World Ages: Rethinking a Demographic Crisis* (Cambridge, MA: Harvard University Press, 2018), 62.

111. MSS Information Corporation, *Dogs and Other Large Mammals in Aging Research: I* (New York: MSS Information Corporation, 1974), 8; Hyung Wook Park, *Old Age, New Science: Gerontologists and Their Biosocial Visions, 1900–1960* (Pittsburgh: University of Pittsburgh Press, 2016), 21.

112. Park, *Old Age, New Science*, 27.

113. Feack, "UCD Scientists Enlist Beagles' Aid in Radiation Studies," 8.

114. For seven to ten years, see William F. Arbogast, "Experimental Dog's Life Big Problem for AEC," *Austin Statesman*, July 24, 1959, 14. For nine to ten, see *Quarterly Report for the Division of Biology and Medicine, US Atomic Energy Commission* (Salt Lake City: Radiobiology Laboratory, University of Utah, June 30, 1953).

115. Allen C. Andersen and Leon S. Rosenblatt, "Survival of Beagles under Natural and Laboratory Conditions," *Experimental Gerontology* 1, no. 3 (1965): 194, https://doi.org/10.1016/0531-5565(65)90003-3. For the thirteen figure, see William Trombley, "Beagles Find Longevity at UC Davis," *Los Angeles Times*, February 4, 1975.

116. Norman M. Sulkin, *The Laboratory Animal in Gerontological Research: Proceedings of a Symposium Conducted by Committee on Animal Research in Gerontology, Institute of Laboratory Animal Resources, National Research Council* (Washington, DC: National Academy of Sciences, 1968), 1.

117. Susan D. Jones, *Valuing Animals: Veterinarians and Their Patients in Modern America* (Baltimore: Johns Hopkins University Press, 2002), 122.

118. Allen C. Andersen, "A Comparison of Beagles Released for Private Ownership with Those Maintained under Kennel Conditions," *Journal of the American Veterinary Medical Association* 132, no. 3 (February 1, 1958): 95.

119. Andersen, 95.

120. Marvin Goldman, "Human Radiation Studies: Remembering the Early Years: Oral History of Radiation Biologist Marvin Goldman, Ph.D.," interview by Lori Hefner

and Karoline Gourley, December 22, 1994, https://ehss.energy.gov/ohre/roadmap/histories/0468/0468toc.html.

121. *Progress Report . . . 1953*, 5.

122. *Progress Report . . . 1953*, 3.

123. "Make Your Own Doghouse from Barrel," *Vancouver Sun*, January 3, 1945.

124. Andersen, "Dog—a Facet of Human Society," 6.

125. Roy C. Thompson and Judy A. Mahaffey, eds., *Life-Span Radiation Effects Studies in Animals: What Can They Tell Us?* (Oak Ridge, TN: Office of Scientific and Technical Information, US Department of Energy, 1986), 102.

126. *Quarterly Report for the Division of Biology and Medicine, US Atomic Energy Commission* (Salt Lake City: Radiobiology Laboratory, University of Utah, June 30, 1952), 1.

127. Allen C. Andersen, *The Effect of X-Radiation on Work Capacity and Longevity of the Dog: Seventh Annual Progress Report* (Davis: School of Veterinary Medicine, University of California, Davis / Atomic Energy Commission, June 1958), 6–7.

128. Quoted in Thompson and Mahaffey, *Life-Span Radiation Effects Studies in Animals*, 102.

129. Otniel E. Dror, "The Affect of Experiment: The Turn to Emotions in Anglo-American Physiology, 1900–1940," *Isis* 90, no. 2 (June 1, 1999): 236, https://doi.org/10/ctwt5t.

130. *Progress Report . . . 1951*, 3.

131. Andersen, *A.E.C. Contract . . . Fourth Annual Progress Report*, 4.

132. Andersen, "Dog—a Facet of Human Society," 6.

133. Dror, "Affect of Experiment," 207.

134. Hyung Wook Park has called this gerontology's "biosocial vision." Park, *Old Age, New Science*, 5.

135. Andersen and Hart, "Kennel Construction and Management"; Allen C. Andersen, "Debarking in a Kennel: Technic and Results," *Veterinary Medicine* 50 (September 1955): 409–11; Allen C. Andersen, "Puppy Production to the Weaning Age," *Journal of the American Medical Association* 130, no. 4 (February 15, 1957): 151–58.

136. Rehfeld, Blomquist, and Taylor, "Beagles," 50.

137. I discuss the RPM further in Brad Bolman, "Strontium," in *An Anthropogenic Table of Elements*, ed. Timothy Neale, Thao Phan, and Courtney Addison (Toronto: University of Toronto Press, 2023).

138. Andersen, *A.E.C. Contract . . . Fourth Annual Progress Report*, 5.

139. *Progress Report . . . 1951*, 3.

140. Georges Canguilhem, *Knowledge of Life*, trans. Stefanos Geroulanos and Daniela Ginsburg, Forms of Living (New York: Fordham University Press, 2008), 128.

141. Tiago Moreira and Paolo Palladino, "Ageing between Gerontology and Biomedicine," *BioSocieties* 4, no. 4 (December 1, 2009): 349–65, https://doi.org/10.1017/S1745855209990305.

142. Allen C. Andersen and Douglas H. McKelvie, "Long-Term Experiments in Relation to Treatment Effects: Radiobiological and Gerontological Problems," *Proceedings of the Animal Care Panel* 12, no. 4 (September 1962): 178.

143. Moreira and Palladino argue that the emphasis on disease treatment was characteristic of England's "social medicine" approach. See Moreira and Palladino, "Ageing between Gerontology and Biomedicine," 356.

144. Sivaramakrishnan, *As the World Ages*, 63.

145. Allen C. Andersen, "Life-Span Studies on Irradiated Beagles: Practical and Academic Contributions," *Small Animal Clinician* 2, no. 9 (1962): 506.

146. Animal Welfare Institute, *Information Report* (New York: Animal Welfare Institute, December 1955).

147. Andersen and McKelvie, "Long-Term Experiments in Relation to Treatment Effects," 179.

148. Andersen, *Effect of X-Radiation on Work Capacity and Longevity of the Dog: Seventh Annual Progress Report*, 8.

149. Allen C. Andersen, D. E. Jasper, and George H. Hart, *A.E.C. Project No. 6 Annual Progress Report (1958)* (Davis: University of California, Davis, 1959), 5.

150. Rehfeld, Blomquist, and Taylor, "Beagles," 56–57.

151. Laura A. Bruno, "The Bequest of the Nuclear Battlefield: Science, Nature, and the Atom during the First Decade of the Cold War," *Historical Studies in the Physical and Biological Sciences* 33, no. 2 (2003): 237, https://doi.org/10.1525/hsps.2003.33.2.237.

152. Joseph Masco, "The Age of Fallout," *History of the Present* 5, no. 2 (Fall 2015): 137–68, https://doi.org/10.5406/historypresent.5.2.0137; Joseph Masco, "The Crisis of Crisis," *Current Anthropology* 58, no. 15 (February 2017): S65–76.

153. Bruno, "Bequest of the Nuclear Battlefield," 240.

154. On the baby tooth survey and surrounding activism, see Kelly Moore, *Disrupting Science: Social Movements, American Scientists, and the Politics of the Military, 1945–1975* (Princeton, NJ: Princeton University Press, 2008); Michael Egan, *Barry Commoner and the Science of Survival: The Remaking of American Environmentalism* (Cambridge, MA: MIT Press, 2007).

155. Walter Schneir, "Strontium-90 in US Children," *Nation*, April 25, 1959, 355.

156. Wright H. Langham to Charles L. Dunham, July 23, 1957, 1, Nuclear Testing Archive.

157. Langham to Dunham, July 23, 1957, 3.

158. Thomas L. Shipman to Norris E. Bradbury, "Proposal for Study of Sr90 in Dogs," May 2, 1957, Nuclear Testing Archive.

159. Austin M. Brues to Robley D. Evans, November 21, 1957, Nuclear Testing Archive.

160. Wright H. Langham to H. D. Bruner, April 9, 1958, Nuclear Testing Archive.

161. Andersen, Jasper, and Hart, *A.E.C. Project No. 6 Annual Progress Report (1958)*, 1.

162. Robley D. Evans to Walter D. Claus, June 14, 1959, Nuclear Testing Archive.

163. Henry A. Blair to Robley D. Evans, October 8, 1957, Nuclear Testing Archive.

164. Blair to Evans, October 8, 1957, 4. Blair appeared to think that the point of the work was to establish a maximum permissible dose for strontium, and thus that a large experiment wasn't worthwhile. Others were interested in more expansive analysis of strontium-90's health risks.

165. Wright H. Langham to Robley D. Evans, January 7, 1958, 1, Nuclear Testing Archive.

166. Langham to Evans, January 7, 1958, 2–3.

167. *Atomic Energy Commission Appropriations for 1960* (Washington, DC: US Government Printing Office, June 1959), 45.

168. *Atomic Energy Commission Appropriations for 1960*, 46.

169. Arbogast, "Experimental Dog's Life Big Problem for AEC," 14.

170. Joseph Masco, *The Theater of Operations: National Security Affect from the Cold War to the War on Terror* (Durham, NC: Duke University Press, 2014), 48. See also Paul Boyer, *By the Bomb's Early Light: American Thought and Culture at the Dawn of the Atomic Age* (Chapel Hill: University of North Carolina Press, 1994).

171. Masco, "Age of Fallout," 140.

172. Sharon Olds, "When," in *Writing in a Nuclear Age*, ed. Jim Schley (Hanover, NH: University Press of New England, 1984), 72.

173. William Dickey, "Armageddon," in Schley, *Writing in a Nuclear Age*, 43.

174. See Radin, *Life on Ice*.

175. Joseph Masco, *The Nuclear Borderlands: The Manhattan Project in Post–Cold War New Mexico* (Princeton, NJ: Princeton University Press, 2006), 310.

176. Masco, 311.

177. See, again, Bolman, "Pig Mentations."

178. Kate Brown has explored the intertwined histories of these places in Kate Brown, *Plutopia: Nuclear Families, Atomic Cities, and the Great Soviet and American Plutonium Disasters* (Oxford: Oxford University Press, 2015).

179. Roger McClellan, personal communication with author, 2020–21.

180. Gail Davies has written extensively on recent efforts to "humanize" mice and their meaning for thinking about the becoming-human of experimental organisms. See, for instance, Gail Davies, "What Is a Humanized Mouse? Remaking the Species and Spaces of Translational Medicine," *Body and Society* 18, nos. 3–4 (September 1, 2012): 126–55; Gail Davies, "Writing Biology with Mutant Mice: The Monstrous Potential of Post Genomic Life," *Geoforum* 48 (August 1, 2013): 268–78, https://doi.org/10.1016/j.geoforum.2011.03.004.

181. "A Dog's Life," *University Explorer* (San Francisco: Columbia Broadcasting System, June 5, 1960), box 23, Marvin D. Goldman / Laboratory for Energy-Related Health Research (LEHR) Collection, Archives and Special Collections, University of California, Davis. On automobility, see John Urry, "The 'System' of Automobility," *Theory, Culture and Society* 21, nos. 4–5 (October 1, 2004): 25–39, https://doi.org/10.1177/0263276404046059.

182. Allen C. Andersen, *The Effect of X-Radiation on the Work Capacity and Longevity of the Dog: Fifth Annual Report* (Davis: School of Veterinary Medicine, University of California, Davis / Atomic Energy Commission, 1956), 3. This comment reflects the complicated position of Asian Americans as "model minorities" in United States scientific contexts. For a contemporary reflection on this, see Roli Varma, "High-Tech Coolies: Asian Immigrants in the US Science and Engineering Workforce," *Science as Culture* 11, no. 3 (September 1, 2002): 337–61, https://doi.org/10.1080/0950543022000005078.

183. In this sense they were also "technoscientific organisms," as Tiago Saraiva terms the agricultural animals that became key elements in the production of fascist modernity in Germany. See Tiago Saraiva, *Fascist Pigs: Technoscientific Organisms and the History of Fascism* (Cambridge, MA: MIT Press, 2016), 9.

184. Rehfeld, Blomquist, and Taylor, "Beagles," 48.

185. See the introduction.

186. Rehfeld, Blomquist, and Taylor, "Beagles," 48.

187. The "beaglers" reference appears, for instance, in a retrospective interview. See J. Newell Stannard, "Interview with Leo K. Bustad," 1981, 24, J. Newell Stannard Papers, Betsey B. Creekmore Special Collections and University Archives, University of Tennessee, Knoxville.

188. See, for instance, Leo K. Bustad to Tom Pigford, "On Gofman and Tamplin vs. Evans," n.d., box 38, "Gofman and Tamplin," Marvin D. Goldman / Laboratory for Energy-Related Health Research (LEHR) Collection, Archives and Special Collections, University of California, Davis.

189. The newsletter appears in box 11, Marvin D. Goldman / Laboratory for Energy-Related Health Research (LEHR) Collection, Archives and Special Collections, University of California, Davis.

190. "Never Bug a Beagle!" (Laboratory for Energy-Related Health Research, University of California, Davis, 1970), Sub 2.5, 23.261, "Never Bug a Beagle," Leo K. Bustad Papers, Manuscripts, Archives, and Special Collections, Washington State University Libraries.

191. Most cartoons appear in "Comics" and "Just for Laughs," box 44, Marvin D. Goldman / Laboratory for Energy-Related Health Research (LEHR) Collection, Archives and Special Collections, University of California, Davis. The specific cartoon referenced here is photocopied in box 11.

192. Arnold Arluke argues that this blurred distinction between "object" and "pet" sits at the heart of debates about the treatment of laboratory animals. For many researchers, experimental animals are both at once. See Arnold Arluke, "Sacrificial Symbolism in Animal Experimentation: Object or Pet?," *Anthrozoös* 2, no. 2 (1988): 98–117.

193. See Marvin Goldman to Doug, January 25, 1980, box 17, "Federal Strategy," Marvin D. Goldman / Laboratory for Energy-Related Health Research (LEHR) Collection, Archives and Special Collections, University of California, Davis. In *Peanuts*, the Head Beagle is sometimes Snoopy himself, but at other times the reference is to an external, higher Head Beagle of ambiguous ontological status.

194. "Man's Best Friend Faces Strontium 90," 26.

195. The process parallels the way in which xenotransplantation researchers have imagined their work fitting within a transhistoric desire for multispecies hybridity. See Brad Bolman, "Farm-to-Operating Table," *Grow by Ginkgo*, October 2022.

196. See Robin Wolfe Scheffler, "Brightening Biochemistry: Humor, Identity, and Scientific Work at the Sir William Dunn Institute of Biochemistry, 1923–1931," *Isis* 111, no. 3 (September 1, 2020): 514, https://doi.org/10.1086/710666. Scheffler shows the critical role of humor in defining workplace anxiety and professional identity among biochemists.

197. Creager, *Life Atomic*, 369–72.

198. William J. Bair, "Oral History of Health Physicist William J. Bair, Ph.D.," interview by David Harrell and Cindy Shindledecker, October 14, 1994, DOE Openness: Human Radiation Experiments, https://ehss.energy.gov/ohre/roadmap/histories/0463/0463toc.html.

199. Harry A. Kornberg to James Newell Stannard, August 9, 1982, 2–4, William Bair Papers, Hanford History Project Collections, Richland, WA. Kornberg's account exemplifies the omnipresence of self-experimentation in midcentury science, yet a case that was minimally valorized. See Rebecca Herzig, *Suffering for Science: Reason and Sacrifice in Modern America* (New Brunswick, NJ: Rutgers University Press, 2005).

200. McClellan, personal communication with author, 2020–21. The quote about Park appears in Kornberg to Stannard, August 9, 1982. McClellan remembered a friendly competition between Bair's and Bustad's groups.

201. M. G. Brown, "Hanford's Dog Colony," July 5, 1962, William Bair Papers, Hanford History Project Collections; H. E. Dziuk, "Outline for Animal Care Procedures in Use at Hanford" (Texas Meeting: Hanford Radiobiology, June 8, 1960), William Bair Papers, Hanford History Project Collections.

202. Clayton Samuel White, October 14, 1985, Oral History Project, University of New Mexico Medical Center Library, https://nmdigital.unm.edu/digital/collection/nmhhc/id/36/.

203. Bruner, interview, November 14, 1979, 3.

204. White, October 14, 1985, Oral History Project, University of New Mexico Medical Center Library. Lovelace later became the Lovelace Respiratory Research Institute. Today it is known as Lovelace Biomedical.

205. For background on Lovelace, see Jake W. Spidle, *The Lovelace Medical Center: Pioneer in American Health Care* (Albuquerque: University of New Mexico Press, 1987).

206. Ivan Oransky, "Clayton Samuel White," *Lancet* 363, no. 9424 (June 5, 2004): 1913, https://doi.org/10.1016/S0140-6736(04)16381-2.

207. McClellan noted White's interest in sustaining the AEC connection in our communications.

208. Bruner, interview, November 14, 1979, 6.

209. White, October 14, 1985, Oral History Project, University of New Mexico Medical Center Library. Mercer's son, Robert, would go on to become CEO of Renaissance Technologies, a principal investor in Cambridge Analytica, and major Republican donor. See Jane Mayer, "The Reclusive Hedge-Fund Tycoon behind the Trump Presidency," *New Yorker*, March 17, 2017, https://www.newyorker.com/magazine/2017/03/27/the-reclusive-hedge-fund-tycoon-behind-the-trump-presidency.

210. William J. Bair, interview by Robert Bauman, video, August 14, 2013, Hanford Oral History Project, Hanford History Project Collections, http://www.hanfordhistory.com/items/show/9.

211. Jerry Smothers, "1,000 Dogs Will Live in New $3 Million Lab," *Albuquerque Tribune*, August 1, 1963, B6.

212. "A.E.C. to Expose 1,000 Dogs to Radioactive Air," *New York Times*, August 3, 1963, 2.

213. Glenn T. Seaborg, *Plutonium Revisited*, ed. Betsy J. Stover and Webster S. S. Jee (Salt Lake City: J.W., 1972).

214. M. E. Wrenn et al., "Summary of Dosimetry, Pathology, and Dose Response for Bone Sarcomas in Beagles Injected with Radium-226," in *Life-Span Radiation Effects Studies in Animals: What Can They Tell Us?* (Oak Ridge, TN: Office of Scientific and Technical Information, US Department of Energy, 1986), 267.

215. "Atom Beagles Love Beer Barrel Beds," *Oakland Tribune*, August 9, 1959, 11. For more on the card system, see Douglas H. McKelvie, T. J. Powel, and Sally Lee, "Programming for Computer Storage, Retrieval, and Analyses of Clinical Chemistry Data in a Life-Span Study on Beagles," *Laboratory Animal Care* 17, no. 5 (1967): 494–500.

216. Allen C. Andersen, *The Beagle as an Experimental Dog* (Ames: Iowa State University Press, 1970).

217. Betsy J. Stover, "*The Beagle as an Experimental Dog* by Allen C. Andersen," *Radiation Research* 45, no. 2 (February 1971): 449.

218. "Contract for Publication 'The Beagle as an Experimental Dog,'" June 1964, box 1, Directors Office Files for LEHR (434-90-191), National Archives at San Francisco.

219. Andersen, *Beagle as an Experimental Dog*, 3.

220. Andersen, 3–4.

221. Andersen, 4.

222. Andersen, "Dog—a Facet of Human Society," 10.

223. Leo K. Bustad et al., "The Choice of the Beagle for Radiobiologic Studies," in *Radiobiology of Plutonium* (Salt Lake City: J.W., 1972), 204.

224. Eva Giraud and Gregory Hollin, "Care, Laboratory Beagles and Affective Utopia," *Theory, Culture and Society* 33, no. 4 (July 2016): 36, https://doi.org/10.1177/0263276415619685.

225. Mariam Motamedi Fraser, *Dog Politics: Species Stories and the Animal Sciences* (Manchester: Manchester University Press, 2024), 182. "Politeness" is a concept Fraser develops from the work of Vinciane Despret. See Vinciane Despret, "The Becomings of Subjectivity in Animal Worlds," *Subjectivity* 23, no. 1 (July 1, 2008): 123–39, https://doi.org/10.1057/sub.2008.15.

226. Andersen, *Beagle as an Experimental Dog*, xi.

227. John Paul Scott, "A Laboratory Breed," *Science* 170, no. 3959 (November 13, 1970): 723.

228. Andersen, "Dog—Facet of Human Society," 9.

229. Scott, "Laboratory Breed," 723.

230. Stover, "*Beagle as an Experimental Dog* by Allen C. Andersen," 449.

231. Andersen, *Beagle as an Experimental Dog*, xiii.

232. Allen C. Andersen and L. S. Good, "The Beagle as an Experimental Dog," October 26, 1965, box 1, Directors Office Files for LEHR (434-90-191), National Archives at San Francisco.

233. Allen C. Andersen to Marvin Goldman, September 20, 1966, box 1, Directors Office Files for LEHR (434-90-191), National Archives at San Francisco.

234. Marcus M. Mason, *Bibliography of the Dog* (Ames: Iowa State University Press, 1959), iii. My copy of *Bibliography* was originally owned by the Pet Food Nutrition Lab of General Mills, one demonstration of interest in its subject. Mason would go on to run his own research laboratory and received significant funding from the tobacco industry for his work. The connection is well documented, but a revealing letter is W. T. Hoyt to Marcus M. Mason, August 13, 1971, Council for Tobacco Research Records, https://www.industrydocuments.ucsf.edu/docs/qkfj0044. As mentioned above, many items cited in connection to tobacco are available at this University of California, San Francisco, tobacco database.

235. Roger O. McClellan, "Working Group—'Status of Use of Large Long-Lived Domestic Animals in Radiobiology,'" April 12, 1972, box 4, folder 21, Dr. Goldman Records (434-90-0192), National Archives at San Francisco.

236. Marvin Goldman to Leo K. Bustad, "Trip Report on Long Lived Domestic Mammals in Radiobiology," December 20, 1971, box 4, folder 21, Dr. Goldman Records (434-90-0192), National Archives at San Francisco.

237. See the conclusion, as well as Bolman, "How Experiments Age," 244–48.

238. Susan E. Lederer, "Political Animals: The Shaping of Biomedical Research Literature in Twentieth-Century America," *Isis* 83, no. 1 (March 1992): 61–79.

239. Thompson, *Life-Span Effects of Ionizing Radiation in the Beagle Dog*, iii.

240. McDonald E. Wrenn, "On the Infeasibility of Combining Long Term Dog Studies at Argonne, Utah, and Davis," March 30, 1982, box 11, folder 5, Directors Office Files for LEHR (434-90-191), National Archives at San Francisco.

241. See K. S. Crump, L. S. Rosenblatt, and M. A. Schneiderman, "Concluding Panel Discussion," in *Life-Span Radiation Effects Studies in Animals*, 713.

242. Wrenn, "On the Infeasibility," 4.

243. Ed Branam, "Medical Impact on LEHR Dog Relocation," March 26, 1982, box 11, folder 5, Directors Office Files for LEHR (434-90-191), National Archives at San Francisco.

244. Wrenn, "On the Infeasibility," 4.

245. Wrenn, 3.

246. Marvin Goldman, "Telephone Contact Report," March 24, 1982, box 11, folder 5, Directors Office Files for LEHR (434-90-191), National Archives at San Francisco.

247. J. W. Overstreet to Richard Nolan, "Transfer of Dogs from LEHR to the University," February 12, 1986, box 14, folder 10, Directors Office Files for LEHR (434-90-191), National Archives at San Francisco.

248. Thompson and Mahaffey, *Life-Span Radiation Effects Studies in Animals*, ix.

249. Thompson and Mahaffey, xi.

250. Thompson, *Life-Span Effects of Ionizing Radiation in the Beagle Dog.*

251. Thompson, 6–7. "The data still to be obtained from animals still alive is, in many respects, the most interesting data of all, since it is usually data from the longest-lived animals," he notes.

252. Stannard, *Radioactivity and Health,* 1800.

253. Cathy Horyn, "Dogs Have Their Own 'Legionnaire's Disease,'" *Sterling Daily Gazette,* May 19, 1979, 2.

254. Crump, Rosenblatt, and Schneiderman, "Concluding Panel Discussion," 713.

255. Associated Press, "New Virus Killing Thousands of Dogs around US," *New York Times,* August 13, 1980, A15.

256. Ellen Bilgore, "Canine Parvovirus Can Be Prevented," *New York Times,* August 21, 1980, C8.

257. *Advancing Veterinary Medicine through Research* (Ithaca, NY: James A. Baker Institute for Animal Health, 2000), 7.

Chapter 3

1. "About," Vanda Pharmaceuticals, https://www.vandapharma.com/about/, accessed March 2024. Gastroparesis is a condition common among diabetics in which the stomach cannot empty itself in expected fashion.

2. "Vanda Sues FDA over Mandate to Test Drug on Dogs," *Pharmaceutical Manufacturing,* February 8, 2019, https://www.pharmamanufacturing.com/industrynews/2019/vanda-sues-fda-over-mandate-to-test-drug-on-dogs/.

3. "Vanda Puts Dog Studies in Spotlight in Suit Challenging FDA Clinical Hold," *Pink Sheet,* February 10, 2019, LexisNexis.

4. Vanda Pharmaceuticals to US Food and Drug Administration, "Open Letter to the Food and Drug Administration," February 5, 2019, 2, https://mma.prnewswire.com/media/818251/Open_Letter.pdf. Vanda researchers followed up with a presentation at the American Society of Human Genetics conference in October 2019, arguing that inbreeding in beagles made them a poor model for drug toxicity predictions. Sandra Smieszek et al., "Using Patterns in Regions of Homozygosity to Evaluate the Use of Dogs as Preclinical Models in Human Drug Development," 2019, https://www.prnewswire.com/news-releases/vanda-announces-new-study-results-showing-a-high-degree-of-inbreeding-in-beagle-dogs-300944277.html#.

5. "Vanda Announces New Study Results Showing a High Degree of Inbreeding in Beagle Dogs," *PR Newswire,* October 23, 2019, https://www.prnewswire.com/news-releases/vanda-announces-new-study-results-showing-a-high-degree-of-inbreeding-in-beagle-dogs-300944277.html.

6. Clair Linzey and Andrew Linzey, "From the Editors: In Praise of Corporate Bravery," *Journal of Animal Ethics* 10, no. 2 (2020): v–vii, https://doi.org/10.5406/janimalethics.10.2.000v.

7. See, e.g., Harold Boxenbaum, "Interspecies Scaling, Allometry, Physiological Time, and the Ground Plan of Pharmacokinetics," *Journal of Pharmacokinetics and Biopharmaceutics* 10, no. 2 (1982): 203–4; R. T. Williams, "Interspecies Scaling," in *Pharmacology and Pharmacokinetics,* ed. Torsten Teorell, Robert L. Dedrick, and Peter G. Condliffe, Fogarty International Center Proceedings 20 (New York: Plenum, 1974), 105.

8. On the "™," Haraway writes, "I am extremely curious about what kinds of bodies, what forms of frozen as well as motile sociotechnical alliances, also called social

relationships, these little ornaments can adorn, at whose cost, and to whose benefit." See Haraway, *Modest_Witness@Second_Millennium.FemaleMan_Meets_OncoMouse,* 7–8.

9. Emily Martin, "The Pharmaceutical Person," *BioSocieties* 1, no. 3 (September 1, 2006): 273–87, https://doi.org/10/frrddh.

10. Joseph Dumit, *Drugs for Life: How Pharmaceutical Companies Define Our Health* (Durham, NC: Duke University Press, 2012), 89.

11. Paul C. Underwood and Charles G. Durbin, "Beagles for Pharmacological Studies," *Laboratory Animal Care* 13, no. 4 (August 1963): 525.

12. J. Stuart Paterson, "The Dog (*Canis familiaris*)," in *The UFAW Handbook on the Care and Management of Laboratory Animals* (London: Universities Federation for Animal Welfare, 1957), 533.

13. Underwood and Durbin, "Beagles for Pharmacological Studies," 525.

14. Bert J. Vos et al., History of the Food and Drug Administration, interview by James Harvey Young et al., June 20, 1980, 119, Oral History, National Library of Medicine, Bethesda, MD, https://www.fda.gov/media/81332/download.

15. Vos et al., 119–20. Dawson would later advise the Jackson Labs dog study in its early years. See Brad Bolman, "Critical Periods in Science and the Science of Critical Periods: Canine Behavior in America," *Berichte zur Wissenschaftsgeschichte* 45, nos. 1–2 (2022): 119.

16. "Surplus" dogs from the study, including collies, a German shepherd, and a chow chow, were sold to the public in 1941. Alfred Friendly, "Dogs Refusing to Bite at Trick Baffle Experts," *Washington Post,* October 29, 1939, X8, ProQuest Historical Newspapers; UP (later UPI), "US Has Dogs for Sale," *New York Times,* November 11, 1941, 25.

17. Underwood and Durbin, "Beagles for Pharmacological Studies," 526.

18. George E. Burch, "Of the Normal Dog," *American Heart Journal* 58, no. 6 (December 1959): 806.

19. Underwood and Durbin, "Beagles for Pharmacological Studies," 525.

20. Stephen Pemberton, "Canine Technologies, Model Patients: The Historical Production of Hemophiliac Dogs in American Biomedicine," in *Industrializing Organisms: Introducing Evolutionary History* (New York: Routledge, 2004), 200. Brinkhous's dogs hailed from Cortland, New York, not far from Cornell, where their illness was diagnosed by researchers who would go on to work at the RLDD. A subsequent laboratory at Oklahoma State University dedicated to blood disorders would use beagles; see Associated Press, "Beagle Colony Aids OSU Blood Research," *Amarillo Globe Times,* January 5, 1966, 37.

21. Leslie Tillotson Webster and Jordi Casals, "A Dog Test for Measuring the Immunizing Potency of Antirabies Vaccines," *Journal of Experimental Medicine* 71, no. 5 (April 30, 1940): 719–30; Leslie Tillotson Webster and Jordi Casals, "The Quantity of Irradiated Non-virulent Rabies Virus Required to Immunize Mice and Dogs," *Journal of Experimental Medicine* 73, no. 5 (May 1, 1941): 601–15; Leslie Tillotson Webster, *Rabies* (New York: Macmillan, 1942).

22. Leslie Tillotson Webster to Wilbur A. Sawyer, June 20, 1938, Correspondence, "Sawyer, Wilbur A.," Leslie Tillotson Webster Papers, American Philosophical Society, Philadelphia.

23. In another example, Wolfgang Felix von Oettingen and his colleagues in the Experimental Biology and Medicine Institute of the National Institutes of Health explored the toxicity of chlorinated methanes on beagles in 1949. See Wolfgang Felix von Oettingen et al., "Relation between the Toxic Action of Chlorinated Methanes and Their Chemical and Physicochemical Properties," *National Institutes of Health Bulletin* 191 (1949).

24. William Layton, interview by Tanya Levin, May 20, 1997, Oral Histories, American Institute of Physics, https://www.aip.org/history-programs/niels-bohr-library/oral-histories/6941.

25. Edmund Mayer, "Inhibition of Thyroid Function in Beagle Puppies by Propylthiouracil without Disturbance of Growth or Health," *Endocrinology* 40, no. 3 (March 1947): 181. A definite date for the founding of Stamford's colony is difficult to pin down, but research at Stamford used beagles as early as 1943 according to Edmund Mayer and Lina A. Ottolenghi, "Protrusion of Tubular Epithelium into the Space of Bowman's Capsule in Kidneys of Dogs and Cats," *Anatomical Record* 99, no. 4 (1947): 477–509, https://doi.org/10.1002/ar.1090990405. The American Cyanamid Company's story is chronicled in Alfred Dupont Chandler, *Shaping the Industrial Century: The Remarkable Story of the Evolution of the Modern Chemical and Pharmaceutical Industries* (Cambridge, MA: Harvard University Press, 2009). Mayer was another German émigré that the Rockefeller Foundation helped relocate during World War II.

26. Frank Blair Hanson, "Diary," 1941, 168, Rockefeller Foundation Records, Officers' Diaries, RG 12, F-L, Hanson, Frank B., Rockefeller Archive Center, Sleepy Hollow, NY.

27. See, for instance, R. H. Silber, E. E. Howe, and C. C. Porter, "The Utilization of Amino Acid Mixtures by Dogs," *Transactions of the New York Academy of Sciences*, ser. 2, 10, no. 8 (1948): 277–80, https://doi.org/10/ggrs46; R. H. Silber et al., "Growth and Maintenance of Dogs Fed Amino Acids as the Source of Dietary Nitrogen: One Figure," *Journal of Nutrition* 37, no. 4 (April 1, 1949): 429–41, https://doi.org/10.1093/jn/37.4.429; G. E. Boxer and J. C. Rickards, "Studies on the Metabolism of the Carbon of Cyanide and Thiocyanate," *Archives of Biochemistry and Biophysics* 39, no. 1 (1952): 7–26, https://doi.org/10.1016/0003-9861(52)90256-7.

28. John E. Gilmartin, "The Establishment of a Dog Breeding Kennel for Pharmaceutical Research," *Proceedings of the Animal Care Panel* 11, no. 4 (September 1961): 222.

29. B. Jackson and V. P. Cappiello, "Ranges of Normal Organ Weights of Dogs," *Toxicology and Applied Pharmacology* 6 (1964): 664. Ellis J. Robinson of American Cyanamid notes their experience raising beagles for research in the discussion following Sidney Farber, "Animal Colony Maintenance—Financing and Budgeting—Viewpoint of the University," *Annals of the New York Academy of Sciences* 46, no. 1 (1945), https://doi.org/10.1111/j.1749-6632.1945.tb36163.x.

30. Victor Heiman, "Dogs for Nutritional Research," *Proceedings of the Animal Care Panel* 11, no. 4 (September 1961): 215.

31. Robert K. Plumb, "Group to Regulate Research Animals," *New York Times*, September 13, 1953, 39.

32. Institute of Animal Resources, *A Handbook of Laboratory Animals* (Washington, DC: National Academy of Sciences, National Research Council, 1954), 50, http://hdl.handle.net/2027/mdp.39015082038525.

33. Both companies were identified with their New Jersey locations. See Institute of Animal Resources, 67, 70.

34. Institute of Laboratory Animal Resources, *Laboratory Animals: II. Animals for Research, a Directory of Commercial Sources of Laboratory Animals and Equipment* (Washington, DC: National Academy of Sciences, National Research Council, 1961), 22–23.

35. A short history of the Iowa dogs appears in Robert Daniel Whiteford, "Distribution of Lipofuscin as Related to Aging in the Canine and Porcine Brain" (PhD diss., Iowa State University, 1964), 17. The Florida glaucoma project was reported in local papers, for instance "Beagle Colony Helps Scientists Find New Glaucoma Information," *Panama City News*, August 31, 1976, 5A.

36. H. Reinert and G. K. A. Smith, "The Establishment of an Experimental Beagle Colony," *Journal of the Animal Technicians Association* 14, no. 2 (1963): 75. The colony itself was founded in 1958, but Reinert and Smith note that beagles were not collected for it until 1959.

37. G. K. A. Smith and L. P. Scammell, "Congenital Abnormalities Occurring in a Beagle Breeding Colony," *Laboratory Animals* 2, no. 1 (April 1, 1968): 84, https://doi.org/10.1258/002367768781035403. The failure to understand older dogs due to the colony's short existence is noted in R. F. Davies and H. Reinert, "Arteriosclerosis in the Young Dog," *Journal of Atherosclerosis Research* 5, no. 2 (March 4, 1965): 187, https://doi.org/10.1016/S0368-1319(65)80060-6.

38. Douglas H. Appleton, *The Beagle Handbook*, Dog Lover's Library (London: Nicholson and Watson, 1959); Douglas H. Appleton and Carol Appleton, "The Production of the Research Beagle," *Journal of the Institute of Animal Technicians* 18, no. 3 (September 1967): 124–30.

39. An example study is R. Heywood, "Drug-Induced Retinopathies in the Beagle Dog," *British Veterinary Journal* 130, no. 6 (1974): 564–69.

40. "Humane Group Assails US Housing of Dogs," *Washington Post*, May 6, 1961, C1.

41. North American Newspaper Alliance, "Dog Lovers Criticize FDA over Locked-Up Beagles," *Hays Daily News*, February 28, 1960, 3.

42. Stefan Timmermans and Valerie Leiter, "The Redemption of Thalidomide: Standardizing the Risk of Birth Defects," *Social Studies of Science* 30, no. 1 (February 1, 2000): 44, https://doi.org/10.1177/030631200030001002.

43. Bustad's speech was published as Leo K. Bustad, "The Experimental Subject—a Choice, Not an Echo," *Perspectives in Biology and Medicine* 14, no. 1 (Autumn 1970): 4.

44. Quoted in Marjorie Hunter, "Drug Curbs, Scoffed at in '60, Are Now Being Sped in Congress," *New York Times*, August 12, 1962, 57.

45. Scott H. Podolsky, *The Antibiotic Era: Reform, Resistance, and the Pursuit of a Rational Therapeutics* (Baltimore: Johns Hopkins University Press, 2014), 73; Jeremy A. Greene and Scott H. Podolsky, "Reform, Regulation, and Pharmaceuticals—the Kefauver–Harris Amendments at 50," *New England Journal of Medicine* 367, no. 16 (October 18, 2012): 50, https://doi.org/10.1056/NEJMp1210007.

46. Hunter, "Drug Curbs," in which Taussig is quoted.

47. Laura E. Bothwell et al., "Assessing the Gold Standard—Lessons from the History of RCTs," *New England Journal of Medicine* 374 (June 2016): 2175. See also Wen-Hua Kuo, "Japan and Taiwan in the Wake of Bio-globalization: Drugs, Race and Standards" (PhD diss., Massachusetts Institute of Technology, 2005).

48. Arnold J. Lehman et al., "Procedures for the Appraisal of the Toxicity of Chemicals in Foods," *Food Drug Cosmetic Law Quarterly*, September 1949, 412–34.

49. Arnold J. Lehman et al., "Procedures for the Appraisal of the Toxicity of Chemicals in Foods, Drugs and Cosmetics," *Food, Drug, Cosmetic Law Journal* 10, no. 10 (October 1955): 709. The procedure description mentions beagles twice but references no other breeds explicitly, a clear suggestion of the FDA's preference.

50. Arnold J. Lehman, "Newer Trends in the Laboratory Evaluation of the Safety of Drugs," *Bulletin of the Parenteral Drug Association* 13 (1959): 3.

51. Victor R. Berliner, "US Food and Drug Administration Requirements for Toxicity Testing of Contraceptive Products," Meeting on Pharmacological Models to Assess Toxicity and Side Effects of Fertility Regulating Agents, Geneva, Switzerland, 1973, 243.

52. Berliner, 244.

53. Morton Mintz and Tim O'Brien, "The Guinea Pigs: Nobody Knows If Drugs Tested on Humans Will Cause Cancer," *Washington Post*, November 22, 1969, A2.

54. Morton Mintz, "FDA Requests Prosecution of Merck and Co.," *Washington Post*, April 16, 1968, A8.

55. Berliner, "US Food and Drug Administration Requirements," 244; Marion J. Finkel and Victor R. Berliner, "The Extrapolation of Experimental Findings (Animal to Man): The Dilemma of the Systemically Administered Contraceptives," 62nd Annual Meeting of the International Academy of Pathology, Washington, DC, 1973, 13.

56. Harold M. Schmeck Jr., "2 Birth Pills Discontinued after Tests Made on Dogs," *New York Times*, October 24, 1970, 1, 23.

57. G. R. Daniel, "Chlormadinone Contraceptive Withdrawn," *British Medical Journal* 1, no. 5691 (January 31, 1970): 303.

58. Schmeck, "2 Birth Pills Discontinued after Tests Made on Dogs," 23.

59. Debates over Depo-Provera would prove far more intense and wide ranging. See, for background, William Green, *Contraceptive Risk: The FDA, Depo-Provera, and the Politics of Experimental Medicine* (New York: New York University Press, 2017), 46–47, http://muse.jhu.edu/book/56545.

60. Finkel quoted in Morton Mintz, "Contraceptive a Carcinogen?," *Washington Post*, May 6, 1973, K8. Marion Finkel is not the Miriam Finkel from chapter 2.

61. Morton Mintz, "Ribicoff Hits FDA for Laxity in Human Testing of Drug Safety," *Washington Post*, September 24, 1973, A8.

62. The conference presentation was published in Robert Hill and Kenneth Dumas, "The Use of Dogs for Studies of Toxicity of Contraceptive Hormones," in *Pharmacological Models in Contraceptive Development: Animal Toxicity and Side-Effects in Man*, ed. M. H. Briggs and E. Diczfalusy (Copenhagen: Bogtrykkeriet Forum, 1974), 83.

63. Finkel and Berliner, "Extrapolation of Experimental Findings," 18.

64. Jane E. Brody, "Contraception Study Slows as Costly Testing Curbs Industry's Prospects of Profits," *New York Times*, March 5, 1975, 21.

65. Michael Briggs, "The Beagle Dog and Contraceptive Steroids," *Life Sciences* 21, no. 3 (August 1, 1977): 275–84, https://doi.org/10/cpvfwx.

66. Radiation studies required a baseline understanding of "normal" tumor tendencies. Although numerous studies were done and multiple large reports written on the subject, the topic remained in dispute for decades.

67. The quote appears in a discussion following the Hill and Dumas paper. See Hill and Dumas, "Use of Dogs for Studies of Toxicity of Contraceptive Hormones," 88.

68. M. Fathy El Etreby and Klaus-Jürgen Gräf, "Effect of Contraceptive Steroids on Mammary Gland of Beagle Dog and Its Relevance to Human Carcinogenicity," in "Pharmacology and Therapeutics," special issue, *Pharmacological Methods in Toxicology* 5, no. 1 (January 1, 1979): 369–402, https://doi.org/10/fr826t.

69. Green, *Contraceptive Risk*, 75–76. Activist Gena Corea later argued that the focus on breast cancers in beagles obscured the threat of uterine cancer posed by Depo-Provera, although subsequent studies have suggested that the drug may lower risks of uterine and other cancers. See Gena Corea, "The Depo-Provera Weapon," in *Birth Control and Controlling Birth: Women-Centered Perspectives*, ed. Helen B. Holmes, Betty B. Hoskins, and Michael Gross (Clifton, NJ: Humana, 1980), 112–13; Carolyn Westhoff, "Depot-Medroxyprogesterone Acetate Injection (Depo-Provera®): A Highly Effective Contraceptive Option with Proven Long-Term Safety," *Contraception* 68, no. 2 (August 1, 2003): 75–87, https://doi.org/10.1016/S0010-7824(03)00136-7.

70. Ryan Noah Shapiro, "Bodies at War: National Security in American Controversies over Animal and Human Experimentation from WWI to the War on Terror" (PhD diss., Massachusetts Institute of Technology, 2018), 404. Shapiro covers the "Battle of the Beagles" in extensive detail and shows convincingly how earlier efforts to paint animal protestors as un-American or dangerous failed amid widespread antimilitarism and government distrust.

71. Associated Press, "Committee Defends Tests on Beagles," *(Washington Court House, OH) Record-Herald*, June 19, 1974.

72. A. C. Jacobs and K. P. Hatfield, "History of Chronic Toxicity and Animal Carcinogenicity Studies for Pharmaceuticals," *Veterinary Pathology* 50, no. 2 (March 1, 2013): 328, https://doi.org/10.1177/0300985812450727.

73. America's Cancer Chemotherapy National Service Center, an early, federally funded institute to investigate and test experimental cancer treatments integrated with the National Cancer Institute, also made beagles a default testing animal during the early 1960s.

74. Carpentier is little remembered today, but along with an extended military career, he helped found the zoo in Meknes, Morocco. See Louis Nicol, "Réception de M. C.-J. Carpentier," *Bulletin de l'Académie vétérinaire de France* 113, no. 8 (1960): 433–37; J. Ladrat, "Éloge funèbre du Vétérinaire-Lieutenant-Colonel Carpentier," *Bulletin de l'Académie vétérinaire de France* 130, no. 2 (1977): 239–40.

75. C.-J. Carpentier, "Le chien de laboratoire," *Bulletin de l'Académie vétérinaire de France* 111, no. 7 (1958): 359.

76. See, for instance, M. Dor and L. Brull, "Consommation d'oxygéne et production d'acide carbonique du rein de chien normal transplanté," *Archives internationales de physiologie* 50, no. 2 (1940): 244–56; M. Adant, "Quelques effets de l'injection intraveineuse de polyvinylpyrrolidone au chien normal ou hépatectomisé," *Archives internationales de physiologie* 62, no. 1 (January 1, 1954): 145–46, https://doi.org/10.3109/13813455409145387.

77. Carpentier, "Le chien de laboratoire," 368.

78. Carpentier, 363.

79. Eric Baratay and Philippe Delisle, eds., *Milou, Idéfix et cie: Le chien en BD* (Paris: Éditions Karthala, 2012), 88.

80. Leon Fradley Whitney, "The Dog," in *The Care and Breeding of Laboratory Animals*, ed. Edmond J. Farris (New York: John Wiley and Sons, 1950), 198. Whitney wrote a popular book on making money with dogs, but selling to laboratories is mentioned as a less desirable pathway, indicating his own challenges. See Leon Fradley Whitney, *Dollars in Dogs* (Orange, CT: Practical Science, 1956). Charles Whitney, Leon's grandson, told me that playing with beagles was among his first memories and that the family kept hundreds of dogs for Leon's work.

81. Carpentier, "Le chien de laboratoire," 365–66.

82. Carpentier, 367.

83. The proceedings are reported in E. Despret, "Élevage collectif du chien," in *Symposium international sur l'avenir des animaux de laboratoires* (Paris: Vigot Frères, 1967), 214.

84. Jean-Pierre Vaissaire and Pierre Escuret, "Le beagle et son élevage: I. Caractéristiques générales," *Expérimentation animale* 2, no. 1 (1969): 25.

85. Jean-Pierre Vaissaire, *Le chien, animal de laboratoire* (Belgium: Editions Vigot Frères, 1972), 176.

86. Pierre Goret, "Review: *Le chien, animal de laboratoire*," *Bulletin de l'Académie vétérinaire de France* 126, no. 9 (1973): 361–62.

87. "See Nicolas Aspiotis, "Le chien animal de laboratoire," *Recueil de médecine vétérinaire* 136, no. 1 (January 1960): 109.

88. Vaissaire and Escuret, "Le beagle et son élevage: I. Caractéristiques générales," 26.

89. The survey is reported in Vaissaire, *Le chien*, 12. The CNRS plans are noted in Michel Sabourdy, *L'animal de laboratoire dans la recherche biologique et médicale* (Paris: Presses Universitaires de France, 1967), vi. Sabourdy explicitly references the growth of pharmacodynamics and the pharmaceutical industry. Although French companies were slow to aggressively target the American market, influence was certainly felt by midcentury. See Viviane Quirke, "Targeting the American Market for Medicines, ca. 1950s–1970s: ICI and Rhône-Poulenc Compared," *Bulletin of the History of Medicine* 88, no. 4 (2014): 654–96.

90. Arthur Daemmrich and Georg Krücken, "Risk versus Risk: Decision-Making Dilemmas of Drug Regulation in the United States and Germany," *Science as Culture* 9, no. 4 (December 1, 2000): 505–34, https://doi.org/10.1080/713695270.

91. Siegfried Kamphans, Wilhelm Schumacher, and Herbert Strasser, "Über Haltung und Zucht von Versuchshunden," *Zeitschrift für Versuchstierk* 3 (1963): 41. The company Farbwerke vorm. Meister Lucius & Brüning AG shortened its name to Farbwerke Hoechst AG before merging into IG Farben in 1925. After World War II, Hoechst became independent once more as IG Farben was dissolved by the Allied Powers. Hoechst then proceeded to eat up smaller competitors, including Behringwerke. "AG" is short for "Aktiengesellschaft."

92. Siegfried Kamphans, "Über unsere Beagle-Hunde," *Die blauen Hefte für den Tierarzt* 3, no. 4 (1964): 31.

93. Hunting with dogs retained an ambiguous valence in Germany. Adolf Hitler's government, with the exception of avid hunter Hermann Göring, sought to curtail the practice on animal welfare grounds, which produced a mixed response. The country's dog fanciers were also more likely to raise prized national breeds such as the German shepherd, popular domestically and abroad despite a lingering identification with the Nazi racial project: Arnold Arluke and Boria Sax, "Understanding Nazi Animal Protection and the Holocaust," *Anthrozoös* 5, no. 1 (1992): 8. See also Thorsten Gieser, "Hunting Wild Animals in Germany: Conflicts between Wildlife Management and 'Traditional' Practices of Hege," in *Managing the Return of the Wild: Human Encounters with Wolves in Europe*, ed. Michaela Fenske and Bernhard Tschofen (New York: Routledge, 2020).

94. Kamphans, "Über unsere Beagle-Hunde," 31.

95. Kamphans, 31. See also Herbert Strasser and Wilhelm Schumacher, "Breeding Dogs for Experimental Purposes: II. Assessment of 8-Year Breeding Records for Two Beagle Strains," *Journal of Small Animal Practice* 9 (1968): 605.

96. H. Oettel and H. Frohberg, "Zum nachweis teratogener Wirkung im Tierversuch," in *Use and Abuse of Drugs*, Fourth International Congress of the International Federation for Hygiene and Preventive Medicine (Vienna: Verlag der Wiener Medizinischen Akademie Wien I, 1965), 343. Baden Aniline and Soda Factory, now BASF, is one of the world's largest chemical producers.

97. J. Grauwiler, "Das normale Elektrokardiogramm des Beagle-Hundes," *Naunyn-Schmiedebergs Archiv für Pharmakologie* 266, no. 4 (August 1970): 337.

98. Wilhelm Schumacher and Herbert Strasser, "Breeding Dogs for Experimental Purposes: I. Breeding Installations, Breeding Processes and Maintenance Conditions," *Journal of Small Animal Practice* 9 (1968): 597.

99. Schumacher and Strasser, 597.

100. Maki Umemura, *The Japanese Pharmaceutical Industry: Its Evolution and Current Challenges* (New York: Routledge, 2011), 13.

101. Umemura, 19.

102. Central Institute for Experimental Animals, “History Museum of CIEA,” n.d., https://www.ciea.or.jp/en/about/history.html, accessed March 2024. On April 1, 2024, CIEA changed its name to CIEM: Central Institute for Experimental Medicine and Life Science.

103. Michiko Nomura, “米国における実験動物飼育管理の現況” (Current status of laboratory animal breeding and management in the United States*), *実験動物彙報* (Bulletin of the experimental animals) 2, no. 4 (1953): 38–44. I have followed J-STAGE, where possible, for journal names and article titles, including rendering “実験” as “experimental” (although “laboratory” is often appropriate). Titles denoted with * are my translations.

104. In a reflective piece in 1954, Koji Ando noted that Britain’s laboratory animal system, set up in the aftermath of the war, offered a path forward for Japan. See Koji Ando, “回顧と今後の方針” (Retrospective and future plans*), *実験動物彙報* (Bulletin of the experimental animals) 3, no. 5 (September 1954): 52–53. The International Committee was announced in *Nature* in February 1957, although discussions began in 1956. See William Lane-Petter, “The International Committee on Laboratory Animals,” *Nature* 179, no. 4553 (February 1, 1957): 240–41, https://doi.org/10.1038/179240a0. For historical background, see Tone Druglitrø and Robert G. W. Kirk in “Building Transnational Bodies: Norway and the International Development of Laboratory Animal Science, ca. 1956–1980,” *Science in Context* 27, no. 2 (2014): 337–38. A letter to Koji Ando about the new initiative was published in *Experimental Animals* later in 1957, encouraging a survey of Japanese laboratory animal production. See Koji Ando, “実験動物に関する国際協力” (International Cooperation on Experimental Animals*), *実験動物* (Experimental animals) 6, no. 3 (1957): 79.

105. Koji Ando, “ユネスコ国際会議の内容並びζ　各国実験動物の現” (Summary of the UNESCO international conference and present state of experimental animals in each country*), *実験動物* (Experimental animals) 8, no. 2 (1959): 41.

106. Junichi Kaneko, “米国の実験動物，とくにその生産を見学して” (Observing experimental animals in the United States, especially their production*), *実験動物* (Experimental animals) 11, no. 3 (1962): 99.

107. Kaneko, 101.

108. Kaneko, 101–2.

109. Koji Ando, “実験動物としてのイヌ　：　わが国の現状と対策” (Dogs as experimental animals: Current situation and measures in Japan*), *実験動物* (Experimental animals) 10, nos. 4–6 (1961): 126. In one example, the cardiovascular effects of varying potassium diets were tested on beagles at the Iwate Medical College in Morioka in 1964. See Ichio Ono, Tokuo Hukuoka, and Isao Onodera, “The Effects of Varying Dietary Potassium on the Electrocardiogram and Blood Electrolytes in Young Dogs,” *Japanese Heart Journal* 5, no. 3 (1964): 272–86.

110. Tatsuji Nomura and Teruhisa Noguchi, “SPF 動物から Gnotobiotesへ” (From SPF animals to gnotobiotes*), *Farumashia* 2, no. 3 (1966): 188–89.

111. A note appears in Rikio Niki, Yoshio Takagaki, and Masanobu Fukui, “The Occurrence of Thyroiditis in Laboratory Dogs: A Study on 124 Beagles,” *Keio Journal of Medicine* 21, no. 3–4 (1972): 171. On Chugai’s drugs, see Heidi Wrightsman and Christopher Kasic, “Chugai Pharmaceutical Co., Ltd.,” in *International Directory of Company Histories*, ed. Karen Hill (New York: St. James, 2013), 109–14.

112. The estimate appears in Experimental Animal Survey Committee, "実験につかわれた動物種ならびにその数" (Species and numbers of animals used in experiments: Results of survey in 1970), *実験動物* (Experimental animals) 22, no. 4 (1973): 307–40. Some of these dogs were likely mixed- or non-beagle breeds, but the survey notes that beagles were the only breed singled out by researchers.

113. The trip is discussed in Chuhei Yamauchi, "米 国における実験用 イヌの状況" (The status of experimental dogs in the United States*), *実験動物* (Experimental animals) 20, no. 3 (1971): 173–77.

114. See, for instance, Nobuo Watanabe, "Commercialized Beagles," a section of "Proceedings of the 19th Symposia on Laboratory Animals, Tokyo," *実験動物* (Experimental animals) 22, no. 1 (1973): 57–115.

115. Kazushige Sakai et al., "Pharmacological Features of Peripheral Vascular Beds of Beagles," *Japanese Journal of Pharmacology* 24 (1974): 659. Other facilities had begun to breed dogs as well. See, for instance, Katsumoto Ueda et al., "Spontaneous *Brucella canis* Infection in Beagles: Bacteriological and Serological Studies," *Japanese Journal of Veterinary Science* 36, no. 5 (1974): 381–89.

116. See John Paul Scott, "Symposium: The Use of Purebred Dogs in Research, Introduction," *Proceedings of the Animal Care Panel* 12, no. 4 (September 1962): 149. A corresponding mimeographed document produced at Jackson in 1960, referred to as "Directory of research workers using purebred dogs and institutions maintaining purebred dog colonies," now appears lost.

117. See Bodil Schmidt-Nielsen, "Choice of Experimental Animals for Research," *Federation Proceedings* 20, no. 4 (December 1961): 902. Trained initially in dentistry, Schmidt-Nielsen was the youngest daughter of August Krogh and worked closely with her husband, the physiologist Knut Schmidt-Nielsen, until they separated in 1962. She later became the first woman to lead the American Physiological Society. See Michel Anctil, *Animal as Machine: The Quest to Understand How Animals Work and Adapt* (Montreal: McGill-Queen's University Press, 2022), 133.

118. Nicolaas Adriaan van der Velden, "De Beagle als laboratorium Hond," *Biotechniek* 7 (1968): 124.

119. Karen A. Rader, *Making Mice: Standardizing Animals for American Biomedical Research, 1900–1955* (Princeton, NJ: Princeton University Press, 2004), 172.

120. Stanley Saxe, personal communication with author, 2019.

121. Paul Daubigne, *Le beagle* (Vesoul, France: Crépin-Leblond, 1953), 14.

122. R. R. Brunk, "Standard Values in the Beagle Dog: Haematology and Clinical Chemistry," *Food and Cosmetics Toxicology* 7 (1969): 141–48.

123. Despret, "Élevage collectif du chien," 215.

124. Vaissaire, *Le chien*, 21; Sakai et al., "Pharmacological Features," 660.

125. Hans Hurni, "L'élevage du chien de laboratoire," in *Symposium international sur l'avenir des animaux de laboratoires*, 205.

126. See E. E. Galal et al., "Invited Comments from Representatives of Other National Drug Regulatory Agencies," in Briggs and Diczfalusy, *Pharmacological Models in Contraceptive Development*, 255.

127. A study by Srivastava et al. in 2000 calls the beagle model "conventional" but difficult to acquire for "economic" reasons. See Pratima Srivastava et al., "A Simple and Rapid Evaluation of Methemoglobin Toxicity of 8-Aminoquinolines and Related Compounds," *Ecotoxicology and Environmental Safety* 45, no. 3 (March 1, 2000): 236–39, https://doi.org/10.1006/eesa.1999.1868.

128. Alexander Jordan, "FDA Requirements for Nonclinical Testing of Contraceptive Steroids," *Contraception* 46 (1992): 499–509. See also Jacobs and Hatfield, "History of Chronic Toxicity and Animal Carcinogenicity Studies for Pharmaceuticals," 328.

129. Environmental Protection Agency, "Toxic Substances Control Act (TSCA): Health Effects Testing Guidelines," *Federal Register* 40, no. 798 (1989): 3320.

130. "To: Faculty Members Using Dogs and Cats," n.d., box 39, folder 20, "Animal Costs 1980 June 27–1988 January 27," A. Clifford Barger Papers, Center for the History of Medicine in the Francis A. Countway Library, Harvard University.

131. On the history of pound seizure and the emergence of "standard" dogs, see Brad Bolman, "In the Animal House: Rabies, Labor, and Salvage Dogs in Jim Crow Birmingham" (in review). See also Shapiro, "Bodies at War," 549.

132. "Young Farmers at State Fair," *Daily Messenger*, August 27, 1935, 3.

133. "Ferrets," *Democrat and Chronicle*, December 27, 1941, 22; Gilman Marshall advertisement, *Field and Stream* 49, no. 8 (December 1944): 118.

134. See Michael Bresalier and Michael Worboys, "'Saving the Lives of Our Dogs': The Development of Canine Distemper Vaccine in Interwar Britain," *British Journal for the History of Science* 47, no. 2 (June 2014): 321.

135. Bill Beeney, "Raising Ferrets Big Business," *Democrat and Chronicle*, March 7, 1960, 15.

136. Norman J. Pyle, "Use of Ferrets in Laboratory Work and Research Investigations," *American Journal of Public Health* 30 (July 1940): 788.

137. Early twentieth-century newspapers contain a multitude of stories and advertisements about them. For one of many examples, see "Ferret Ranch Located Near Heart of City," *Dubuque Times Journal*, September 14, 1922, 3.

138. Gilman Marshall advertisement, *Science* 109 (June 24, 1949): 17. He would, however, donate a supposedly "bloodthirsty pair" to Rochester's Seneca Park Zoo in 1955. See "Bloodthirsty Pair Checked in at Zoo," *Democrat and Chronicle*, January 1955, 15.

139. Beeney, "Raising Ferrets Big Business," 15.

140. The self-published pamphlet remains scarce in library collections but appears, for instance, in "Publications," *ILAR News* 9, no. 1 (October 1965): 9.

141. Lynn Steinberg, "Ferret Fanciers," *Seattle Post-Intelligencer*, April 26, 1993, C1.

142. Quoted in Peter Lovenheim, "The Puppy Farm," *Plain Dealer Magazine*, June 10, 1979, 21.

143. "Dogs—Pets—Supplies 96," *Post-Standard*, February 20, 1964, sec. Classifieds, 35.

144. "Area Beagles to Be Used in Research," *Post-Standard*, November 4, 1964, 6.

145. As noted above, many documents concerning tobacco research are available at the University of California, San Francisco, tobacco database; here see "Research Contracts Executed during July, August and September, 1966" (National Institute of Health, Supply Management Branch, 1967), Tobacco Institute Records, University of California, San Francisco, tobacco database, https://www.industrydocuments.ucsf.edu/docs/kfjm0027.

146. Gilman Marshall advertisement, *Federation Proceedings* 25, no. 2 (April 1966).

147. "Area Beagles to Be Used in Research," 6.

148. Sarah Franklin, "Dolly: A New Form of Transgenic Breedwealth," *Environmental Values* 6 (1997): 432.

149. Terry Culvin, "Army Plans to Buy 350 Beagle Puppies for Use in Edgewood Arsenal Chemical Experiments," *Cumberland News*, June 5, 1975, 6.

150. Lovenheim, "Puppy Farm," 22.

151. Edna Goldberg, "Price on Head of Test Puppy Now Tops $100, Army Finds," *Sun*, June 5, 1975, C16.

152. Marshall Research Animals, Inc., Marshall advertisement, *Laboratory Animal Science*, February 1978, 27, HathiTrust.

153. Donna J. Haraway, *When Species Meet* (Minneapolis: University of Minnesota Press, 2008), 65.

154. Lovenheim, "Puppy Farm," 21.

155. Marshall Research Animals, Inc., advertisement, 27.

156. Marshall Research Animals, Inc., "The Marshall Beagle," *Laboratory Animal Science*, 1981, HathiTrust, n.p.

157. My analysis here benefited from a conversation with Arnold Arluke.

158. Marshall Farms, "Building a Better Beagle," *Laboratory Animal Science*, 1990, n.p., HathiTrust.

159. Marshall Farms, n.p.

160. Gene Palmer, "Shipment of Dogs Prompts Charges," *Post-Standard*, January 16, 1992, A1.

161. Gene Palmer, "Puppy Breeder Avoids $220,000 Fine," *Syracuse Herald-Journal*, January 17, 1992, B5.

162. Marshall Farms, "A Global Standard," *Lab Animal*, June 1998, n.p., HathiTrust.

163. "Trademarks," *Official Gazette of the United States Patent and Trademark Office* 1238, no. 1 (September 5, 2000): 351.

164. Noah Phillips, "Get Your Research Dogs Here," *Isthmus*, March 5, 2015, 15-17.

165. Utopia/Marshall were accused of importing dogs into France without paying customs fees in 2014. See "Case C-40/14: Request for a Preliminary Ruling from the Cour de Cassation (France) Lodged on 27 January 2014," Court of Justice of the European Union, January 27, 2014, https://op.europa.eu/en/publication-detail/-/publication/727cebb6-be2a-11e3-86f9-01aa75ed71a1.

166. Meredith Wadman, "Lab-Animal Flights Squeezed," *Nature* 489 (2012): 344–45, https://doi.org/10.1038/489344a.

167. Meredith Wadman, "Research Dogs Shipped to India under Airline's Radar," *Nature News* (blog), November 21, 2012, http://blogs.nature.com/news/2012/11/research-dogs-shipped-to-india-under-airlines-radar.html.

168. "Marshall to Sell Green Hill Lab-Dog Breeding Facility; Says Italy Too Restrictive on Scientific Use of Animals," *ANSA English Media Service*, November 22, 2016, Factiva.

169. "Humane Society of the United States Undercover Investigation Shows Plight of Dogs in a Laboratory Being Dosed with Pesticides and Drugs," *Global Newswire*, March 12, 2019.

170. "Biomolecular Research and Development—Laboratory Animals," *Medical and Healthcare Marketplace Guide* (Philadelphia: Dorland's Biomedical Publications, 1999).

171. Karl Johan Öbrink, "A Short Survey of the Scientific Work of Torsten Teorell," *Upsala Journal of Medical Sciences* 100, no. 1 (1995): 11.

172. Martin Henriksson Holmdahl, "Torsten Teorell, the Teacher and Researcher: Address of Welcome," *Upsala Journal of Medical Sciences* 100, no. 1 (1995): 6.

173. Upon arrival that September, Teorell learned that Cannon was headed to China, so the vast majority of his fellowship was spent studying the permeability of cell membranes and "the relations between the nervous system and internal secretion" with Osterhout. "Teorell, Dr. E. Torsten A.," n.d., RG 10.2, series 9 (Medical and Natural Sciences), "Sweden," Rockefeller Archive Center.

174. "Teorell, Dr. E. Torsten A.," n.d., RG 10.2, series 9 (Medical and Natural Sciences), "Sweden," Rockefeller Archive Center.

175. Teorell returned to Sweden aboard the *Gripsholm* in February 1936. The money was dispersed through his supervisor, Gunnar Blix.

176. Torsten Teorell, "An Attempt to Formulate a Quantitative Theory of Membrane Permeability," *Proceedings of the Society for Experimental Biology and Medicine* 33, no. 2 (1935): 282–85.

177. Torsten Teorell, "Kinetics of Distribution of Substances Administered to the Body: I. The Extravascular (*) Modes of Administration," *Archives internationales de pharmacodynamie et de thérapie* 57 (1937): 205.

178. Teorell, 206, emphasis in the original.

179. Teorell, 210. Teorell termed the two principal "driving forces" moving particles across these boundaries the "concentration gradient" (the tendency for concentrations of particles to equalize) and the "electrical potential gradient" (causing electrically charged particles such as ions or colloid particles to be unequally distributed). To simplify, he proposed ignoring the latter in favor of former (Teorell, 207–8). Because any measure would have to be relative, he proposed using percentage "amounts" and generalized "time units" to build up equations for the "Typical Case" of drug administration (Teorell, 216). The term "depot fat" emerged, in the mid-1800s, seemingly in parallel with the rapid construction of railroads and their storage sites, although further historical work on the subject would be illuminating.

180. Teorell, 223.

181. Walther Gehlen, "Wirkungsstärke intravenös aerabreichter Arzneimittel als Zeitfunktion: Ein Beitrag zur mathematischen Behandlung pharmakolischer Probleme," *Archiv für experimentelle Pathologie und Pharmakologie* 171, no. 1 (December 1933): 541.

182. Emilio Beccari, "Distribuzione dei farmaci nell'organismo: Teorie e controlli sperimentali biochimici e farmacologici," *Archives internationales de pharmacodynamie et de thérapie* 58 (1938): 453.

183. Öbrink, "Short Survey of the Scientific Work of Torsten Teorell," 18.

184. "Teorell, Dr. E. Torsten A.," n.d., RG 10.2, series 9 (Medical and Natural Sciences), "Sweden," Rockefeller Archive Center.

185. Friedrich Hartmut Dost, *Der Blutspiegel: Kinetik der Konzentrationsabläufe in der Kreislaufflüssigkeit* (Leipzig: Georg Thieme Verlag, 1953), 244.

186. Dost's involvement with Nazism appeared to impact his subsequent career minimally. See Ernst Klee, *Das Personenlexikon zum Dritten Reich: Wer war was vor und nach 1945?* (Frankfurt am Main: S. Fischer, 2003), 117.

187. Friedrich Hartmut Dost, "Die Clearance," *Klinische wochenschrift* 27, nos. 15–16 (April 1949): 257–64. The paper was requested by a colleague in pediatrics.

188. Dost, *Der Blutspiegel*, 15.

189. Dost, 7.

190. Dost, 251, 264.

191. Widmark's breakthrough came in a reliable approach for quantitatively analyzing blood alcohol, known as the "Widmark micromethod," which required only a small quantity of blood from the fingertip or earlobe. See Rune Andréasson, "Erik Matteo Prochet Widmark: Widmarks Mikrometod och Trafiknykterhetslagen," *Medical History Society of Southern Sweden*, supp. 5 (1985). Widmark's approach was so simple and successful that, by the mid-1940s, Sweden, Norway, and Germany were all using it to test drunk drivers. See Rune Andreasson and A. Wayne Jones, "The Life and Work of Erik M. P. Widmark," *American Journal of Forensic Medicine and Pathology* 17, no. 3 (1996): 177–90. In 1932, Widmark published a monograph in German summarizing his work, arguing that the body's

metabolism of alcohol could be understood in terms of the "velocity of elimination." The book's impact in Europe was significant, although a translation remained unavailable to American readers until 1981. See Erik M. P. Widmark, *Die Theoretischen Grundlagen und die Praktische Verwendbarkeit der Gerichtlich-Medizinischen Alkoholbestimmung* (Berlin: Urban und Schwarzenberg, 1932); Erik M. P. Widmark, Principles and Applications of Medicolegal Alcohol Determination, trans. R. C. Baselt (Davis, CA: Biomedical Publications, 1981). Teorell read and cited it.

192. Dost, *Der Blutspiegel*, 16. On a broader holistic approach to the organism, see Kurt Goldstein, *The Organism: A Holistic Approach to Biology Derived from Pathological Data in Man* (New York: Zone Books, 1995).

193. Dost, *Der Blutspiegel*, 244.

194. The story appears in Erich Gladtke, "History of Pharmacokinetics," in *Pharmacokinetics: Mathematical and Statistical Approaches to Metabolism and Distribution of Chemicals and Drugs*, ed. A. Pecile and A. Rescigno (New York: Plenum, 1988), 1.

195. H. Gibian, "Introduction," in *Schering Workshop on Pharmacokinetics: Berlin, May 8 and 9, 1969*, ed. Gerhard Raspé, Advances in the Biosciences 5 (Oxford: Pergamon, 1970), 1.

196. See Enno Freerksen, "Introduction," in *Antibiotica et chemotherapia*, ed. Freerksen, vol. 12, *Pharmakokinetik und Arzneimitteldosierung* (Basel, Switzerland: S. Karger, 1964), vi–vii. For "multinational and multidisciplinary," see John Wagner, "Biopharmaceutics: Gastrointestinal Absorption Aspects," in Freerksen, *Antibiotica et chemotherapia*, 12:61.

197. Eino Nelson and Ekkehard Krüger-Thiemer, "Abstract of Pharmacokinetic Models and Symbols: Übersicht der pharmakokinetischen Modelle und Formelzeichen," in Freerksen, *Antibiotica et chemotherapia*, 12: viii.

198. In 1920, A. Richard Bliss Jr. had differentiated pharmacology and pharmacodynamics by noting that "pharmacodynamics"—"so frequently called 'Pharmacology'"—ultimately designated "the science which treats of the action of drugs on the living organisms." See A. Richard Bliss Jr., "Pharmacodynamics in the Schools and Colleges of Pharmacy," *Journal of the American Pharmaceutical Association* 9, no. 4 (April 1920): 380.

199. Wagner, "Biopharmaceutics: Gastrointestinal Absorption Aspects," 12:60.

200. John G. Wagner, "Biopharmaceutics: Absorption Aspects," *Journal of Pharmaceutical Sciences* 50, no. 5 (1961): 360, https://doi.org/10.1002/jps.2600500502.

201. Gibian, "Introduction," 2.

202. Erwin Schrödinger, *What Is Life? The Physical Aspect of the Living Cell* (Cambridge: Cambridge University Press, 1944).

203. James B. Conant, *On Understanding Science* (New York: New American Library, 1951). The lecture reflected Conant's advocacy for and teaching of "history of science" as a field. Among Conant's teaching assistants was Thomas Kuhn. See James Hershberg, *James B. Conant: Harvard to Hiroshima and the Making of the Nuclear Age* (New York: Knopf, 1993), 409–10; Christopher Hamlin, "The Pedagogical Roots of the History of Science: Revisiting the Vision of James Bryant Conant," *Isis* 107, no. 2 (June 2016): 282–308, https://doi.org/10.1086/687217.

204. Wagner, "Biopharmaceutics: Absorption Aspects," 384.

205. He did so two decades apart. See Giorgio Segre, "Mathematical Models in the Study of Drug Kinetics," in *Mathematical Models in Medicine* (Berlin: Springer-Verlag, 1976), 204; Giorgio Segre, "Pharmacokinetics—Compartmental Representation," *Pharmaceutical Therapy* 17 (1982): 111.

206. Gehlen, "Wirkungsstärke intravenös verabreichter Arzneimittel als Zeitfunktion," 542–43.

207. David Kaiser, *Drawing Theories Apart: The Dispersion of Feynman Diagrams in Postwar Physics* (Chicago: University of Chicago Press, 2009), 8–9.

208. Stefan Helmreich, *A Book of Waves* (Durham, NC: Duke University Press, 2023), 36.

209. Torsten Teorell, "Closing Address," in Raspé, *Schering Workshop on Pharmacokinetics*, 265.

210. Friedrich Hartmut Dost, "Opening," in Raspé, *Schering Workshop on Pharmacokinetics*, 3.

211. Dost, 4.

212. Segre, "Pharmacokinetics—Compartmental Representation," 113. If none of the major precursors of pharmacokinetics spoke in terms of "compartments," a more abstract "compartment" metaphor floated around the academic world prior to pharmacokinetic application. On the one hand, railroad and ship "compartments" were common objects of analysis. On the other hand, George Trumbull Ladd, a professor of philosophy at Yale, published *The Philosophy of Mind: An Essay in the Metaphysics of Psychology* in 1895 (London: Longmans, Green), which Edward T. Dixon heralded in *Nature* for its "energetic and effective protest" against the "water-tight compartment theory" of the sciences. See Edward T. Dixon, "Review *The Philosophy of Mind*," *Nature* 52, no. 1338 (1895): 172. In the 1920s, James H. Ryan wrote in defense of a Catholic view of public education that refused the "compartment theory" of man, which considered the total human being "a composite made up of parts which are separable." See James H. Ryan, "The Catholic View," *Religious Education* 22, no. 6 (1927): 583. These compartment theories were different from what Dost, Segre, or Wagner meant, but the longer history reveals interest and resistance to scientific approaches that broke up the human being into abstract, separable "compartments."

213. Richard E. Bellman, "Use of Digital Computers in Defining Pharmacokinetic Parameters," in Raspé, *Schering Workshop on Pharmacokinetics*, 80.

214. William Ogilvy Kermack and Anderson Gray McKendrick, "A Contribution to the Mathematical Theory of Epidemics," *Proceedings of the Royal Society of London A: Mathematical, Physical and Engineering Sciences* 115, no. 772 (August 1, 1927): 700–721. Mathias Grote has shown that the language of "compartments" later took hold in electrochemical studies of cells and membranes in the 1960s and 1970s, while Andrew S. Reynolds has argued that the compartmentalizing view of modern cell biology was connected to a conception of the cell-as-factory. Mathias Grote, "Surfaces of Action: Cells and Membranes in Electrochemistry and the Life Sciences," *Studies in History and Philosophy of Science Part C: Studies in History and Philosophy of Biological and Biomedical Sciences*, 41, no. 3 (September 1, 2010): 183–93, https://doi.org/10.1016/j.shpsc.2010.07.007; Andrew S. Reynolds, *The Third Lens: Metaphor and the Creation of Modern Cell Biology* (Chicago: University of Chicago Press, 2018).

215. Angela N. H. Creager, *Life Atomic: A History of Radioisotopes in Science and Medicine* (Chicago: University of Chicago Press, 2013), 141. See also Jacob Hamblin, *The Wretched Atom: America's Global Gamble with Peaceful Nuclear Technology* (Oxford: Oxford University Press, 2021).

216. Charles Wilcox Sheppard, "The Theory of the Study of Transfers within a Multi-compartment System Using Isotopic Tracers," *Journal of Applied Physics* 19, no. 1 (January 1, 1948): 70–76, https://doi.org/10.1063/1.1697874.

217. David Hawkins, "Some Conditions of Macroeconomic Stability," *Econometrica* 16, no. 4 (1948): 309–22, https://doi.org/10.2307/1909272.

218. Charles Wilcox Sheppard and W. R. Martin, "Cation Exchange between Cells and Plasma of Mammalian Blood: I. Methods and Application to Potassium

Exchange in Human Blood," *Journal of General Physiology* 33, no. 6 (July 20, 1950): 703–22, https://doi.org/10.1085/jgp.33.6.703. See also the discussion of compartments in H. Quastler and F. G. Sherman, "Cell Population Kinetics in the Intestinal Epithelium of the Mouse," *Experimental Cell Research* 17, no. 3 (June 1, 1959): 420, https://doi.org/10.1016/0014-4827(59)90063-1.

219. Sheppard and Martin, "Cation Exchange between Cells and Plasma," 711.

220. Aldo Rescigno, "A Contribution to the Theory of Tracer Methods: Part II," *Biochimica et biophysica acta* 21 (1956): 111–16.

221. The textbook was first published in 1961 and translated into English in 1966 as Aldo Rescigno and Giorgio Segre, *Drug and Tracer Kinetics*, trans. Piero Ariotti (Waltham, MA: Blaisdell, 1966).

222. Friedrich Hartmut Dost, *Grundlagen der Pharmakokinetik* (Stuttgart: Georg Thieme Verlag, 1968), v.

223. I prefer the literal translation of "Wortbildung."

224. Dost, "Opening," 3.

225. The 1962 paper was later published as Wagner, "Biopharmaceutics: Gastrointestinal Absorption Aspects"; see p. 12:61.

226. Wagner, 12:83.

227. Edward R. Garrett, "Basic Concepts and Experimental Methods of Pharmacokinetics," in Raspé, *Schering Workshop on Pharmacokinetics*, 9.

228. Garrett, 10.

229. Torsten Teorell, "General Physico-chemical Aspects of Drug Distribution," in Raspé, *Schering Workshop on Pharmacokinetics*, 23.

230. Dost, "Opening," 4. See, in addition, Peter Galison, "The Ontology of the Enemy: Norbert Wiener and the Cybernetic Vision," *Critical Inquiry* 21, no. 1 (Autumn 1994): 228–66; Ronald R. Kline, *The Cybernetics Moment: Or Why We Call Our Age the Information Age* (Baltimore: Johns Hopkins University Press, 2015).

231. Peter Galison, "The Many Faces of Big Science," in *Big Science: The Growth of Large-Scale Research* (Stanford, CA: Stanford University Press, 1992); Joseph Masco, *The Nuclear Borderlands: The Manhattan Project in Post–Cold War New Mexico* (Princeton, NJ: Princeton University Press, 2006), 56–58.

232. Peter Galison, *Image and Logic: A Material Culture of Microphysics* (Chicago: University of Chicago Press, 1997), 803–4.

233. Sheppard, "Theory of the Study of Transfers," 76.

234. Dost, *Grundlagen der Pharmakokinetik*, v.

235. UPI, "Electronic Dog Tests Drugs to Determine Dosages for Humans," *Paterson Evening News*, February 18, 1961, 3.

236. Edward R. Garrett et al., "Psicofuranine: Kinetics and Mechanisms *In Vivo* with the Application of the Analog Computer," *Journal of Pharmacology and Experimental Therapeutics* 130, no. 1 (1960): 117.

237. Gibian, "Introduction," 2.

238. Edward R. Garrett, Richard L. Johnston, and Elliott J. Collins, "Kinetics of Steroid Effects on Ca47 Dynamics in Dogs with the Analog Computer II," *Journal of Pharmaceutical Sciences* 52, no. 7 (July 1, 1963): 677, https://doi.org/10.1002/jps.2600520715.

239. Horst Röpke and Jürgen Riemann, "The Application of the Analog Computer in Pharmacokinetics," in Raspé, *Schering Workshop on Pharmacokinetics*, 76.

240. Richard Murray, "Dr. Garrett's Electronic Dog," *Pensacola News-Journal*, October 11, 1964, AF9.

241. Edward R. Garrett and Howard J. Lambert, "Analog Computer in Drug Dosage and Formulation Design," *Journal of Pharmaceutical Sciences* 55, no. 6 (June 1, 1966): 626, https://doi.org/10.1002/jps.2600550621.

242. Garrett et al., "Psicofuranine: Kinetics and Mechanisms," 107.

243. See Jean Baudrillard, *Simulacra and Simulation*, trans. Sheila Faria Glaser (Ann Arbor: University of Michigan Press, 1994).

244. UPI, "Electronic Dog Tests Drugs to Determine Dosages for Humans," 3.

245. "Electronic Dogs Used," *Science News-Letter* 83, no. 3 (January 19, 1963): 36.

246. Murray, "Dr. Garrett's Electronic Dog," AF8.

247. On the changing role of computers in the biomedical sciences, see Joseph A. November, *Biomedical Computing: Digitizing Life in the United States* (Baltimore: Johns Hopkins University Press, 2012), https://doi.org/10.1353/book.14634.

248. John G. Wagner, "Use of Computers in Pharmacokinetics," *Clinical Pharmacology and Therapeutics* 8, no. 1, part 2 (1967): 201, https://doi.org/10.1002/cpt196781part2201.

249. To emphasize how computer simulations might change pharmaceutical testing, Wagner offered an example from his own research on an antibiotic developed by Upjohn under the name Lincocin. When administered intramuscularly, Lincocin moved through the body in two ways: two-thirds of the dose appeared to flow quickly into the blood, functioning just as an intravenous injection might; one-third, however, formed a depot that released the drug over ten to twelve hours. A simulation model could enable researchers to test multiple dose quantities and analyze whether it was possible to lower the peak blood level of the drug while raising its serum concentration. The simulations showed alternative, profitable possibilities. Similar digital computer projects (he listed seven useful programs) were, Wagner thought, the future of pharmacokinetics. See Wagner, 213.

250. Röpke and Riemann, "Application of the Analog Computer in Pharmacokinetics," 56.

251. On computers and the Cold War, see Atsushi Akera, *Calculating a Natural World: Scientists, Engineers, and Computers during the Rise of US Cold War Research* (Cambridge, MA: MIT Press, 2006). Programming also required teams of trained technicians, which many researchers and companies had little access to. For the broader transition from, and debates over, analog and digital computers, see James S. Small, *The Analogue Alternative: The Electronic Analogue Computer in Britain and the USA, 1930–1975* (New York: Routledge, 2013). On computerization, see Nathan L. Ensmenger, *The Computer Boys Take Over: Computers, Programmers, and the Politics of Technical Expertise* (Cambridge, MA: MIT Press, 2012), 113.

252. Bellman, "Use of Digital Computers," 79. See also David Sankoff, "The Early Introduction of Dynamic Programming into Computational Biology," in *Advances in the Mathematical Sciences: CRM's 25 Years* (Providence, RI: American Mathematical Society, 1997), 403–13. Familiarity was also a problem: once "a computer was a pretty, young girl who worked with a Frieden calculator all day long," Bellman joked ("Use of Digital Computers," 88). On this, see also Jennifer S. Light, "When Computers Were Women," *Technology and Culture* 40, no. 3 (July 1, 1999): 455–83. Mar Hicks has also shown how female computers played a central role in wartime projects, despite being ejected from the workforce afterward in England; see Mar Hicks, *Programmed Inequality: How Britain Discarded Women Technologists and Lost Its Edge in Computing* (Cambridge, MA: MIT Press, 2017).

253. Bellman, "Use of Digital Computers," 85. "The probability is practically one that if an untutored person tries to do any kind of difficult problem on a computer,"

responded Bellman, "the result would be meaningless." Researchers attuned to analog methods had to learn how to port their models over to a novel work environment and make sense of the endless errors spit back by the devices. Hallam Stevens has argued that the development of digital computers revealed a continuity with earlier practices but also entirely novel data techniques. See Hallam Stevens, "A Feeling for the Algorithm: Working Knowledge and Big Data in Biology," *Osiris* 32, no. 1 (September 1, 2017): 151–74, https://doi.org/10/gg4bjg.

254. Teorell's question appears in the discussion section of Bellman, "Use of Digital Computers," 88.

255. Editors, "Friedrich Hartmut Dost," *Chemotherapy* 22, no. 2 (1976): 73–74.

256. Guenther Hochhaus, Jeffrey S. Barrett, and Hartmut Derendorf, "Evolution of Pharmacokinetics and Pharmacokinetic/Dynamic Correlations during the 20th Century," *Journal of Clinical Pharmacology* 40, no. 9 (2000): 908, https://doi.org/10.1177/00912700022009648.

257. Ronald Alsop, "Scientists Are Turning to Computers in Search for New Chemicals, Drugs," *Wall Street Journal*, August 23, 1983.

258. Teorell, "Closing Address," 265.

259. Harold Boxenbaum, "Interspecies Scaling, Allometry, Physiological Time, and the Ground Plan of Pharmacokinetics," *Journal of Pharmacokinetics and Biopharmaceutics* 10, no. 2 (1982): 203–4.

260. Harold Boxenbaum, "Time Concepts in Physics, Biology, and Pharmacokinetics," *Journal of Pharmaceutical Sciences* 75, no. 11 (November 1986): 1056. Hartmut Rosa notes that "measurements of time, perceptions of time, and time horizons are highly culturally dependent and change with the social structure of societies." See Hartmut Rosa, *Social Acceleration: A New Theory of Modernity*, trans. Jonathan Trejo-Mathys (New York: Columbia University Press, 2015), 5.

261. Alexis Carrel, "Physiological Time," *Science* 74, no. 1929 (December 1931): 619. Carrel was a subject of frequent protest from the antivivisection movement, and his legacy was tarnished by association with Nazism.

262. On Carrel's theory of time, see also Hannah Landecker, *Culturing Life: How Cells Became Technologies* (Cambridge, MA: Harvard University Press, 2010); Rosine Kelz, "Tissue Culture and Biological Time: Alexis Carrel, Henri Bergson and the Plasticity of Living Matter," *BioSocieties* 17, 2021: 442–60.

263. Boxenbaum, "Interspecies Scaling," 202.

264. Boxenbaum, 204.

265. Peter Galison, "Einstein's Clocks: The Place of Time," *Critical Inquiry* 26 (Winter 2000): 360–61.

266. Boxenbaum, "Time Concepts in Physics, Biology, and Pharmacokinetics," 1057.

267. Boxenbaum, 1058.

268. Kenneth B. Bischoff, "Physiological Pharmacokinetics," *Bulletin of Mathematical Biology* 48, no. 3/4 (1986): 309.

269. Wagner, "Biopharmaceutics: Absorption Aspects."

270. The 1972 paper was published as R. T. Williams, "Interspecies Scaling," in *Pharmacology and Pharmacokinetics*, ed. Torsten Teorell, Robert L. Dedrick, and Peter G. Condliffe, Fogarty International Center Proceedings 20 (New York: Plenum, 1974); see p. 105. One example of the complexity of cross-species comparison concerns aminotriazole, an herbicide contaminating the 1959 Thanksgiving cranberry harvest, which FDA tests found to be carcinogenic in rats. The finding devastated farmers' profits and ultimately

pushed Ocean Spray into producing cranberry juice to alleviate the company's holiday dependence, but dog studies initially found no danger from ingesting the chemical. See Michael Tortorello, "The Great Cranberry Scare of 1959," *New Yorker*, November 24, 2015, https://www.newyorker.com/tech/annals-of-technology/the-great-cranberry-scare.

271. Robert L. Dedrick, "Animal Scale-Up," *Journal of Pharmacokinetics and Biopharmaceutics* 1, no. 5 (1973): 435.

272. Dedrick, 436.

273. Dedrick, 458.

274. Alan Poole and George B. Leslie, *A Practical Approach to Toxicological Investigations* (Cambridge: Cambridge University Press, 1989), 39.

275. See, for instance, Renpei Nagashima, Gerhard Levy, and Nathan Back, "Comparative Pharmacokinetics of Coumarin Anticoagulants II: Pharmacokinetics of Bishydroxycoumarin Elimination in the Rat, Guinea Pig, Dog, and Rhesus Monkey," *Journal of Pharmaceutical Sciences* 57, no. 1 (January 1968): 68–71; Anthony R. DiSanto and John G. Wagner, "Pharmacokinetics of Highly Ionized Drugs II: Methylene Blue—Absorption, Metabolism, and Excretion in Man and Dog after Oral Administration," *Journal of Pharmaceutical Sciences* 61, no. 7 (July 1972): 1086–90; Karl Olof Borg et al., "Pharmacokinetic Studies of Metoprolol-(3H) in the Rat and the Dog," *Acta Pharmologica et toxicologica* 36, supp. 5 (1975): 104–15.

276. Dumit, *Drugs for Life*, 8.

277. T. G. Hiebert, "Panel Discussion," in Raspé, *Schering Workshop on Pharmacokinetics*, 257.

Chapter 4

1. Quoted in Sally O'Reilly, *The Ambivalents* (New York: Cabinet Books, 2017), 21.

2. Allan Brandt, *The Cigarette Century: The Rise, Fall, and Deadly Persistence of the Product That Defined America* (New York: Basic Books, 2009); Robert N. Proctor, *Golden Holocaust: Origins of the Cigarette Catastrophe and the Case for Abolition* (Berkeley: University of California Press, 2012); Sarah Milov, *The Cigarette: A Political History* (Cambridge, MA: Harvard University Press, 2019). One of the first presentations of Auerbach's beagle findings came at a news conference for the 1966 AMA meeting in Chicago. See American Medical Association, "Schedule of News Conferences," Chicago, 1966, 2, Tobacco Institute Records, https://www.industrydocuments.ucsf.edu/docs/txvm0027. In 1970, further findings from the beagle research appeared in major newspapers across the United States.

3. "Smoking and Health: Report of the Advisory Committee to the Surgeon General of the Public Health Service" (Washington, DC: Public Health Service, US Department of Health, Education, and Welfare, 1964).

4. Auerbach's story contrasts in important ways with that of Ernst L. Wynder, who received extensive support from the tobacco industry as part of efforts to court his research. See, for instance, N. Fields and S. Chapman, "Chasing Ernst L. Wynder: 40 Years of Philip Morris' Efforts to Influence a Leading Scientist," *Journal of Epidemiology and Community Health* 57, no. 8 (August 1, 2003): 571–78, https://doi.org/10.1136/jech.57.8.571.

5. See Peter Galison and Robert N. Proctor, "Agnotology in Action: A Dialogue," in *Science and the Production of Ignorance: When the Quest for Knowledge Is Thwarted*, ed. Janet Kourany and Martin Carrier (Cambridge, MA: MIT Press, 2020), 32.

6. Quoted in "Tobacco Medicare Lawsuit [6]," n.d., 90, WJC-DPC: Records of the Domestic Policy Council (Clinton Administration), ca. 1992–1/20/2001, Bruce Reed's Admin Files, 1993–2001, William J. Clinton Library, Little Rock, AR, https://catalog.archives.gov/id/26413634.

7. Brandt mentions Auerbach twice in *Cigarette Century* (pp. 187, 236), and Proctor emphasizes Auerbach's epidemiological work. Both note the beagle work mostly in passing. For Proctor, the beagle studies appear more as "history of farce" than as "history of science" (personal communication with author). The most detailed attention to Auerbach is in Richard Kluger's *Ashes to Ashes: America's Hundred-Year Cigarette War, the Public Health, and the Unabashed Triumph of Philip Morris* (New York: Vintage Books, 1997), but Kluger's account focuses on the person of Auerbach rather than on the dogs. Kluger's interview notes, preserved at Yale, include information that never made his book, and Kluger also lacked some recently available industry documents that this chapter draws on.

8. Quoted in William Kloepfer Jr., "Untitled" (Tobacco Institute, January 3, 1971), Philip Morris Records, https://www.industrydocuments.ucsf.edu/docs/ftvd0122.

9. Jonathan Miles, "Tobacco Road," *New York Times Book Review*, May 6, 2007, https://www.nytimes.com/2007/05/06/books/review/Miles.t.html.

10. Donna J. Haraway, *Primate Visions: Gender, Race, and Nature in the World of Modern Science* (New York: Routledge, 1989), 243. Unlike "sadism," which suggests a conscious enjoyment of inflicting pain on beagles, most researchers discussed here actively protested the need for dog experiments. *Voyeurism* gets at the curious pleasure, in many descriptions of beagles smoking, of seeing dogs doing something they normally could not, or should not, do. On Harlow, see also Marga Vicedo, "Mothers, Machines, and Morals: Harry Harlow's Work on Primate Love from Lab to Legend," *Journal of the History of the Behavioral Sciences* 45, no. 3 (Summer 2009): 193–218; Marga Vicedo, *The Nature and Nurture of Love: From Imprinting to Attachment in Cold War America* (Chicago: University of Chicago Press, 2013).

11. Lawrence K. Altman, "12 Dogs Develop Lung Cancer in Group of 86 Taught to Smoke," *New York Times*, February 6, 1970, Late City edition, 1. The "Veterans Administration" became the "Department of Veterans Affairs" in the late 1980s. I have used "VA" throughout to refer to both entities.

12. Delos Smith, "Scientists Prove Cigarette-Cancer Cause and Effect," *Columbia Missourian*, February 8, 1970, sec. Science Page, 22; Washington Post Service, "Smoking Dogs May Be Final Cancer Link," *Arizona Republic*, February 6, 1970, 1.

13. Walter Sullivan, "The Beagle Smoked; the Beagle Got Lung Cancer," *New York Times*, February 8, 1970, sec. The Week in Review, E9.

14. See Robert N. Proctor, "The History of the Discovery of the Cigarette–Lung Cancer Link: Evidentiary Traditions, Corporate Denial, Global Toll," *Tobacco Control* 21, no. 2 (March 1, 2012): 87–91, https://doi.org/10.1136/tobaccocontrol-2011-050338.

15. The anecdote was relayed by a contemporary of Auerbach's. For an example discussion of his work ethic, see Elizabeth McFadden, "Pathologist Seeks to Expand Cancer-Smoking Link Study," *Newark Sunday News*, September 25, 1960, 20.

16. Tuberculosis was "the captain of all these men of death," in the words of Puritan preacher John Bunyan. See John Bunyan, *The Life and Death of Mr. Badman*, ed. James F. Forrest and Roger Sharrock (Oxford: Clarendon, 1988). On the history of tuberculosis, see also Helen Bynum, *Spitting Blood: The History of Tuberculosis* (Oxford: Oxford University Press, 2012); Christian W. McMillen, *Discovering Tuberculosis: A Global History,*

1900 to the Present (New Haven, CT: Yale University Press, 2015); Janina Kehr and Flurin Condrau, "Recurring Revolutions? Tuberculosis Treatments in the Era of Antibiotics," in *Therapeutic Revolutions: Pharmaceuticals and Social Change in the Twentieth Century*, ed. Jeremy A. Greene, Flurin Condrau, and Elizabeth Siegel Watkins (Chicago: University of Chicago Press, 2016). The quote on Seaview comes from "'Hospital on Trolleys' for the Tuberculous," *Commercial Tribune*, May 12, 1913, 12.

17. Gordon L. Snider, "Tuberculosis Then and Now: A Personal Perspective on the Last 50 Years," *Annals of Internal Medicine* 126, no. 3 (February 1, 1997): 237.

18. Virginia Cameron and Esmond R. Long, *Tuberculosis Medical Research: National Tuberculosis Association 1904–1955* (New York: National Tuberculosis Association, 1959), 80–81.

19. See Barron H. Lerner, "Once upon a Time, a Plague Was Vanquished," *New York Times*, October 14, 2003, https://www.nytimes.com/2003/10/14/health/once-upon-a-time-a-plague-was-vanquished.html.

20. Edward Pilley, "New Hope in Fight to Control TB Cited," *Birmingham Post-Herald*, October 5, 1954, 1. For a then-contemporary discussion of the gradual decline in TB cases, see Theodore C. Doege, "Tuberculosis Mortality in the United States, 1900 to 1960," *JAMA* 192, no. 12 (June 21, 1965): 1045–48, https://doi.org/10.1001/jama.1965.03080250023005.

21. Lawrence Garfinkel, "Oscar Auerbach, M.D.," *Cancer* 79, no. 10 (May 15, 1997): 1854.

22. Earl Ubell, "Overgrowth Is Found in Smokers' Windpipes," *New York Herald Tribune*, June 3, 1955.

23. *Cigarette Labeling and Advertising* (Washington, DC: Hearing of the Committee on Commerce, US Senate, March 23, 1965), 168.

24. The "E.," for Edward, was nearly always left as an initial.

25. E. Cuyler Hammond and Daniel Horn, "The Relationship between Human Smoking Habits and Death Rates: A Follow-Up Study of 187,766 Men," *Journal of the American Medical Association* 155, no. 15 (August 7, 1954): 1316–28, https://doi.org/10.1001/jama.1954.03690330020006.

26. Cited in Proctor, "History of the Discovery of the Cigarette–Lung Cancer Link," 88.

27. Brandt, *Cigarette Century*, 141.

28. Proctor, *Golden Holocaust*, 152.

29. Proctor, 291.

30. *False and Misleading Advertising (Filter-Tip Cigarettes)* (Washington, DC: Legal and Monetary Affairs Subcommittee of the Committee on Government Operations, US Congress, July 19, 1957).

31. Pat McGrady, *Cigarettes and Health*, Public Affairs Pamphlet, no. 220A (New York: Public Affairs Committee, March 1960), 11.

32. Quoted in *Cigarette Labeling and Advertising*, 169.

33. Alton Ochsner to Oscar Auerbach, January 8, 1960, Alton Ochsner Papers, box 26, folder 1, Historical New Orleans Collection, New Orleans, LA.

34. "A Double Barrel," (*York*) *Gazette and Daily*, November 12, 1963, clippings, Oscar Auerbach Papers, Special Collections in the History of Medicine, George F. Smith Library of the Health Sciences, Rutgers University, Newark, NJ. The claim of disappearing lung growths was a slight exaggeration of Auerbach's findings.

35. Brandt, *Cigarette Century*, 156.

36. Ruth Winter, "Lung Cancer Researcher Gets $60,000 Grant," *Syracuse-Herald-Journal*, May 14, 1964.

37. Robert Goldenstein (Associated Press), "Researchers Tell of Producing Lung Disease in Dogs," *Star-Gazette*, June 27, 1966, 2.

38. Malcolm M. Manber, "Cigarettes Are Seen Leading to 'Nation of Lung Cripples,'" *Newark Evening News*, May 10, 1963, 23, Newark Public Library.

39. In 1930, the institution received a $250,000 donation from Minnie Hayward, the wife of deceased New York theater impresario Al Hayward. See "Jewess Leaves $250,000 to Tuberculosis Sanatorium," *Jewish Daily Bulletin*, November 13, 1930, 3. Rockey received a $10,000 grant.

40. "Seen and Heard: MASH," *Pfizer Spectrum, Journal of the American Medical Association* 152, no. 11 (July 11, 1953): 27–28; Edward Ernest Rockey, "Management of Thoracoabdominal Injuries at a Mobile Army Surgical Hospital Level in Korea," *American Journal of Surgery* 85, no. 6 (June 1, 1953): 738–46, https://doi.org/10.1016/0002-9610(53)90560-1. Descriptions of Rockey build on notes from a public presentation in Janet C. Brown, "Report on Panel Discussion of 'The Tobacco Cancer Problem' Held under the Auspices of the Society of Medical Jurisprudence at the New York Academy of Medicine, Monday, May 9, 1955," May 9, 1955, Tobacco Products Liability Project Collection, https://www.industrydocuments.ucsf.edu/docs/tmdy0050.

41. Edward Ernest Rockey et al., "The Effect of Tobacco Tar on the Bronchial Mucosa of Dogs," *Cancer* 11 (June 1958): 466.

42. Rockey et al., 466.

43. J. A. Hernandez et al., "Pulmonary Parenchymal Defects in Dogs following Prolonged Cigarette Smoke Exposure," *American Review of Respiratory Disease* 93, no. 1 (January 1, 1966): 78.

44. Ernst L. Wynder, Evarts A. Graham, and Adele B. Croninger, "Experimental Production of Carcinoma with Cigarette Tar," *Cancer Research* 13, no. 12 (December 1953): 855–64. Brandt explores the response to Wynder and Graham's animal studies in Brandt, *Cigarette Century*, 160–62.

45. See Edward Ernest Rockey, Edgar Mayer, and Israel Rappaport, "Tracheal Fenestration: Experimental Aspect," *Diseases of the Chest* 30, no. 2 (August 1, 1956): 224–28, https://doi.org/10.1378/chest.30.2.224; Edward Ernest Rockey, "Surgical Technic of Tracheal Fenestration," *American Journal of Surgery* 94 (September 1957): 486–89.

46. Rockey, Mayer, and Rappaport, "Tracheal Fenestration," 227. Direct application of tobacco tar to the bronchial mucosa, Rockey later noted, "was made possible by the development of the technique of tracheal fenestration." Rockey et al., "Effect of Tobacco Tar on the Bronchial Mucosa of Dogs," 466.

47. Edward Ernest Rockey to Robert C. Hockett, July 24, 1956, Council for Tobacco Research Records, https://www.industrydocuments.ucsf.edu/docs/xkjp0216.

48. Robert C. Hockett to Clarence Cook Little et al., "Study by Drs. Edgar Mayer and Ernest E. Rockey," June 14, 1956, Council for Tobacco Research Records, https://www.industrydocuments.ucsf.edu/docs/nkjp0216. Little's prominent support for the industry is detailed further in Brandt, *Cigarette Century*, 175–76. In a demonstration of the interconnectedness of all things, Little received a letter in 1965 supporting the CTR grant application of Ivan Parfentjev from his old friend Leon Whitney. See Leon Fradley

Whitney to Clarence Cook Little, May 27, 1965, Council for Tobacco Research Records, https://www.industrydocuments.ucsf.edu/docs/qqdv0217.

49. Hockett to Little et al., June 14, 1956.

50. Robert C. Hockett to Paul Kotin, June 5, 1956, Council for Tobacco Research Records, https://www.industrydocuments.ucsf.edu/docs/pgvw0216.

51. Edward Ernest Rockey, "Application for Research Grant: A Study of the Effect of Tobacco Tar on the Bronchial Mucosa of Dogs," July 2, 1956, 2, Council for Tobacco Research Records, https://www.industrydocuments.ucsf.edu/docs/jkjp0216.

52. W. T. Hoyt to Edward Ernest Rockey, September 4, 1956, Council for Tobacco Research Records, https://www.industrydocuments.ucsf.edu/docs/gkjp0216.

53. Edward Ernest Rockey to Robert C. Hockett, September 24, 1956, Council for Tobacco Research Records, https://www.industrydocuments.ucsf.edu/docs/zjjp0216.

54. Edward Ernest Rockey et al., "The Effect of Cigarette Smoke Condensate on the Bronchial Mucosa of Dogs," *Cancer* 15, no. 6 (December 1962): 1100; see also "Dogs to Undergo Cancer Tar Test," *New York World-Telegram and Sun*, February 26, 1958; "$968,105 Is Granted for Cancer Studies," *New York Times*, February 26, 1958.

55. Rockey et al., "Effect of Cigarette Smoke Condensate on the Bronchial Mucosa of Dogs," 1100–1101.

56. E. Cotchin, "Some Tumours of Dogs and Cats of Comparative Veterinary and Human Interest," *Veterinary Record* 71, no. 45 (December 26, 1959): 1046.

57. Robert N. DuPuis to Paul Kotin, February 4, 1958, Philip Morris Records, https://www.industrydocuments.ucsf.edu/docs/zmgm0109.

58. Rockey et al., "Effect of Tobacco Tar on the Bronchial Mucosa of Dogs."

59. Robert C. Hockett to Edward Ernest Rockey, June 2, 1958, Council for Tobacco Research Records, https://www.industrydocuments.ucsf.edu/docs/qjjp0216.

60. See Leonard S. Zahn to Carl Thompson, "Dr. Edward Rockey," September 16, 1959, Council for Tobacco Research Records, https://www.industrydocuments.ucsf.edu/docs/yjjp0216.

61. Rockey et al., "Effect of Cigarette Smoke Condensate on the Bronchial Mucosa of Dogs," 1114.

62. United Press International, "Link Tobacco with Cancer, Heart Disease," *Elwood Call-Leader*, June 26, 1962, 1.

63. Rockey et al., "Effect of Cigarette Smoke Condensate on the Bronchial Mucosa of Dogs," 1115.

64. Carl C. Seltzer to Robert C. Hockett, January 5, 1963, Council for Tobacco Research Records, https://www.industrydocuments.ucsf.edu/docs/gggh0216. It is now well documented that Seltzer received hundreds of thousands of dollars from industry sources during his career. See, for instance, "List of Special Projects Administered by the Council for Tobacco Research—USA, Inc.," September 27, 1994, Depositions and Trial Testimony (DATTA), https://www.industrydocuments.ucsf.edu/docs/kghp0034.

65. Israel Rappaport to Robert C. Hockett, January 29, 1963, Tobacco Institute Records, https://www.industrydocuments.ucsf.edu/docs/zhkh0046.

66. T. D. Day, "Comments on the Paper: 'The Effect of Cigarette Smoke Condensate on the Bronchial Mucosa of Dogs,'" March 20, 1963, British American Tobacco Records, https://www.industrydocuments.ucsf.edu/docs/lxpf0203.

67. Edward Ernest Rockey et al., "Experimental Study on Effect of Cigarette Smoke on Bronchial Mucosa," *Journal of the American Medical Association* 182, no. 11 (December 15,

1962): 1098. The large number of dogs required off-site keeping, so space was borrowed from the Coler Memorial Hospital and Home on Welfare (today's Roosevelt) Island. Commissioners Morris A. Jacobs and Roy I. Trussell of the Department of Hospitals aided the group in acquiring these facilities without which "it would not have been possible to conduct such a large-scale study."

68. Rockey et al., 1097. See also R. M. Mulligan, *Neoplasms of the Dog* (Baltimore: Williams and Wilkins, 1949).

69. Rockey et al., "Experimental Study on Effect of Cigarette Smoke on Bronchial Mucosa," 1097.

70. Russell W. Weller to Robert C. Hockett, "Confidential Comments on the Paper Entitled 'Experimental Study on the Effect of Cigarette Smoke Condensate on Bronchial Mucosa' by E. E. Rockey, et al., *JAMA*, December 15, 1962," February 18, 1963, Council for Tobacco Research Records, https://www.industrydocuments.ucsf.edu/docs/lqgh0216.

71. Dale L. Tipton and T. Timothy Crocker, "Duration of Bronchial Squamous Metaplasia Produced in Dogs by Cigarette Smoke," *JNCI: Journal of the National Cancer Institute* 33, no. 3 (1964): 488.

72. "Topics for Discussion," ca. 1968, 7, Ness Motley Law Firm Documents, https://www.industrydocuments.ucsf.edu/docs/ytlc0040. The preserved version of the memo lacks an author, but it likely came, at least in part, from Little, given the cited studies.

73. Day, "Comments on the Paper."

74. Edward Ernest Rockey, "Evolution of Cigarette Smoking Technics in Dogs," *International Surgery* 46, no. 5 (1966): 409.

75. Rockey, 412.

76. A report on Rockey's public comments from 1970 includes his note that the meetings took place on July 30 and August 1, but his published articles on the techniques give July 20. See "Questions following Papers Given by E. C. Hammond and Oscar Auerbach, American Medical Association Convention June 24, 1970," n.d., 2, box 45, Tobacco Institute Records, https://www.industrydocuments.ucsf.edu/docs/zhfx0117.

77. Richard Kluger, "Oscar Auerbach Notes," n.d., 2, box 17, folder 186, Richard Kluger Papers (MS 1443), Manuscripts and Archives, Yale University Library.

78. "Questions following Papers Given," 2.

79. Joan Lee Faust, "He Picks His Apples on the 21st Floor," *New York Times*, March 17, 1977, sec. Gardening, 86.

80. William G. Cahan and David Kirman, "An Effective System and Procedure for Cigarette Smoking by Dogs," *Journal of Surgical Research* 8, no. 12 (December 1968): 567.

81. Oscar Auerbach, *Report of Research Project: Production of Carcinoma of Tracheobronchial Tree in Dogs* (East Orange, NJ: Veterans Administration Hospital, May 25, 1964), Brown and Williamson Records, https://www.industrydocuments.ucsf.edu/docs/ssbh0138.

82. "Of Beagles, Smoking, and Cancer," *Roche Medical Image and Commentary*, September 1970, 11.

83. In one recent exception, Nicole C. Nelson and Kaitlin Stack Whitney offer parallel lab narratives from the perspective of experimental mice. An older mouse with a startlingly nuanced understanding of his workplace warns a new arrival: "You may be nervous on the day of your first experiment, but don't worry—remember, you've been bred for this!" See Nicole C. Nelson and Kaitlin Stack Whitney, "Becoming a Research Rodent," in *Living with Animals: Bonds across Species*, ed. Natalie Porter and Ilana Gershon

(Ithaca, NY: Cornell University Press, 2018), 202. The account, which is focused more on describing lab protocols, remains an outlier. See also Otniel E. Dror, "The Affect of Experiment: The Turn to Emotions in Anglo-American Physiology, 1900–1940," *Isis* 90, no. 2 (June 1, 1999): 236, https://doi.org/10/ctwt5t; Otniel E. Dror, "Afterword: A Reflection on Feelings and the History of Science," *Isis* 100, no. 4 (December 1, 2009): 848–51, https://doi.org/10/cj4dr5.

84. To take one example, PETA campaign coordinator Jay M. Kelly criticized R. J. Reynolds, in 2000, for continuing to "subject animals to horrific pain and suffering" through animal experimentation, even as the group "takes no position for or against the use of tobacco products." See Jay M. Kelly to Andy Schindler, July 18, 2000, R. J. Reynolds Records, https://www.industrydocuments.ucsf.edu/docs/knfn0099. (The R. J. Reynolds company name has appeared with different punctuation and spacing over the years; I am using the spelling shown here throughout the book, including in citations.) Notable in the latter category is the work of Larry Carbone; see, for instance, Larry Carbone, *What Animals Want: Expertise and Advocacy in Laboratory Animal Welfare Policy* (Oxford: Oxford University Press, 2004); Larry Carbone and Jamie Austin, "Pain and Laboratory Animals: Publication Practices for Better Data Reproducibility and Better Animal Welfare," *PLoS One* 11, no. 5 (May 12, 2016): e0155001, https://doi.org/10.1371/journal.pone.0155001.

85. Another way of analyzing resistance and subjectivity might be to study conditions in which devices or laboratory setups are designed to minimize resistance in the first place, such as Seligman's learned helplessness studies. See Mariam Motamedi Fraser, *Dog Politics: Species Stories and the Animal Sciences* (Manchester: Manchester University Press, 2024), 181.

86. Michel Callon, "Some Elements of a Sociology of Translation: Domestication of the Scallops and the Fishermen of St Brieuc Bay," *Sociological Review* 32, no. 1 supp. (May 1, 1984): 196–233, https://doi.org/10.1111/j.1467-954X.1984.tb00113.x.

87. Bruno Latour, *Politics of Nature: How to Bring the Sciences into Democracy*, trans. Catherine Porter (Cambridge, MA: Harvard University Press, 2004), 76.

88. Such episodes are explored in Marian Stamp Dawkins, *Through Our Eyes Only? The Search for Animal Consciousness* (Oxford: Oxford University Press, 1998), 124–25. See also Brad Bolman, "Parroting Patriots: Interspecies Trauma and Becoming-Well-Together," *Medical Humanities* 45, no. 3 (2019): 305–12.

89. Amanda Rees, "Animal Agents? Historiography, Theory and the History of Science in the Anthropocene," *British Journal for the History of Science: Themes* 2 (2017): 2–3.

90. In one of many examples of the resistance-as-agency theme, the theme for the *Journal for Critical Animal Studies* in December 2018 was "nonhuman animals as agents of resistance." See Amber E. George, "Issue Introduction: Nonhuman Animals Agents of Resistance," *Journal for Critical Animal Studies* 15, no. 6 (December 2018): 3.

91. Justyna Włodarczyk, *Genealogy of Obedience: Reading North American Dog Training Literature, 1850s–2000s* (Leiden: Brill, 2018), x.

92. Chris Pearson calls for looking beyond the "animal as resistor" frame, noting that "obstruction is only one type of nonhuman agency." Other creatures, he emphasizes, "also enable and allow human activities." See Chris Pearson, "Beyond 'Resistance': Rethinking Nonhuman Agency for a 'More-Than-Human' World," *European Review of History / Revue Européenne d'histoire* 22, no. 5 (September 3, 2015): 710, https://doi.org/10.1080/13507486.2015.1070122. At the end of a series of questions about nonhuman agency, Rees wonders, "Can agency be identified in cooperation with authority as well as recognized in resistance?" Rees, "Animal Agents?," 2. Pearson additionally

stresses the importance of including nonhuman animals in scholarship on the history of emotions, noting that "we gain a better understanding of how animals have influenced human emotional states and how human emotions, such as fear, disgust, and compassion, have transformed animal lives, thereby highlighting the material as well as the representational dimensions of emotions." Chris Pearson, *Dogopolis: How Dogs and Humans Made Modern New York, London, and Paris* (Chicago: University of Chicago Press, 2021), 191.

93. Cahan and Kirman, "Effective System and Procedure for Cigarette Smoking by Dogs," 569.

94. Rappaport to Hockett, January 29, 1963.

95. The paper and news conference are referenced in American Medical Association, "Schedule of News Conferences," Chicago, 1966, 2, Tobacco Institute Records, https://www.industrydocuments.ucsf.edu/docs/txvm0027. Cahan would fall out of the picture, gradually, after Hammond gave Auerbach a mild ultimatum about working with either himself or Cahan. See Kluger, "Oscar Auerbach Notes," 2.

96. See for instance, Associated Press, "Cigarette Smoking Gives Dogs Disease," *Atlanta Journal*, June 27, 1966, 3.

97. Associated Press, 3.

98. Quoted in Frank E. Carey, *Emphysema: The Battle to Breathe* (Washington, DC: Public Health Service, 1968), 23, https://www.industrydocuments.ucsf.edu/tobacco/docs/nrgy0145.

99. "Sen. Soaper Says," *Spokesman-Review*, September 24, 1966, 4.

100. "Special News Summary on Auerbach's Paper" (Hill and Knowlton, Inc., August 5, 1966), Tobacco Institute Records, https://www.industrydocuments.ucsf.edu/docs/hxmn0146. The place of Hill and Knowlton in this work is covered by Naomi Oreskes and Erik Conway in *Merchants of Doubt: How a Handful of Scientists Obscured the Truth on Issues from Tobacco Smoke to Global Warming* (New York: Bloomsbury, 2010), 10–35.

101. "The Dog Experiment," *Salt Lake City Tribune*, July 30, 1966, sec. Church News Editorial Page, Tobacco Institute Records, https://www.industrydocuments.ucsf.edu/docs/yjmn0146.

102. "Comments on the Work of Oscar Auerbach Published or Presented since 1963," 1967, Philip Morris Records, https://www.industrydocuments.ucsf.edu/docs/ltgh0119.

103. Hiram T. Langston to Edwin J. Jacob, "Re: Work of Oscar Auerbach," July 21, 1966, Ness Motley Law Firm Documents, https://www.industrydocuments.ucsf.edu/docs/hfyw0040. Langston was another researcher whose funding from the industry was revealed only much later. See Stanton A. Glantz et al., eds., *The Cigarette Papers* (Berkeley: University of California Press, 1996), http://ark.cdlib.org/ark:/13030/ft8489p25j/.

104. "Two Appearances by Dr. Oscar Auerbach" (Hill and Knowlton, Inc., November 10, 1966), American Tobacco Records, https://www.industrydocuments.ucsf.edu/docs/lmmf0177.

105. Robert Goldenstein (Associated Press), "Cigarette-Smoking Dogs Aid Emphysema Research," *Terra Haute Tribune*, June 27, 1966, 2.

106. Quoted in Ann Sullivan, "Scientist Claims Smoking Beagles Develop Emphysema Just as Humans," *Sunday Oregonian*, November 12, 1967, Oscar Auerbach Papers, Special Collections in the History of Medicine, George F. Smith Library of the Health Sciences, Rutgers University. Never one to miss a speaking opportunity, Auerbach had discussed the work while visiting his son, Bruce, a junior at Reed College, in November 1967, at a small press conference sponsored by the Oregon Tuberculosis and Health Association.

107. Kluger, "Oscar Auerbach Notes," 3.

108. The first appears above Miriam Kass, "Smoking Problem Also Hounds Dogs," *Houston Post*, December 13, 1969. The second was sent to Oscar Auerbach along with a clipping of DC's *Evening Star* from February 12, 1970. It is preserved with his other clippings in the Oscar Auerbach Papers, Special Collections in the History of Medicine, George F. Smith Library of the Health Sciences, Rutgers University.

109. One early example appears in "Cigarette-Smoking Dog Makes Ideal 'Nurse Maid,'" *Washington Post*, January 21, 1923, 2.

110. "Puffing Pup," *Philadelphia Daily News*, February 9, 1976, 1.

111. William Weiss, "Undermining the Subsidy for Premature Death," *Archives of Environmental Health* 19, no. 2 (August 1969): 231.

112. Oscar Auerbach et al., "Emphysema Produced in Dogs by Cigarette Smoking," *Journal of the American Medical Association* 199, no. 4 (January 23, 1967): 241–46.

113. Oreskes and Conway, *Merchants of Doubt*, 15.

114. *Cigarette Labeling and Advertising*, 210.

115. *Cigarette Labeling and Advertising*, 216–17.

116. On Koch's postulates and the cigarette debates, see Proctor, *Golden Holocaust*, 424, 597.

117. *Cigarette Labeling and Advertising*, 218.

118. Kluger, *Ashes to Ashes*, 350.

119. A representative example is M. Campbell to Harrison A. Williams, October 10, 1967, Harrison A. Williams Jr. Papers, box 524, folder 28, Special Collections and University Archives, Rutgers University, New Brunswick, NJ. Letters were not nearly as voluminous or organized as those concerning the beagle chemical weapons tests a few years later, but Auerbach's research may well have primed people to be critical of subsequent studies.

120. Proctor, *Golden Holocaust*, 191–92.

121. Auerbach et al., "Emphysema Produced in Dogs by Cigarette Smoking," 241–46.

122. Oscar Auerbach et al., "Histologic Changes in Bronchial Tubes of Cigarette-Smoking Dogs," *Cancer* 20 (December 1967): 2055–66; Julio M. Frasca et al., "Electron Microscopic Observations of the Bronchial Epithelium of Dogs: I. Control Dogs," *Experimental and Molecular Pathology* 9, no. 3 (December 1, 1968): 363–79, https://doi.org/10.1016/0014-4800(68)90026-9.

123. "Of Beagles, Smoking, and Cancer," 13.

124. William J. Bair to George V. Allen, July 15, 1966, R. J. Reynolds Records, https://www.industrydocuments.ucsf.edu/docs/xjxl0056. Pacific Northwest Laboratory was renamed the Pacific Northwest National Laboratory in 1995. PNNL is now the common acronym, but PNL was current at the time.

125. Bair to Allen, July 15, 1966.

126. Associated Press, "Lung Cancer, Uranium Mining Link 'Alarming,'" *San Antonio Express*, May 22, 1967.

127. Associated Press, "Uranium Miners Rate High in Lung Cancer Deaths," *Albuquerque Journal*, May 28, 1961, 10A.

128. Technician Vanis Daniels remembered that the explanation for using such a particular breed of dog was "that the closest thing to a humans' physique was the beagle." Vanis Daniels, interview by Laura Arata, video, November 14, 2013, Hanford Oral History Project, Hanford History Project Collections, http://www.hanfordhistory.com/items/show/202. The differences in beagle weights (ranging anywhere from fifteen to forty-seven pounds) correlated well with the wide divergences in human weight, making a fifteen-pound dog roughly equivalent to a 130-pound

man and therefore close to the "standard man" of toxicological research. See Brad Bolman, "Women Radiobiologists and 'Standard Man,'" *Lady Science*, May 17, 2018, https://thenewinquiry.com/blog/women-radiobiologists-and-standard-man/; Maria Rentetzi, "Gendering Things," in *The Gender of Things: How Epistemic and Technological Objects Become Gendered* (London: Routledge, 2024), 4–6.

129. William J. Bair, interview by Robert Bauman, video, August 14, 2013, Hanford Oral History Project, Hanford History Project Collections, http://www.hanfordhistory.com/items/show/9.

130. Bair to Allen, July 15, 1966.

131. John Walsh, "Battelle: New Contractor's Role at AEC Lab Means Diversification for Hanford, Growth for Institute," *Science* 145, no. 3635 (1964): 905–7, https://doi.org/10/c6d2zx.

132. H. J. Kinkel, "Investigations Aimed at the Development of a Method for Individual Inhalation of Measurable Quantities of Fresh Tobacco Smoke by Animals: Part I—Theoretical Studies Based on Preliminary Experiments" (Frankfurt am Main: Battelle-Institut E.V., April 10, 1964), box 902, file b4911, British American Tobacco Records, https://www.industrydocuments.ucsf.edu/docs/lxlb0208.

133. Edward H. DeHart to Frederick Haas et al., July 21, 1966, Ness Motley Law Firm Documents, https://www.industrydocuments.ucsf.edu/docs/rrgc0040.

134. "Proposed Research Program on One Year Study of Dogs Inhaling Cigarette Smoke (BNPI-195)" (Inhalation Toxicology, Pacific Northwest Laboratories, Battelle Memorial Institute, November 1966), Ness Motley Law Firm Documents, https://www.industrydocuments.ucsf.edu/docs/qrgc0040. For the difficulty of getting to New York, see William J. Bair to Robert C. Hockett, August 15, 1966, Council for Tobacco Research Records, https://www.industrydocuments.ucsf.edu/docs/gqkl0216.

135. Daniels, interview by Laura Arata.

136. "Proposed Research Program on One Year Study of Dogs Inhaling Cigarette Smoke (BNPI-195)," 10.

137. David R. Hardy to Philip Grant, April 21, 1967, Lorillard Records, https://www.industrydocuments.ucsf.edu/docs/gndy0104.

138. "Memorandum of Conference on March 23, 1966 at the New Research Facility of the American Tobacco Company" (American Tobacco Company, March 23, 1966), Tobacco Products Liability Project Collection, https://www.industrydocuments.ucsf.edu/docs/ffwy0050.

139. The beagle studies reveal "agnotology," the production of ignorance, in the action. See Robert N. Proctor and Londa Schiebinger, eds., *Agnotology: The Making and Unmaking of Ignorance* (Stanford, CA: Stanford University Press, 2008).

140. The network quote comes from Brandt, *Cigarette Century*, 195.

141. JBL to EDH, February 8, 1967, Ness Motley Law Firm Documents, https://www.industrydocuments.ucsf.edu/docs/gslm0042.

142. Brandt, *Cigarette Century*, 195.

143. Tobacco Institute, Inc., "Special to the Journal of the American Medical Association," May 10, 1967, Tobacco Institute Records, https://www.industrydocuments.ucsf.edu/docs/ghmn0146.

144. The industry received an extensive report on the announcement and subsequent discussion: "'Effects of Cigarette Smoking on Dogs': Reports Read by Dr. E. Cuyler Hammond (Part I) and Dr. Oscar Auerbach (Part II)," February 5, 1970, Philip Morris Records, https://www.industrydocuments.ucsf.edu/docs/txpg0111.

145. "'Effects of Cigarette Smoking on Dogs': Reports Read by Dr. E. Cuyler Hammond (Part I) and Dr. Oscar Auerbach (Part II)," 38.

146. "Backgrounder: 'The Effects of Cigarette Smoking upon Dogs,'" American Cancer Society News Service, February 1970, Lorillard Records, https://www.industrydocuments.ucsf.edu/docs/zqdc0124.

147. "Experiments with Dogs," Radio TV Reports, Inc., February 5, 1970, Tobacco Institute Records, https://www.industrydocuments.ucsf.edu/docs/yjgb0002.

148. Dwight E. Harken to Oscar Auerbach, February 10, 1970, box 81, folder 22, Dwight E. Harken Papers, Center for the History of Medicine in the Francis A. Countway Library, Harvard University.

149. "Questions Following Papers Given," 2–3.

150. Steven M. Spencer, "Last Gasp for Cigarettes?," *Reader's Digest*, April 1970.

151. Brian R. Davis, "The Effects of Cigarette Smoking on Dogs and How These Experiments Relate to the Work at Harrogate," March 23, 1970, British American Tobacco Records, https://www.industrydocuments.ucsf.edu/docs/mspv0208.

152. V. D. Tughan to Managing Director, Gallaher Limited, "Auerbach/Hammond Beagle Experiment," April 3, 1970, American Tobacco Records, https://www.industrydocuments.ucsf.edu/docs/kldg0001.

153. Tughan to Managing Director, Gallaher Limited, 4.

154. See Milov, *Cigarette*, 136–43.

155. UPI, "Tobacco Stock Hit by 'Cancer Scare,'" *Birmingham Post-Herald*, February 6, 1970, 17.

156. Frederick Panzer, "Informational Memorandum" (Tobacco Institute, February 5, 1970), R. J. Reynolds Records, https://www.industrydocuments.ucsf.edu/docs/ffgk0056.

157. Leonard S. Zahn to William Kloepfer Jr., February 6, 1970, Tobacco Institute Records, https://www.industrydocuments.ucsf.edu/docs/fhby0059.

158. Quoted in Carey, *Emphysema*, 24.

159. Frederick Panzer to William Kloepfer Jr., February 6, 1970, Tobacco Institute Records, https://www.industrydocuments.ucsf.edu/docs/fsxl0045.

160. Lincoln P. Brower, "Smoking and Cancer," *New York Times*, February 15, 1970, sec. Letters to the Editor of the Times.

161. In a 1998 deposition for *Washington v. American Tobacco Co.*, legal counsel for the tobacco companies cross-examined witnesses about Brower's article as proof of scientific critique of Auerbach. See "Deposition of John Mercer Pinney, May 13, 1998, *Washington v. American Tobacco Co.*," May 13, 1998, 76–78, Depositions and Trial Testimony (DATTA), https://www.industrydocuments.ucsf.edu/docs/gqkl0001.

162. Brower, "Smoking and Cancer."

163. A letter from Frederick Haas on February 19, 1970, discussing the *Times* letter notes, "I understand Ed Cooke has already seen Professor Brower." In previous years, Ed Cooke had put together a list of useful scientists for future litigation testimony. He commonly met with or made initial contact with these individuals. See Frederick Haas to Donald J. Cohn, February 19, 1970, Liggett and Myers Records, https://www.industrydocuments.ucsf.edu/docs/stpw0014. An industry research librarian had assembled a biography and bibliography for Brower, to better defend his bona fides, by February 18, 1970.

164. Lincoln Pierson Brower, "Sodium Cyclamate and Bladder Carcinoma," *Science* 170, no. 3957 (October 30, 1970): 558–558, https://doi.org/10.1126/science.170.3957.558.

165. Alexander Holtzman, "Comments on the Report of Lung Cancer in Smoking Beagles," February 16, 1970, 6–7, American Tobacco Records, https://www.industrydocuments.ucsf.edu/docs/hfpn0174.

166. Earle C. Clements to Joseph F. Cullman III, February 25, 1970, Philip Morris Records, https://www.industrydocuments.ucsf.edu/docs/fkpk0112.

167. Joseph F. Cullman III to William B. Lewis, February 27, 1970, Council for Tobacco Research Records, https://www.industrydocuments.ucsf.edu/docs/fxcv0215.

168. William Kloepfer Jr., "Memorandum," March 2, 1970, Tobacco Research Records, https://www.industrydocuments.ucsf.edu/docs/nkmd0047.

169. William B. Lewis to Joseph F. Cullman III, March 12, 1970, Philip Morris Records, https://www.industrydocuments.ucsf.edu/docs/kjnk0000.

170. Joseph F. Cullman III to William B. Lewis, March 20, 1970, Tobacco Institute Records, https://www.industrydocuments.ucsf.edu/docs/xjjm0049.

171. William Kloepfer Jr., "Notes on Communications Committee Meeting" (March 17, 1970), Tobacco Institute Records, https://www.industrydocuments.ucsf.edu/docs/hyfx0030. The suggestion about expressing "disappointment" also appears in these meeting notes.

172. John F. Banzhaf III to Dwight E. Harken, May 14, 1970, box 89, folder 25, Dwight E. Harken Papers, Center for the History of Medicine in the Francis A. Countway Library, Harvard University.

173. Bowling quoted in Robert K. Heimann to Robert B. Walker, "Memorandum," April 20, 1970, Tobacco Products Liability Project Collection, https://www.industrydocuments.ucsf.edu/docs/fkvy0050.

174. For example, see Tobacco Institute, "The Tobacco Institute believes," *Chicago Tribune*, May 6, 1970, B28.

175. "The Cigarette-Cancer Dispute," *New York Times*, May 9, 1970, 24.

176. William Kloepfer Jr., "Informational Memorandum" (Tobacco Institute, June 19, 1970), Liggett and Myers Records, https://www.industrydocuments.ucsf.edu/docs/ssph0009.

177. Brian R. Davis, "Report on Visit to the United States of America," June 2, 1967, 11, British American Tobacco Records, https://www.industrydocuments.ucsf.edu/docs/sndb0208.

178. Dwight E. Harken to Oscar Auerbach, March 16, 1970, box 81, folder 22, Dwight E. Harken Papers, Center for the History of Medicine in the Francis A. Countway Library, Harvard University.

179. Dorothy Gordon to Oscar Auerbach, June 11, 1970, Lorillard Records, https://www.industrydocuments.ucsf.edu/docs/fjml0123.

180. Alexander Holtzman to Sheldon C. Sommers, Telegram, 1970, Philip Morris Records, https://www.industrydocuments.ucsf.edu/docs/ghmd0111.

181. Sheldon C. Sommers to Oscar Auerbach, telegram, June 16, 1970, Council for Tobacco Research Records, https://www.industrydocuments.ucsf.edu/docs/rkxh0067.

182. D. G. Felton, "Report on Visit to USA by Dr. F. J. C. Roe," June 5, 1972, American Tobacco Records, https://www.industrydocuments.ucsf.edu/docs/zmmk0204.

183. "Dr. Auerbach's Smoking Beagles," *Tobacco Reporter*, March 1970, 30–31.

184. Horace R. Kornegay to Members of Congress, October 9, 1970, Ness Motley Law Firm Documents, https://www.industrydocuments.ucsf.edu/docs/sjnp0042.

185. "Strategy" (Tobacco Institute, May 1970), Philip Morris Records, https://www.industrydocuments.ucsf.edu/docs/rhgh0119.

186. 1492 Productions, Inc., "Script No. 1 'The Future Smoking Beagle Inquest,'" July 1970, 2, Philip Morris Records, https://www.industrydocuments.ucsf.edu/docs/xkwm0116.

187. 1492 Productions, Inc., "Script No. 2 Proposed for An Audio-Visual Presentation for the Tobacco Institute, Inc.," July 1970, 6, Philip Morris Records, https://www.industrydocuments.ucsf.edu/docs/jkwm0116.

188. William Kloepfer Jr. to Horace R. Kornegay, "The ACS/TI Dispute—Status and Recommendations," June 12, 1970, Council for Tobacco Research Records, https://www.industrydocuments.ucsf.edu/docs/xrbg0215.

189. Alexander Holtzman and Paul D. Smith, "SECRET," March 13, 1970, Philip Morris Records, https://www.industrydocuments.ucsf.edu/docs/hzvp0101. Ackerman was one of many scientists to receive funding from the industry in a covert manner; see Lopez, Edwards, Frank and Co., "Lauterstein and Lauterstein Special Account Number 4, Statement as at February 28, 1971," March 29, 1971, Philip Morris Records, https://www.industrydocuments.ucsf.edu/docs/hlhm0116.

190. Oscar Auerbach to Franz J. Ingelfinger, February 13, 1970, box 17, folder 186, Richard Kluger Papers (MS 1443), Manuscripts and Archives, Yale University Library.

191. Franz J. Ingelfinger to Oscar Auerbach, February 27, 1970, box 17, folder 186, Richard Kluger Papers (MS 1443), Manuscripts and Archives, Yale University Library. "This was much to the sorrow of the editor and staff," Harken later summarized. Dwight E. Harken to Richard Stanton, March 23, 1971, box 89, folder 26, Dwight E. Harken Papers, Center for the History of Medicine in the Francis A. Countway Library, Harvard University. See also "Smoking and Health: Missing Link," *Medical Tribune*, February 1970, Council for Tobacco Research Records, https://www.industrydocuments.ucsf.edu/docs/hjff0215.

192. Thomas Blair Carleton, "To the Editor," *New England Journal of Medicine* 283, no. 4 (July 1970): 207. The "sole" acceptance note appears in the above letter from Kloepfer to Kornegay, "ACS/TI Dispute—Status and Recommendations," June 12, 1970.

193. She jokingly hoped that he would not "be 'hounded' too much by the country's so-called science writers." See M. Therese Southgate to Oscar Auerbach, March 13, 1970, box 17, folder 186, Richard Kluger Papers (MS 1443), Manuscripts and Archives, Yale University Library.

194. Alexander Holtzman, "Auerbach/Hammond Manuscript and JAMA," June 3, 1970, Philip Morris Records, https://www.industrydocuments.ucsf.edu/docs/kqwf0115.

195. James C. Bowling, "Memo to Files," June 24, 1970, Council for Tobacco Research Records, https://www.industrydocuments.ucsf.edu/docs/ggjg0215.

196. "Press Conference, during American Medical Association Convention, June 23, 1970, in Chicago, Featuring Dr. E. Cuyler Hammond and Dr. Oscar Auerbach," transcript, June 23, 1970, 3–4, Philip Morris Records, https://www.industrydocuments.ucsf.edu/docs/fhyv0023.

197. M. Therese Southgate to Oscar Auerbach, July 2, 1970, box 17, folder 186, Richard Kluger Papers (MS 1443), Manuscripts and Archives, Yale University Library.

198. Oscar Auerbach and E. Cuyler Hammond to M. Therese Southgate, July 16, 1970, box 17, folder 186, Richard Kluger Papers (MS 1443), Manuscripts and Archives, Yale University Library.

199. Quoted in I. W. Hughes, "Auerbach Paper," November 19, 1970, British American Tobacco Records, https://www.industrydocuments.ucsf.edu/docs/ylhf0212.

200. The article was eventually published in 1970 in two parts: E. Cuyler Hammond, Oscar Auerbach, et al., "Effects of Cigarette Smoking on Dogs: I. Design of Experiment, Mortality, and Findings in Lung Parenchyma," *Archives of Environmental Health* 21, no. 6 (1970): 740–53 (according to which the article was submitted on July 20 and accepted on

July 29; see p. 740); and Oscar Auerbach et al., "Effects of Cigarette Smoking on Dogs: II. Pulmonary Neoplasms," *Archives of Environmental Health* 21, no. 6 (1970): 754–68.

201. "Beagles, Smoking Article to Be Printed," *Winston-Salem Journal*, September 16, 1970; Horace R. Kornegay to Joseph F. Cullman III, September 18, 1970, Ness Motley Law Firm Documents, https://www.industrydocuments.ucsf.edu/docs/rjnp0042.

202. James Jackson Kilpatrick to Leonard S. Zahn, August 31, 1970, Council for Tobacco Research Records, https://www.industrydocuments.ucsf.edu/docs/gnhg0215.

203. James Jackson Kilpatrick, "What Became of 'Landmark' Cancer Study?," *Washington Evening Star*, September 15, 1970, A11.

204. Frederick Panzer to Leonard S. Zahn, September 17, 1970, Council for Tobacco Research Records, https://www.industrydocuments.ucsf.edu/docs/lzmf0215.

205. See James J. Kilpatrick, "Cigarette Report One Year Later," *(Philadelphia) Evening Bulletin*, February 4, 1971, Council for Tobacco Research Records, https://www.industrydocuments.ucsf.edu/docs/nmhg0215.

206. J. V. Blalock to Addison Y. Yeaman, November 13, 1970, Tobacco Products Liability Project Collection, https://www.industrydocuments.ucsf.edu/docs/rsxn0050.

207. Helmut Wakeham, "Visit of Research Directors to Dr. Gori's Office for Discussion of the Repeat of the Auerbach Smoking Dogs Experiment" (Philip Morris USA, February 23, 1972), Philip Morris Records, https://www.industrydocuments.ucsf.edu/docs/xpjw0181.

208. Oscar Auerbach, "Effects of High and Low Nicotine Cigarettes on Male Beagle Dogs" (East Orange, NJ: Veterans Administration Hospital, December 1978), Philip Morris Records, https://www.industrydocuments.ucsf.edu/docs/kjhc0123. Auerbach would later explain to Richard Kluger that the studies had not "reached their goal" and "high and low nicotine cigarette findings showed very little difference." See Oscar Auerbach to Richard Kluger, January 19, 1989, box 17, folder 188, Richard Kluger Papers (MS 1443), Manuscripts and Archives, Yale University Library.

209. Gio Batta Gori, *An Operational Manual for the Beagle Dog Inhalation Model* (Bethesda, MD: Smoking and Health Program, National Cancer Institute, July 1978), 15.

210. Milov, *Cigarette*, 266–67.

211. John A. Bukowski, Daniel Wartenberg, and Michael Goldschmidt, "Environmental Causes for Sinonasal Cancers in Pet Dogs, and Their Usefulness as Sentinels of Indoor Cancer Risk," *Journal of Toxicology and Environmental Health, Part A* 54, no. 7 (August 1, 1998): 579–91, https://doi.org/10/c4k83z.

212. John S. Reif et al., "Passive Smoking and Canine Lung Cancer Risk," *American Journal of Epidemiology* 135, no. 3 (February 1, 1992): 234–39, https://doi.org/10.1093/oxfordjournals.aje.a116276; John S. Reif, Christa Bruns, and Kimberty S. Lower, "Cancer of the Nasal Cavity and Paranasal Sinuses and Exposure to Environmental Tobacco Smoke in Pet Dogs," *American Journal of Epidemiology* 147, no. 5 (March 1, 1998): 488–92, https://doi.org/10/gg67hn.

213. Kass, "Smoking Problem Also Hounds Dogs."

214. Felton, "Report on Visit to USA by Dr. F. J. C. Roe," 8.

215. On this, see also Brad Bolman, "In the Animal House: Rabies, Labor, and Salvage Dogs in Jim Crow Birmingham," *Historical Studies in the Natural Sciences* 52, no. 5 (November 2022): 589–628.

216. Felton, "Report on Visit to USA by Dr. F. J. C. Roe," 8.

217. Kluger, "Oscar Auerbach Notes," 3.

218. Felton, "Report on Visit to USA by Dr. F. J. C. Roe," 8.

219. This number comes from a comparison of citation counts on Google Scholar.

220. Dave Dryden, "Author Says Industry Put Spotlight On," *Oklahoman*, November 14, 1970.

221. The quote appears in "Raymond Yesner," May 9, 1988, Lorillard Records, https://www.industrydocuments.ucsf.edu/docs/skxp0060.

222. "Premature Puff for Smoking Beagles," *Nature* 230, no. 30 (April 30, 1971): 548.

223. Ernst L. Wynder and Dietrich Hoffmann, "Obituary Listing: Oscar Auerbach, MD," *JAMA* 277, no. 13 (April 2, 1997): 1088, https://doi.org/10/dwzmp3; "The Man Who Helped Indict Smoking," *New York Times*, January 18, 1997, 22.

224. Roy Greenslade, "Mary Beith, the Journalist Who Broke the 'Smoking Beagles' Story," *Guardian*, May 20, 2012, http://www.theguardian.com/media/greenslade/2012/may/20/thepeople-investigative-journalism.

225. "Research," *Tobacco Institute Newsletter*, April 8, 1975, Tobacco Institute Records, https://www.industrydocuments.ucsf.edu/docs/ltgl0004.

226. Patrick Barkham, "The Failed Bomb Plot Has Given Lucozade a PR Problem," *Guardian*, September 8, 2009, https://www.theguardian.com/media/2009/sep/09/lucozade-bombers-brand-terrorism.

227. Robert S. Hirth and Girard Harold Hottendorf, "Lesions Produced by a New Lungworm in Beagle Dogs," *Veterinary Pathology* 10 (1973): 386.

228. Girard Harold Hottendorf and Robert S. Hirth, "Lesions of Spontaneous Subclinical Disease in Beagle Dogs," *Veterinary Pathology* 11 (1974): 240–58.

229. Robert S. Hirth and Girard Harold Hottendorf, "A Spontaneous Pulmonary Disease in Purebred Beagle Dogs Used in Toxicity Studies," in *Abstracts of Papers for the Eight Annual Meeting of the Society of Toxicology* (Williamsburg, VA: Eighth Annual Meeting of the Society of Toxicology, 1969), 639.

230. Hirth and Hottendorf, "Lesions Produced by a New Lungworm in Beagle Dogs," 405.

231. E. Cuyler Hammond, Oscar Auerbach, et al., "Effects of Cigarette Smoking on Dogs: I. Design of Experiment, Mortality, and Findings in Lung Parenchyma," *Archives of Environmental Health* 21, no. 6 (1970): 740–53; Oscar Auerbach et al., "Effects of Cigarette Smoking on Dogs: II. Pulmonary Neoplasms," *Archives of Environmental Health* 21, no. 6 (1970): 754–68.

232. Ivor Wallace Hughes to David R. Hardy, November 20, 1974, Tobacco Products Liability Project Collection, https://www.industrydocuments.ucsf.edu/docs/jngn0050; Ivor Wallace Hughes to William U. Gardner, November 20, 1974, Ness Motley Law Firm Documents, https://www.industrydocuments.ucsf.edu/docs/gpjy0042.

233. Donald K. Hoel to William Kloepfer Jr., December 19, 1974, Tobacco Institute Records, https://www.industrydocuments.ucsf.edu/docs/tmpl0049.

234. Alexander Holtzman, "Lungworm in Beagle Dogs," February 5, 1975, Philip Morris Records, https://www.industrydocuments.ucsf.edu/docs/mzwh0119.

235. Christopher R. E. Coggins, head of Carson Watts Consulting and formerly employed by R. J. Reynolds, would publish multiple reviews critical of Auerbach's work in the early 2000s, for instance. See, e.g., Christopher R. E. Coggins, "A Review of Chronic Inhalation Studies with Mainstream Cigarette Smoke, in Hamsters, Dogs, and Nonhuman Primates," *Toxicologic Pathology* 29, no. 5 (2001): 550–57.

236. On the search for cancer viruses, see Robin Wolfe Scheffler, *A Contagious Cause: The American Hunt for Cancer Viruses and the Rise of Molecular Medicine* (Chicago: University of Chicago Press, 2019).

237. See Arthur W. Burke Jr., "Oscar Auerbach: Report of a Lecture Delivered at the Medical College of Virginia," June 5, 1970, American Tobacco Records, https://www.industrydocuments.ucsf.edu/docs/jrjw0005.

238. See Julio M. Frasca, Oscar Auerbach, et al., "Alveolar Cell Hyperplasia in the Lungs of Smoking Dogs," *Experimental and Molecular Pathology* 21 (1974): 309.

239. Auerbach's report is summarized in R. D. Carpenter to T. S. Osdene, "Tobacco Working Group, September 9 and 10, 1975," September 16, 1975, 5, Ness Motley Law Firm Documents, https://www.industrydocuments.ucsf.edu/docs/jjfc0040.

240. Richard J. Hickey, of the University of Pennsylvania's Wharton School, made the argument in letters to the editor of *JAMA* in the 1980s. See Richard J. Hickey, "In Reply," *JAMA* 255, no. 8 (1986): 1016. He shared reprints of these letters directly to industry contacts. See Richard J. Hickey to Robert F. Gertenbach, February 28, 1986, Council for Tobacco Research Records, https://www.industrydocuments.ucsf.edu/docs/qkhm0041. Hickey received extensive funding from the industry during his career. See Glantz et al., *Cigarette Papers.*

241. "Deposition of Michele Helene Bloch, M.D., Ph.D., October 4, 2001, *United States of America v. Philip Morris USA Inc.*," October 4, 2001, 1109, Depositions and Trial Testimony (DATTA), https://www.industrydocuments.ucsf.edu/docs/zydy0019.

242. Noboru Kagei et al., "Lungworm, *Filaroides hirthi* Infection in Imported Beagle Dogs," *Experimental Animals* 25, no. 3 (1976): 141–48.

243. I. Beveridge et al., "*Filaroides hirthi* in Dogs," *Australian Veterinary Journal* 60 (February 1983): 59.

244. See T. Waner, M. Pirak, and A. Nyska, "Lungworm (*Filaroides hirthi*) Infestation in a Batch of Beagle Dogs: A Case Report and Review of the Literature," *Israel Journal of Veterinary Medicine* 46, no. 3 (1991): 106–9.

245. R. Bahnemann and C. Bauer, "Lungworm Infection in a Beagle Colony: *Filaroides hirthi*, a Common but Not Well-Known Companion," *Experimental and Toxicologic Pathology* 46, no. 1 (March 1, 1994): 55–62, https://doi.org/10.1016/S0940-2993(11)80017-5.

246. Lungworm cases are documented chronologically in L. Crippa, "Lungworm Infection in Laboratory Dogs Reared in Italy," *Parassitologia* 37 (1995): 83–85.

247. Georgi is quoted in "Research on Parasite Found in Lab Dogs Will Aid Scientific Studies," *NIH Record*, September 5, 1979, 5.

248. Andrew Fenelon and Samuel H. Preston, "Estimating Smoking-Attributable Mortality in the United States," *Demography* 49, no. 3 (May 18, 2012): 797–818, https://doi.org/10.1007/s13524-012-0108-x.

249. Samuel H. Preston et al., "Projecting the Effect of Changes in Smoking and Obesity on Future Life Expectancy in the United States," *Demography* 51, no. 1 (November 23, 2013): 27–49, https://doi.org/10.1007/s13524-013-0246-9; Joseph T. Lariscy, Robert A. Hummer, and Richard G. Rogers, "Cigarette Smoking and All-Cause and Cause-Specific Adult Mortality in the United States," *Demography* 55, no. 5 (September 19, 2018): 1855–85, https://doi.org/10.1007/s13524-018-0707-2.

Chapter 5

1. Kathryn Armstrong, "Bobi, the World's Oldest Dog Ever, Dies Aged 31," BBC News, October 23, 2023, https://www.bbc.com/news/world-europe-67194721.

2. Emily Anthes, "Could a Drug Give Your Pet More Dog Years?," *New York Times*, November 28, 2023, sec. Science, https://www.nytimes.com/2023/11/28/science/longevity-drugs-dogs.html.

3. US Food and Drug Administration, "Eligibility Criteria for Expanded Conditional Approval of New Animal Drugs: Guidance for Industry," July 2021, https://www.fda.gov/regulatory-information/search-fda-guidance-documents/cvm-gfi-261-eligibility-criteria-expanded-conditional-approval-new-animal-drugs.

4. Emily Longeretta, "'Beef' Wins Limited Series Emmy, Creator Lee Sung Jin Thanks Fans Who Have 'Reached Out about Their Own Personal Struggles,'" *Variety*, January 15, 2024, https://variety.com/2024/tv/awards/beef-limited-series-emmy-1235869303/.

5. Giovanni Vitale et al., "Role of IGF-1 System in the Modulation of Longevity: Controversies and New Insights from a Centenarians' Perspective," *Frontiers in Endocrinology* 10 (2019), https://www.frontiersin.org/articles/10.3389/fendo.2019.00027.

6. Halioua, quoted in Anthes, "Could a Drug Give Your Pet More Dog Years?"

7. This slogan appeared atop Loyal's webpage, Loyalfordogs.com, when it was accessed in March 2024.

8. This suggested what might be called an "environmentalization" of Alzheimer's. See Etienne S. Benson, *Surroundings: A History of Environments and Environmentalisms* (Chicago: University of Chicago Press, 2020), 11.

9. See Hannah Landecker, "Antibiotic Resistance and the Biology of History," *Body and Society* 22, no. 4 (December 1, 2016): 36, https://doi.org/10.1177/1357034X14561341.

10. While many have analyzed the cultural resonance of the growing market for pet drugs, with the storied "Prozac for dogs" a paradigmatic example, there is minimal history of pet pharma or the intersections between human and pet markets. See Donna J. Haraway, "Value-Added Dogs and Lively Capital," in *Lively Capital: Biotechnologies, Ethics, and Governance in Global Markets* (Durham, NC: Duke University Press, 2012), 93–120; Donna J. Haraway, "For the Love of a Good Dog: Webs of Action in the World of Dog Genetics," in *Genetic Nature/Culture: Anthropology and Science beyond the Two-Culture Divide*, ed. Alan H. Goodman, Deborah Heath, and M. Susan Lindee (Berkeley: University of California Press, 2003); Laurel Braitman, *Animal Madness: Inside Their Minds* (New York: Simon and Schuster, 2015). Controversy about Covid-19 and ivermectin brought some attention to these questions: see Erin Woo, "How Covid Misinformation Created a Run on Animal Medicine," *New York Times*, September 28, 2021, https://www.nytimes.com/2021/09/28/technology/ivermectin-animal-medicine-shortage.html.

11. Kaushik Sunder Rajan's suggestion that biocapitalism involves a dynamic of scientific fact production and innovation finds unique fulfilment here, as scientific experiments were justified by the possibility (and later necessity) of profiting from existing pharmaceutical products that, in turn, generated new avenues for experimentation. See Kaushik Sunder Rajan, *Biocapital: The Constitution of Postgenomic Life* (Durham, NC: Duke University Press, 2006), 135–37.

12. Bill Milgram, personal communication with author, 2019.

13. It followed a line of prior hydrazine-derivative drugs, such as the now defunct pargyline. See M. J. Parnham, "The History of L-deprenyl," in *Inhibitors of Monoamine Oxidase B: Pharmacology and Clinical Use in Neurodegenerative Disorders*, ed. I. Szelenyi (Basel: Birkhauser Verlag, 1993), 238. Miller's contribution is less frequently discussed, but covered in Alastair Dow, *The Deprenyl Story* (Toronto: Stoddart, 1990), 75–76.

14. Quoted in Parnham, History of L-deprenyl," 239. Knoll's first name, József, is typically anglicized.

15. Thomas A. Ban and Fridolin Sulser, eds., *An Oral History of Neuropsychopharmacology: The First Fifty Years, Peer Interviews*, vol. 3, *Neuropharmacology* (Brentwood, TN: American College of Neuropsychopharmacology, 2011).

16. Further background on Knoll appears in Dow, *Deprenyl Story*, 25.

17. Joseph Knoll et al., "Phenylisopropylmethylpropinylamine: HCl (E-250) egy új hatásspektrumú pszichoenergetikum," *Magyar Tudományos Akadémia V. Orvosi Tudományok Osztályának Közleményei* 15 (1964): 231–39.

18. At the time, monoamine oxidase was not yet divided into MAO-A and MAO-B. See Joseph Knoll et al., "Phenylisopropylmethylpropinylamine (E-250): A New Spectrum Psychic Energizer," *Archives internationales de pharmacodynamie et de thérapie* 155, no. 1 (May 1965): 154–64.

19. Francisco Lopez-Munoz and Cecilio Alamo, "Monoaminergic Neurotransmission: The History of the Discovery of Antidepressants from 1950s until Today," *Current Pharmaceutical Design* 15, no. 14 (May 1, 2009): 1563–86, https://doi.org/10.2174/138161209788168001.

20. Ban and Sulser, *An Oral History of Neuropsychopharmacology*. On Hungary's pharmaceutical industry, see Katalin György and Janos Vincze, "Privatization and Innovation in the Pharmaceutical Industry in a Post-socialist Economy," *Rivista internazionale di scienze sociali* 100, no. 3 (September 1992): 403–17; Sandor Kerpel-Fronius, "The Development of Pharmaceutical Medicine in Hungary," *Pharm Med* 27 (2013): 289–95.

21. J. P. Johnston, "Some Observations upon a New Inhibitor of Monoamine Oxidase in Brain Tissue," *Biochemical Pharmacology* 17, no. 7 (July 1, 1968): 1285–97, https://doi.org/10/d3x5s2. In 1969, Knoll's University Medical School was renamed Semmelweis University.

22. Parnham, "History of L-deprenyl," 242. L-deprenyl is one of two enantiomers of the racemic compound deprenyl; R-deprenyl is the other. Both stereoisomers have pharmacological activity, but L-deprenyl has received the lion's share of attention.

23. Parnham, 242.

24. Oleh Hornykiewicz, "L-Dopa," *Journal of Parkinson's Disease* 7, no. s1 (2017): S3–10.

25. Walther Birkmayer et al., "Implications of Combined Treatment with 'Madopar' and L-Deprenil in Parkinson's Disease: A Long-Term Study," *Lancet* 309, no. 8009 (February 26, 1977): 439–43, https://doi.org/10/bgvj64.

26. Ensuing controversies over who was responsible for the discovery would end Birkmayer and Hornykiewicz's friendship. Ironically (given Knoll's own early life), Birkmayer had been an active member of the SS and later sought to play down his Nazi past. See Herwig Czech and Lawrence A. Zeidman, "Walther Birkmayer, Co-describer of L-Dopa, and His Nazi Connections: Victim or Perpetrator?," *Journal of the History of the Neurosciences* 23, no. 2 (April 3, 2014): 160–91, https://doi.org/10.1080/0964704X.2013.865427.

27. Suzanne McGee, "Shulman Builds Thriving Drug Concern, but His Style Alienates Some Investors," *Wall Street Journal*, September 19, 1991, B1.

28. Quoted in Deborah Charles, "Ailing Doctor Turns Self-Help into Million-Dollar Business Venture," *Reuters News*, September 3, 1991.

29. Michael Waldholz, "Medical Odyssey: Drug for Ills of the Old Is Drawing Attention after 30-Year Struggle," *Wall Street Journal*, November 27, 1990, A1.

30. Parnham, "History of L-deprenyl," 246.

31. Patricia Lush, "Morty Turns Adversity into a New Enterprise," *Globe and Mail*, February 10, 1988, B1.

32. Dianne Maley, "Cure Gives Hope," *Brandon (Manitoba) Sun*, November 20, 1989.

33. "Somerset Eldepryl June 5 Approval Clears Way For Mylan/Bolar Acquisition," *Pink Sheet*, June 12, 1989, https://pink.citeline.com/ps015763/somerset-eldepryl-june-5-approval-clears-way-for-mylanbolar-acquisition.

34. Joseph Knoll, J. Dallo, and T. T. Yen, "Striatal Dopamine, Sexual Activity and Lifespan. Longevity of Rats Treated with (−)DEPRENYL," *Life Sciences* 45, no. 6 (January 1, 1989): 525–31, https://doi.org/10/bx6zj9.

35. Quote in McGee, "Shulman Builds Thriving Drug Concern, but His Style Alienates Some Investors," *Wall Street Journal*, September 19, 1991, B1.

36. Norton William Milgram et al., "Maintenance on L-deprenyl Prolongs Life in Aged Male Rats," *Life Sciences* 47, no. 5 (1990): 415–20.

37. Milgram, personal communication with author, 2019.

38. "Pet Food Maker to Consider Use of Deprenyl Drug," *Globe and Mail*, November 2, 1988, B10; Allan Robinson, "Deprenyl Plans to Market Eldepryl and Other Drugs," *Globe and Mail*, April 1, 1989, B4.

39. Jennifer Greer, "Animal Health Company to Sell Stock to Public to Fund Research," *Kansas City Star*, December 27, 1990, 16.

40. "Deprenyl Animal Health, Inc. Files Preliminary Prospectus," *Canada News-Wire*, December 24, 1990, Factiva.

41. Thomas N. Cochran, "Offerings in the Offing," *Barron's*, January 7, 1991, Factiva.

42. Along with those at the University of Toronto, the company would go on to work with scientists at the University of Missouri–Kansas City, UC Davis, the University of Iowa, and Purdue University. See Stephen Buttry, "Stock Offering Has Raised $6 Million for Research Firm," *Kansas City Star*, March 26, 1991, 33.

43. Cristiana Franco, "Dogs and Humans in Ancient Greece and Rome: Towards a Definition of Extended Appropriate Interaction," in *Dog's Best Friend? Rethinking Canid-Human Relations*, ed. John Sorenson and Atsuko Matsuoka (Montreal: McGill-Queen's University Press, 2019).

44. Quoted in Paul Friedland, "Friends for Dinner: The Early Modern Roots of Modern Carnivorous Sensibilities," *History of the Present* 1, no. 1 (2011): 91, https://doi.org/10.5406/historypresent.1.1.0084. See also Raimond Gaita, *The Philosopher's Dog* (New York: Routledge, 2003).

45. See Emma Townshend, *Darwin's Dogs: How Darwin's Pets Helped Form a World-Changing Theory of Evolution* (London: Frances Lincoln, 2009).

46. Charles Darwin, *The Descent of Man, and Selection in Relation to Sex*, 2nd ed., 1874, Project Gutenberg, http://www.gutenberg.org/files/2300/2300-h/2300-h.htm.

47. Michael Worboys, *Doggy People: The Victorians Who Made the Modern Dog* (Manchester: Manchester University Press, 2023), 175.

48. Conwy Lloyd Morgan quoted in Alan Costall, "How Lloyd Morgan's Canon Backfired," *Journal of the History of the Behavioral Sciences* 29, no. 2 (April 1993): 116, https://doi.org/10/b7v34w.

49. Conwy Lloyd Morgan, *Animal Behaviour* (London: Edward Arnold, 1900), 144. For contextualization of Morgan's thinking to the emergence of dog breeds, see Michael Worboys, Julie-Marie Strange, and Neil Pemberton, *The Invention of the Modern Dog: Breed and Blood in Victorian Britain* (Baltimore: Johns Hopkins University Press, 2018), 168.

50. See Gregory Radick, "Morgan's Canon, Garner's Phonograph, and the Evolutionary Origins of Language and Reason," *British Journal for the History of Science* 33, no. 1 (March 2000): 4, https://doi.org/10/cm2xt9. See also Elliott Sober, "Comparative Psychology Meets Evolutionary Biology: Morgan's Canon and Cladistic Parsimony," in *Thinking with Animals: New Perspectives on Anthropomorphism*, ed. Lorraine Daston and Gregg Mitman (New York: Columbia University Press, 2005). Fraser critiques Morgan for turning "Tony the individual into an ambassador for his species"; see Mariam

Motamedi Fraser, *Dog Politics: Species Stories and the Animal Sciences* (Manchester: Manchester University Press, 2024), 105–9.

51. Quoted in Harriet Ritvo, "Animal Consciousness: Some Historical Perspective," *American Zoologist* 40, no. 6 (December 1, 2000): 849–50, https://doi.org/10.1093/icb/40.6.847.

52. Radick, "Morgan's Canon," 21.

53. On Thorndike's puzzle boxes, see Robert Boakes, *From Darwin to Behaviourism: Psychology and the Minds of Animals* (Cambridge: Cambridge University Press, 1984), 68–72.

54. Edward Lee Thorndike, *Animal Intelligence: Experimental Studies* (New Brunswick, NJ: Transaction, 1970), 110.

55. Edmund Ramsden, "A Neurotic Dog's Life: Experimental Psychiatry and the Conditional Reflex Method in the Work of W. Horsley Gantt," *Isis* 109, no. 2 (June 1, 2018): 276–301, https://doi.org/10/ggr7gk.

56. John Paul Scott, John Paul Scott Oral History, August 22, 1986, 12, Oral History Collection, Jackson Laboratory Historical Archives, https://mouseion.jax.org/oral_history/3/.

57. John Langworthy Fuller, "Purebred Dogs in Psychological and Mental Health Research," *Proceedings of the Animal Care Panel* 12, no. 4 (September 1962): 183.

58. Alexandra Rutherford, *Beyond the Box: B. F. Skinner's Technology of Behaviour from Laboratory to Life, 1950s–1970s* (Toronto: Toronto University Press, 2009); Erica N. Feuerbacher and C. D. L. Wynne, "A History of Dogs as Subjects in North American Experimental Psychological Research," *Comparative Cognition and Behavior Reviews* 6 (2011): 54, https://doi.org/10.3819/ccbr.2011.60001. The project was led by Walter Jetter of the Boston University School of Medicine.

59. Walter W. Jetter, Ogden R. Lindsley, and Frederick J. Wohlwill, *The Effects of X-Irradiation on Physical Exercise and Behavior in the Dog: Related Hematological and Pathological Control Studies* (Boston: Boston University School of Medicine, 1953), 7.

60. Ogden R. Lindsley, "Our Harvard Pigeon, Rat, Dog, and Human Lab," *Journal of the Experimental Analysis of Behavior* 77, no. 3 (May 2002): 386; Ogden R. Lindsley, "Studies in Behavior Therapy and Behavior Research Laboratory: June 1953–1965," in *A History of the Behavioral Therapies: Founders' Personal Histories*, ed. William O'Donohue et al. (Oakland, CA: New Harbinger, 2001), 137.

61. Ogden R. Lindsley, "Conditioned Suppression of Behavior in the Dog and Some Sodium Pentobarbital Effects" (PhD diss., Harvard University, 1956).

62. Richard L. Solomon, Leon J. Kamin, and Lyman C. Wynne, "Traumatic Avoidance Learning: The Outcomes of Several Extinction Procedures with Dogs," *Journal of Abnormal and Social Psychology* 48, no. 2 (1953): 291–302.

63. Martin E. P. Seligman and Steven F. Maier, "Failure to Escape Traumatic Shock," *Journal of Experimental Psychology* 74, no. 1 (May 1967): 1–9; J. Bruce Overmier and Martin E. P. Seligman, "Effects of Inescapable Shock upon Subsequent Escape and Avoidance Responding," *Journal of Comparative and Physiological Psychology* 63, no. 1 (1967): 28–33.

64. Martin E. P. Seligman and Dennis P. Groves, "Nontransient Learned Helplessness," *Psychonomic Science* 19, no. 3 (1970): 191–92. Seligman confirmed the importance of availability and known genetic background in personal correspondence but did not state the provenance of the dogs. Fraser analyzes Seligman's experiments in the context of debates over agency and intersubjectivity in Fraser, *Dog Politics*, 177.

65. Scott, John Paul Scott Oral History, 12.

66. Victor Horsley and Robert Henry Clarke, "The Structure and Functions of the Cerebellum Examined by a New Method," *Brain* 31, no. 1 (May 1, 1908): 74, https://doi.org/10/cfnmdz.

67. Michael J. Aminoff, *Victor Horsley: The World's First Neurosurgeon and His Conscience* (Cambridge: Cambridge University Press, 2022), 65.

68. William F. Windle, foreword to *A Stereotaxic Atlas of the Dog's Brain*, by Robert K. S. Lim, Chan-Nao Liu, and Robert L. Moffitt (Springfield, IL: Charles C. Thomas, 1960), v.

69. Windle, vi.

70. Lim, Liu, and Moffitt, *Stereotaxic Atlas of the Dog's Brain*.

71. Kao Liang Chow, "*A Stereotaxic Atlas of the Dog's Brain* by Robert K. S. Lim, Chan-Nao Liu, and Robert L. Moffitt," *Perspectives in Biology and Medicine* 3, no. 4 (1960): 571.

72. Władysław Traczyk, "The Stereotaxic Instrument for the Dog and Its Appliance at Operations on the Animals Used for Chronic Experiments: Part I. The Stereotaxic Instrument," *Acta Physiologica Polonica* 10 (1959): 407–13.

73. Lim, Liu, and Moffitt, *Stereotaxic Atlas of the Dog's Brain*, 3.

74. Edward F. Domino, "A Simple Dog Head Holder for Use with the Lab Tronics Stereotaxic Instrument," *Electroencephalography and Clinical Neurophysiology* 12, no. 2 (May 1, 1960): 521–22, https://doi.org/10.1016/0013-4694(60)90037-7.

75. Associated Press, "A Senator and Ungulates," *San Francisco Examiner*, August 16, 1962, 5.

76. Edward F. MacNichol Jr., "Letters to the Editor," *Baltimore Sun*, August 21, 1962, 12.

77. Work by Traczyk and others was included in the translation and scientific abstracting programs. A later, representative document is "Assessment of Soviet Electrical Brain Stimulation Research and Applications" (Life Sciences Division, Central Intelligence Agency, July 1975), STARGATE Collection, CIA FOIA Electronic Reading Room, https://www.cia.gov/readingroom/document/cia-rdp96-00792r000600340001-4. For context, see Joel E. Dimsdale, *Dark Persuasion: A History of Brainwashing from Pavlov to Social Media* (New Haven, CT: Yale University Press, 2021).

78. Stanley Coren, *The Intelligence of Dogs: Canine Consciousness and Capabilities* (New York: Free Press, 1994), 45.

79. Only very recently have neuroanatomical studies of differences in breed behavior been undertaken. See, for instance, Erin E. Hecht et al., "Significant Neuroanatomical Variation among Domestic Dog Breeds," *Journal of Neuroscience*, September 2, 2019, 0303–19, https://doi.org/10.1523/JNEUROSCI.0303-19.2019. An overview and critical analysis of more recent canine cognition studies appears in Fraser, *Dog Politics*, 157–59. Fraser also explores Coren's ranking system in *Dog Politics*, 29–35.

80. Leo K. Bustad to Michael W. Fox, October 22, 1968, Directors Office Files for LEHR (434-90-191), "Fox, Michael" (60.56), National Archives at San Francisco.

81. Estrella Triana and Robert Pasnak, "Object Permanence in Cats and Dogs," *Animal Learning and Behavior* 9, no. 1 (March 1, 1981): 135–39, https://doi.org/10/bbpzr5.

82. Such was the case, for instance, in Andersen's *Beagle as an Experimental Dog*: Andersen promised a "psychology" section of the book in 1966, but the closest thing in the final text is a chapter on "behavior" heavily influenced by the Jackson Labs studies. See Allen C. Andersen, *The Beagle as an Experimental Dog* (Ames: Iowa State University Press, 1970). The Franks' research is explored in Feuerbacher and Wynne, "History of Dogs as Subjects in North American Experimental Psychological Research," 63; and Harry Frank et al., "Motivation and Insight in Wolf (*Canis lupus*) and Alaskan Malamute (*Canis familiaris*): Visual Discrimination Learning," *Bulletin of the Psychonomic Society* 27, no. 5 (May 1, 1989): 455–58, https://doi.org/10/ghbsms.

83. Sylvain Gagnon and François Y. Doré, "Search Behavior in Various Breeds of Adult Dogs (*Canis familiaris*): Object Permanence and Olfactory Cues," *Journal of Comparative*

Psychology 106, no. 1 (1992): 66. A useful overview of this and other research appears in Miles K. Bensky, Samuel D. Gosling, and David L. Sinn, *The World from a Dog's Point of View: A Review and Synthesis of Dog Cognition Research,*" Advances in the Study of Behavior 45 (Boston: Academic, 2013): 209–406.

84. Elizabeth Head and Norton William Milgram, "Changes in Spontaneous Behavior in the Dog Following Oral Administration of L-deprenyl," *Pharmacology Biochemistry and Behavior* 43, no. 3 (November 1, 1992): 749, https://doi.org/10/fk8cfw.

85. Norton William Milgram et al., "The Effect of L-deprenyl on Behavior, Cognitive Function, and Biogenic Amines in the Dog," *Neurochemical Research* 18, no. 12 (December 1, 1993): 1211–19, https://doi.org/10/cnw5sn.

86. P. N. Tariot et al., "Cognitive Effects of L-deprenyl in Alzheimer's Disease," *Psychopharmacology* 91, no. 4 (April 1, 1987): 489–95, https://doi.org/10/b5xmkp; T. Sunderland et al., "Cognitive Effects of Long-Term Deprenyl and Hydergine in Alzheimer's Disease Patients," *Clinical Neuropharmacology* 15, supp. 1, part B (1992): 161B.

87. Anton von Braunmühl, "'Kongophile Angiopathie' und 'senile Plaques' bei greisen Hunden," *Archiv für Psychiatrie und Zeitschrift Neurologie* 194 (1956): 396–414. For the latter, see MSS Information Corporation, *Dogs and Other Large Mammals in Aging Research: I* (New York: MSS Information Corporation, 1974).

88. John A. Hardy and Gerald A. Higgins, "Alzheimer's Disease: The Amyloid Cascade Hypothesis," *Science* 256, no. 5054 (April 10, 1992): 184. One group argued for little similarity. See Melvyn J. Ball et al., "Paucity of Morphological Changes in the Brains of Ageing Beagle Dogs: Further Evidence That Alzheimer Lesions Are Unique for Primate Central Nervous System," *Neurobiology of Aging* 4, no. 2 (June 1, 1983): 127–31, https://doi.org/10.1016/0197-4580(83)90036-2; Edythe D. London et al., "Regional Cerebral Metabolic Rate for Glucose in Beagle Dogs of Different Ages," *Neurobiology of Aging* 4 (1983): 121–26. Another, however, found more similarity in amyloid deposition than did previous researchers. See H. M. Wisniewski et al., "Aged Dogs: An Animal Model to Study Beta-Protein Amyloidogenesis," in *Alzheimer's Disease: Epidemiology, Neuropathology, Neurochemistry, and Clinics*, ed. K. Maurer, P. Riederer, and H. Beckmann (Vienna: Springer-Verlag Wien, 1990).

89. Norton William Milgram et al., "Cognitive Functions and Aging in the Dog: Acquisition of Nonspatial Visual Tasks," *Behavioral Neuroscience* 108, no. 1 (1994): 57, https://doi.org/10.1037/0735-7044.108.1.57.

90. Head and Milgram, "Changes in Spontaneous Behavior in the Dog," 756.

91. Dow Jones News Service, "Deprenyl Animal Health in Research and Development Pact," October 15, 1991, Factiva.

92. Michael J. Russell et al., "Familial Influence on Plaque Formation in the Beagle Brain," *NeuroReport* 3 (1992): 1093–96.

93. Brian J. Cummings et al., "β-Amyloid Accumulation in Aged Canine Brain: A Model of Early Plaque Formation in Alzheimer's Disease," *Neurobiology of Aging* 14, no. 6 (November 1, 1993): 248, https://doi.org/10/bnj9kp. The remaining dogs were adopted out by the Marin Humane Society. See "Historical Beagle Colony Officially Closed," *UC Davis News*, October 1, 1993, https://www.ucdavis.edu/news/historical-beagle-colony-officially-closed/.

94. William W. Ruehl et al., "Canine Cognitive Dysfunction as a Model for Human Age-Related Cognitive Decline, Dementia and Alzheimer's Disease: Clinical Presentation, Cognitive Testing, Pathology and Response to L-deprenyl Therapy," *Progress in Brain Research* 106 (1995): 217–25.

95. William W. Ruehl, A. C. DePaoli, and D. S. Bruyette, "L-deprenyl for Treatment of Behavioral and Cognitive Problems in Dogs: Preliminary Report of an Open Label Trial," in *Annual Meeting of the American Veterinary Society of Animal Behavior*, vol. 39.2 (Minneapolis, MN: Applied Animal Behaviour Science, 1993), 191.

96. Ruehl et al., "Canine Cognitive Dysfunction as a Model for Human Age-Related Cognitive Decline, Dementia and Alzheimer's Disease," 223–24.

97. Multiple participants in this early research noted the general perception during interviews for this book.

98. Dow Jones News Service, "Company Plants to Expand Trials," March 27, 1992, Factiva. Cushing syndrome affects humans, but the canine variant is usually called "Cushing's disease in dogs."

99. "Deprenyl Animal Health Granted Patent for L-deprenyl (Anipryl) as Treatment of Cognitive Dysfunction in Dogs," *PR Newswire*, July 6, 1992, sec. Financial News.

100. Terry Raffensperger, "Phoenix Scientific Joins Forces with Kansas Firm," *St. Joseph News-Press*, June 30, 1992. The new patent allowance is reported in "Deprenyl Animal Health, Inc. Reports US Patent Allowance for Use of Anipryl in Maintaining Weight During Mammalian Aging," *PR Newswire*, December 17, 1992.

101. NY Times News Service, "Helping Old Dogs Remember," *New Mexican*, October 4, 1992, C8.

102. "Deprenyl Animal Health, Inc. Reports Preliminary Positive Results in Use of Anipryl for Treating Cognitive Dysfunction (Dementia) in Pet Dogs," *PR Newswire*, January 14, 1993, Factiva.

103. Bruce Owen, "PAL Addresses Unwanted Animal Problem," *Winnipeg Free Press*, June 14, 1994, C1.

104. Ray Moynihan and David Henry, "The Fight against Disease Mongering: Generating Knowledge for Action," *PLoS Med* 3, no. 4 (2006): e191.

105. Elizabeth Head, personal communication, 2019.

106. Owen, "PAL Addresses Unwanted Animal Problem," C1.

107. "Senile Dogs Can Bark Again, Thanks to K-State Researcher," *Hutchinson (Kansas) News*, October 23, 1992, 3.

108. Owen, "PAL Addresses Unwanted Animal Problem," C1.

109. Kane Race, *Pleasure Consuming Medicine: The Queer Politics of Drugs* (Durham, NC: Duke University Press, 2009), 16.

110. Alastair Dow, "Morty's Latest Stock Venture Is No Dog," *Toronto Star*, March 23, 1991, C2.

111. Andrew Leckey, "Not Time to Shop Your Dayton Stock," *Chicago Tribune*, November 18, 1991, B3.

112. Kathleen McDermott, "Parasite Drug on Sale after Years of Work," *Central New Jersey Home News*, July 28, 1987, C7.

113. Associated Press, "It's a Dog's Life: New Pill to Fight Fleas," February 16, 1995, Factiva.

114. Pablo Junquera, "Lufenuron and Program: Episodes on Their Discovery, Development and Marketing," *Parasitipedia*, 2014, https://parasitipedia.net/index.php?option=com_content&view=article&id=3059&Itemid=2953.

115. Farrell Crook, "New Once-a-Month Tablet Knocks Out Cat, Dog Fleas," *Toronto Star*, June 16, 1994, Factiva.

116. Junquera, "Lufenuron and Program."

117. Dow Jones News Service, "Deprenyl Animal to Take Part in Vet Medicine Rules Review," September 24, 1991, Factiva.

118. Canada NewsWire, "Deprenyl Animal Health, Inc. Files Phase III Equivalent Clinical Trial Report with the US FDA," September 28, 1995, Factiva.

119. Reuters, "Deprenyl, Draxis Sign Agreement," December 19, 1995, Factiva.

120. Stephen Northfield, "Deprenyl Stock Rises on Hopes for Dog Drug," *Globe and Mail*, May 8, 1996, B16.

121. Business Wire, "DAHI Submits Canine Cognitive Dysfunction Claim for Anipryl to Canada's Bureau of Veterinary Drugs," June 10, 1996, Factiva.

122. See Joseph Dumit, *Drugs for Life: How Pharmaceutical Companies Define Our Health* (Durham, NC: Duke University Press, 2012). Eric Cazdyn has theorized the emergence of a "new chronic" paradigm of treatment and time to describe a parallel transformation. See Eric Cazdyn, *The Already Dead: The New Time of Politics, Culture, and Illness* (Durham, NC: Duke University Press, 2012), 18.

123. Elaine Carey, "Parkinson's Drug May Perk Up Aging Pooches: Draxis Health Could Have a Booming Market on Its Hands," *Toronto Star*, June 26, 1996, C3.

124. Sharon Voas, "When Old Dogs Forget Old Tricks: New Drug Offers Hope," *Pittsburgh Post-Gazette*, January 5, 1999, D2.

125. Paul Taylor, "Pampered Pets a Marketer's Dream," *Globe and Mail*, December 24, 1997, A9.

126. Taylor, A9.

127. Junquera, "Lufenuron and Program."

128. Stephen Northfield, "Draxis, Pfizer Do Deal on Dotty-Dog Drug," *Globe and Mail*, November 14, 1997, B12.

129. The ad ran widely. See, for instance, Pfizer Animal Health, advertisement for Anipryl, *Vancouver Sun*, June 2, 1998, B6.

130. Pfizer Animal Health, B6.

131. Lawrence Cohen, *No Aging in India: Alzheimer's, the Bad Family, and Other Modern Things* (Berkeley: University of California Press, 2000).

132. References to "signs" of aging appear throughout Anipryl ads, including the one previously cited. See also Stephen Katz and Barbara L. Marshall, "Is the Functional 'Normal'? Aging, Sexuality and the Bio-marking of Successful Living," *History of the Human Sciences* 17, no. 1 (February 1, 2004): 53–75, https://doi.org/10.1177/0952695104043584.

133. Dumit, *Drugs for Life*, 55–56.

134. The reference here is to Louis Althusser's concept of "hailing" (interpellation): Louis Althusser, "Ideology and Ideological State Apparatuses (Notes towards an Investigation)," *Lenin and Philosophy, and Other Essays*, trans. Ben Brewster (New York: Monthly Review Press, 2001): 117–18. For another application in contemporary biomedicine, see Mette N. Svendsen and Lene Koch, "Between Neutrality and Engagement: A Case Study of Recruitment to Pharmacogenomic Research in Denmark," *BioSocieties* 3, no. 4 (December 1, 2008): 399–418, https://doi.org/10.1017/S1745855208006315.

135. "Spending Surges behind Pet Drugs," *Adweek*, March 8, 1999. See also Marlene Cimons, "Animal Drugs Become Big Pet Project for Industry," *Los Angeles Times*, October 12, 1999, A1, A8.

136. "Canine Cognitive Dysfunction Syndrome Can Be Controlled; Take the CDS Signs Quiz," *PR Newswire*, April 15, 1999. The Purina Institute now offers a more complicated evaluation tool, called DISHAA (for "Disorientation, interactions, sleep/wake cycles, housesoiling, learning and memory, activity, anxiety"), to evaluate the presence of CDS in dogs. For "responsibilization," see Nikolas Rose, "The Death of the Social? Re-figuring the Territory of Government," *Economy and Society* 25, no. 3 (1996): 327–56, https://doi.org/10.1080/03085149600000018; Susanna

Trnka and Catherine Trundle, "Competing Responsibilities: Moving beyond Neoliberal Responsibilisation," *Anthropological Forum* 24, no. 2 (April 3, 2014): 136–53, https://doi.org/10.1080/00664677.2013.879051.

137. An expanding literature in the sociology of veterinary practice has emphasized the peculiarity of the "triad" of pet/owner/veterinarian. See, for an overview, Pru Hobson-West and Annemarie Jutel, "Animals, Veterinarians and the Sociology of Diagnosis," *Sociology of Health and Illness* 42, no. 2 (2020): 393–406, https://doi.org/10.1111/1467-9566.13017. For recent investments into alternative diagnostic tools, see Ian Thomas, "The pet drugs vets are now prescribing look a lot more like human medications," CNBC, February 28, 2024, https://www.cnbc.com/2024/02/28/new-pet-drugs-are-looking-a-lot-more-like-human-medications.html.

138. Haraway, "Value-Added Dogs and Lively Capital," 98.

139. Denise Grady, "Human Drugs Approved for Mental Problems in Dogs," *New York Times*, January 6, 1999, 15.

140. Haraway, "Value-Added Dogs and Lively Capital," 100.

141. Naomi Aoki, "Pet Rx Drug Firms Are Coming Out with Dog Versions of Human Drugs to Feed a Healthcare Spending Boom," *Boston Globe*, May 11, 2003, C1.

142. Elyse Tanouye, "The Ow in Bowwow: With Growing Market in Pet Drugs, Makers Revamp Clinical Trials," *Wall Street Journal*, April 13, 1999, A1.

143. Cristine Russell, "Odds of a Vaccine," *Washington Post*, July 14, 1992, 15. For the longer history of controversy over the human vaccine, see Robert A. Aronowitz, "The Rise and Fall of the Lyme Disease Vaccines: A Cautionary Tale for Risk Interventions in American Medicine and Public Health," *Milbank Quarterly* 90, no. 2 (June 2012): 250–77, https://doi.org/10.1111/j.1468-0009.2012.00663.x.

144. Aoki, "Pet Rx Drug Firms," C1.

145. Ben Steelman, "Depressed Dogs and Compulsive Cats," *Morning Star*, January 28, 1999, D1–2.

146. Braitman, *Animal Madness*, 225.

147. Grady, "Human Drugs Approved for Mental Problems in Dogs," 15.

148. Chris Adams, "Most Arthritic Dogs Do Very Well on This Pill, Except Ones That Die," *Wall Street Journal*, March 13, 2000, sec. Front Section, https://www.wsj.com/articles/SB952905343216013738.

149. "Spending Surges behind Pet Drugs."

150. Pfizer Animal Health, advertisement for Anipryl, 1999, Rebecca's TV Commercial Archive, https://www.youtube.com/watch?v=cIhLopYvmBw.

151. Thomas, "Pet Drugs."

152. Persistence Market Research, "Companion Animal Health Market to Reach a Valuation of US$61.6 Billion by 2033," *Globe Newswire*, January 16, 2024.

153. Ken Silverstein, "Millions for Viagra, Pennies for Diseases of the Poor: Research Money Goes to Profitable Lifestyle Drugs," *Nation*, July 19, 1999, 13.

154. Michael Meyer, "When Pets Pop Pills: The Big Drugmakers Are Racing to Develop Products for the Lucrative 'Companion Animal' Market," *Newsweek*, October 11, 1999, 60.

155. Silverstein, "Millions for Viagra," 13.

156. Peter Kuitenbrouwer, "Morty Shulman's Last Battle: Weakened by Parkinson's, the Legendary Doctor Still Rages at the Men Who Cut Him Out of Draxis Health's Big Payoff," *Financial Post*, March 28, 1998, 8.

157. Leonard Zehr, "Draxis Defeats Shulman Challenge: Dissident Group, Aided by Former Founder, Fails in Bid to Replace Four Board Members," *Globe and Mail*, June 26, 1998, B5.

158. Dora Games et al., "Alzheimer-Type Neuropathology in Transgenic Mice Overexpressing V717F β-Amyloid Precursor Protein," *Nature* 373, no. 6514 (February 1995): 523–27, https://doi.org/10.1038/373523a0.

159. Philippa J. Johnson et al., "Stereotactic Cortical Atlas of the Domestic Canine Brain," *Scientific Reports* 10, no. 1 (March 16, 2020): 4781, https://doi.org/10.1038/s41598-020-61665-0.

160. Brian J. Cummings et al., "The Canine as an Animal Model of Human Aging and Dementia," *Neurobiology of Aging* 17, no. 2 (March 1, 1996): 259, https://doi.org/10/d2fhcd.

161. Milgram et al., "Cognitive Functions and Aging in the Dog," 58.

162. Margaret Lock, *The Alzheimer Conundrum: Entanglements of Dementia and Aging* (Princeton, NJ: Princeton University Press, 2013), 5.

163. Elizabeth Head, "The Canine as an Animal Model of Human Aging and Dementia: A Behavioral and Neurobiological Analysis" (PhD diss., University of Toronto, 1997), 33.

164. Elizabeth Head, personal communication. See references in Head, "Canine as an Animal Model of Human Aging and Dementia," 28.

165. Teofil Simchowicz, "Histologische Studien über die senile Demenz," in *Histologische und histopathologische Arbeiten über die Großhirnrinde mit besonderer Berücksichtigung der pathologischen Anatomie der Geisteskrankheiten*, ed. Franz Nissl and Alois Alzheimer, vol. 4 (Jena: Gustav Fischer, 1911), 356–57. For more on Simchowicz, see Andrzej Grzybowski, Aleksandra Pięta, and Marta Pugaczewska, "Teofil Simchowicz (1879–1957)," *Journal of Neurology* 264, no. 8 (2017): 1831–32, https://doi.org/10.1007/s00415-017-8460-9.

166. Gonzalo Rodríguez Lafora, "Neoformaciones dendríticas en las neuronas y alteraciones de la neuroglia en el perro senil," in *Trabajos del laboratorio de investigaciones biológicas de la Universidad de Madrid*, ed. Santiago Ramón y Cajal (Madrid: Hijos de Nicolás Moya, 1914).

167. Breed appeared irrelevant to von Braunmühl; see Braunmühl, "'Kongophile Angiopathie' und 'senile Plaques' bei greisen Hunden," 402. Not to be confused with Anton von Braunmühl the mathematician, von Braunmühl the psychiatrist played a prominent role in popularizing electroshock therapy during the Third Reich. See Lara Rzesnitzek and Sascha Lang, "'Electroshock Therapy' in the Third Reich," *Medical History* 61, no. 1 (January 2017): 66–88, https://doi.org/10.1017/mdh.2016.101.

168. Henryk M. Wiśniewski et al., "Senile Plaques and Cerebral Amyloidosis in Aged Dogs: A Histochemical and Ultrastructural Study," *Laboratory Investigation* 23 (1970): 287–96; Henryk M. Wiśniewski et al., "Aged Dogs: An Animal Model to Study Beta-Protein Amyloidogenesis," in *Alzheimer's Disease: Epidemiology, Neuropathology, Neurochemistry, and Clinics*, ed. K. Maurer, P. Riederer, and H. Beckmann (Vienna: Springer-Verlag Wien, 1990).

169. Betsy J. Stover, "*The Beagle as an Experimental Dog* by Allen C. Andersen," *Radiation Research* 45, no. 2 (February 1971): 450.

170. Cummings et al., "Canine as an Animal Model of Human Aging and Dementia," 261.

171. Cummings et al., 261.

172. *DSM-IV: Diagnostic and Statistical Manual of Mental Disorders* (Washington, DC: American Psychiatric Association, 1994), 142.

173. Head, "Canine as an Animal Model of Human Aging and Dementia," 19.

174. Cummings et al., "Canine as an Animal Model of Human Aging and Dementia," 264.

175. Quoted in Donna J. Haraway, *Primate Visions: Gender, Race, and Nature in the World of Modern Science* (New York: Routledge, 1989), 237.

176. Haraway, 237.

177. Cummings et al., "Canine as an Animal Model of Human Aging and Dementia," 265.

178. Elizabeth Head et al., "Cognitive Function and Alzheimer's-Like Pathology in the Aged Canine: I. Spatial Learning and Memory," in *Abstracts of the Society for Neuroscience*, vol. part 2 (Washington, DC: Society for Neuroscience 23rd Annual Meeting 1993), 1046.

179. See Sarah-Elizabeth Byosiere et al., "What Do Dogs (*Canis familiaris*) See? A Review of Vision in Dogs and Implications for Cognition Research," *Psychonomic Bulletin and Review* 25, no. 5 (October 1, 2018): 1798–813, https://doi.org/10.3758/s13423-017-1404-7; Niranjana Rajalakshmi, "What Colors Do Dogs See?," *Scientific American*, October 2, 2023, https://www.scientificamerican.com/article/what-colors-do-dogs-see/.

180. H. Kalmus, "The Discrimination by the Nose of the Dog of Individual Human Odours and in Particular of the Odours of Twins," *British Journal of Animal Behaviour* 3, no. 1 (January 1, 1955): 25–31, https://doi.org/10/bvvfkz. More recent research suggests the process is aided by unique nasal airflow patterns. See Brent A. Craven, Eric G. Paterson, and Gary S. Settles, "The Fluid Dynamics of Canine Olfaction: Unique Nasal Airflow Patterns as an Explanation of Macrosmia," *Journal of the Royal Society Interface* 7, no. 47 (June 6, 2010): 933–43, https://doi.org/10/dt6b3p.

181. Alexandra Horowitz, "Smelling Themselves: Dogs Investigate Their Own Odours Longer When Modified in an 'Olfactory Mirror' Test," *Behavioural Processes* 143 (October 1, 2017): 17–24, https://doi.org/10.1016/j.beproc.2017.08.001.

182. Head et al., "Cognitive Function and Alzheimer's-Like Pathology in the Aged Canine," 1046.

183. Ira T. Lott and Elizabeth Head, "Dementia in Down Syndrome: Unique Insights for Alzheimer Disease Research," *Nature Reviews: Neurology* 15, no. 3 (March 2019): 135–47, https://doi.org/10.1038/s41582-018-0132-6.

184. J. M. Delabar et al., "Beta Amyloid Gene Duplication in Alzheimer's Disease and Karyotypically Normal Down Syndrome," *Science* 235, no. 4794 (March 13, 1987): 1390–92, https://doi.org/10.1126/science.2950593. For a representative news report, see Gina Kolata, "Gene Mutation That Causes Alzheimer's Is Found," *New York Times*, February 16, 1991, 1, 14.

185. See Lily E. Kay, *Who Wrote the Book of Life? A History of the Genetic Code* (Stanford, CA: Stanford University Press, 2000): xv–xvi; see also Evelyn Fox Keller, *The Century of the Gene* (Cambridge, MA: Harvard University Press, 2009).

186. In 2001, Lott and Head summarized that Alzheimer's brain lesions in individuals with Down syndrome were similar, if not identical, to Alzheimer's lesions in the general population when visualized by light and electron microscopy. See Ira T. Lott and Elizabeth Head, "Down Syndrome and Alzheimer's Disease: A Link between Development and Aging," *Mental Retardation and Developmental Disabilities Research Reviews* 7, no. 3 (2001): 172, https://doi.org/10.1002/mrdd.1025.

187. Bassem Y. Azizeh et al., "Molecular Dating of Senile Plaques in the Brains of Individuals with Down Syndrome and in Aged Dogs," *Experimental Neurology* 163, no. 1 (May 1, 2000): 113, https://doi.org/10.1006/exnr.2000.7359.

188. Marlene Oscar-Berman and Stuart M. Zola-Morgan, "Comparative Neuropsychology and Korsakoff's Syndrome. I—Spatial and Visual Reversal Learning," *Neuropsychologia* 18, no. 4–5 (1980): 499–512, https://doi.org/10.1016/0028-3932(80)90152-9. Oscar-Berman worked at the Boston VA Medical Center and Boston University, while Zola-Morgan worked at the San Diego VA and UC San Diego.

189. Oscar-Berman and Zola-Morgan, 500.

190. Studies of alcoholism in experimental animals had adopted a "situated" perspective of their organisms, wherein the environment became a key part of the modeling process. See Sabina Leonelli et al., "Making Organisms Model Human Behavior: Situated Models in North-American Alcohol Research, 1950–Onwards," *Sci Context* 27, no. 3 (September 2014): 485–509.

191. Marlene Oscar-Berman, "Comparative Neuropsychology: Brain Functions in Nonhuman Primates and Human Neurobehavioral Disorders," in *Neuropsychological Explorations of Memory and Cognition: Essays in Honor of Nelson Butters*, ed. Laird S. Cermak (New York: Springer Science and Business Media, 1994), 18.

192. Morris Freedman and Marlene Oscar-Berman, "Spatial and Visual Learning Deficits in Alzheimer's and Parkinson's Disease," *Brain and Cognition* 11, no. 1 (September 1, 1989): 114–26, https://doi.org/10.1016/0278-2626(89)90009-2.

193. Linda Nelson et al., "Learning and Memory as a Function of Age in Down Syndrome: A Study Using Animal-Based Tasks," in "Canine Model of Cognitive Aging: Further Developments and Practical Applications," ed. Candace J. Ikeda-Douglas, Christina de Rivera, and Norton W. Milgram, special issue, *Progress in Neuro-psychopharmacology and Biological Psychiatry* 29, no. 3 (March 1, 2005): 444, https://doi.org/10.1016/j.pnpbp.2004.12.009.

194. For the diagnostic, previously the Dementia Questionnaire for Mentally Retarded Persons (DMR), see H. M. Evenhuis, "Evaluation of a Screening Instrument for Dementia in Ageing Mentally Retarded Persons," *Journal of Intellectual Disability Research: JIDR* 36, part 4 (August 1992): 337–47, https://doi.org/10.1111/j.1365-2788.1992.tb00532.x. The institute's full name was the Hooge Burch Institute for Mentally Retarded People.

195. Nelson et al., "Learning and Memory as a Function of Age in Down Syndrome," 451.

196. David Wright, *Downs: The History of a Disability* (Oxford: Oxford University Press, 2010), 95–96.

197. Jeff McMahan, *The Ethics of Killing: Problems at the Margins of Life* (Oxford: Oxford University Press, 2002), 146.

198. Michael Lundblad, "Disanimality: Disability Studies and Animal Advocacy," *New Literary History* 51, no. 4 (2020): 765–95, https://doi.org/10.1353/nlh.2020.0048. See also Eva Feder Kittay, "At the Margins of Moral Personhood," *Ethics* 116, no. 1 (October 2005): 100–131, https://doi.org/10.1086/454366.

199. Fraser, *Dog Politics*, 133.

200. Kittay, "At the Margins of Moral Personhood," 130.

201. Environmenticity, which I mean to resonate with "historicity," relates to what Lucie Gerber calls the "eco-zootechnical acquisition" of mouse lemurs and what Leonelli et al. call the "situatedness" of a model organism. Both terms, however, relate to the control and manipulation of an animal's environment in laboratory settings: the sense that the environment must be stabilized as part of the construction of a model organism. With environmenticity, on the other hand, the point is that the dog's supposedly "natural" trait of having an environment shared with humans, and perhaps one that is not

standardizable, became central to their experimental value. See Lucie Gerber, "The Art of Growing Old: Environmental Manipulation, Physiological Rhythms, and the Advent of Microcebus Murinus as a Primate Model of Aging," *History and Philosophy of the Life Sciences* 42, no. 2 (June 11, 2020): 26, https://doi.org/10/gg2hm6; Leonelli et al., "Making Organisms Model Human Behavior."

202. Vadim N. Gladyshev, "The Free Radical Theory of Aging Is Dead. Long Live the Damage Theory!," *Antioxidants and Redox Signaling* 20, no. 4 (February 1, 2014): 727–31, https://doi.org/10.1089/ars.2013.5228.

203. Julia Mcnamee Neenan, "Vitamins, Fruit, Veg Curb Aging's Effect: Dogs Fed Diet Rich in Antioxidants Learn New Tricks," *Hamilton Spectator*, November 28, 2000, D4.

204. Jane E. Brody, "Making a Case for Antioxidants," *New York Times*, April 20, 1994, 51.

205. Norton William Milgram et al., "Landmark Discrimination Learning in the Dog: Effects of Age, an Antioxidant Fortified Food, and Cognitive Strategy," *Neuroscience and Biobehavioral Reviews* 26, no. 6 (October 1, 2002): 681, https://doi.org/10/fc7w7p.

206. Jane E. Brody, "Studies of the Infirmities of Aging Dogs Offer Insights for Humans," *New York Times*, February 2, 2002, sec. Science, F4.

207. Frazier Moore, "Animal's Best Friends," *Fort Myers News-Press*, May 4, 1977, D1.

208. Jeanette Branin, "Pet Doctor Creates Specialized Diets," *Advocate*, July 21, 1979, 7.

209. Marty Becker, "New dog food shows promise in prolonging mental capacity," *Clarion-Ledger*, January 31, 2002, 6E.

210. Chris Duke, "Antioxidants May Help Fight Behaviour Changes in Aging Dogs," *Hamilton Spectator*, December 22, 2001, E6.

211. Norton William Milgram et al., "Dietary Enrichment Counteracts Age-Associated Cognitive Dysfunction in Canines," *Neurobiology of Aging: Brain Aging; Identifying the Brakes and Accelerators* 23, no. 5 (September 1, 2002): 737–45, https://doi.org/10.1016/S0197-4580(02)00020-9.

212. A report on the Army grant was published in 2002; see Bruce A. Muggenburg and Elizabeth Head, *The Effects of Antioxidants and Experience on the Development of Age Dependent Cognitive Dysfunction and Neuropathology in Canines* (Fort Detrick, MD: US Army Medical Research and Materiel Command, October 2002).

213. Norton William Milgram et al., "Learning Ability in Aged Beagle Dogs Is Preserved by Behavioral Enrichment and Dietary Fortification: A Two-Year Longitudinal Study," *Neurobiology of Aging* 26, no. 1 (January 1, 2005): 77–90, https://doi.org/10.1016/j.neurobiolaging.2004.02.014.

214. Margaret Fahnestock et al., "BDNF Increases with Behavioral Enrichment and an Antioxidant Diet in the Aged Dog," *Neurobiology of Aging* 33, no. 3 (March 1, 2012): 546–54, https://doi.org/10/b4jvr5.

215. Steven C. Zicker, "Cognitive and Behavioral Assessment in Dogs and Pet Food Market Applications," in Ikeda-Douglas, de Rivera, and Milgram, "Canine Model of Cognitive Aging," 455–59, https://doi.org/10.1016/j.pnpbp.2004.12.010.

216. National Institute on Aging, "Diet, Exercise, Stimulating Environment Helps Old Dogs Learn," *NIH News*, January 18, 2005, Wayback Machine.

217. Lock, *Alzheimer Conundrum*, 97.

218. CanCog, "Company," https://www.cancog.com/company, accessed March 2024.

219. Candace J. Ikeda-Douglas, Christina de Rivera, and Norton William Milgram, "Pharmaceutical and Other Commercial Uses of the Dog Model," in Ikeda-Douglas, de Rivera, and Milgram, "Canine Model of Cognitive Aging," 355–60, https://doi.org/10.1016/j.pnpbp.2004.12.001.

220. Carl W. Cotman and Elizabeth Head, "The Canine (Dog) Model of Human Aging and Disease: Dietary, Environmental and Immunotherapy Approaches," *Journal of Alzheimer's Disease* 15, no. 4 (January 1, 2008): 686, https://doi.org/10.3233/JAD-2008-15413.

221. Scott, John Paul Scott Oral History, 13.

222. "Alzheimer's 'Epidemic' Now a Deadlier Threat to Elderly," *Morning Edition* (NPR, March 19, 2013), https://www.npr.org/sections/health-shots/2013/03/19/174651566/alzheimers-epidemic-now-a-deadlier-threat-to-elderly. Between 2000 and 2010, it became the sixth leading cause of death in the United States, although historians of medicine David Jones and Jeremy Greene have argued, on the basis of findings that dementia is declining, that "our burden of disease is malleable." See David S. Jones and Jeremy A. Greene, "Is Dementia in Decline? Historical Trends and Future Trajectories," *New England Journal of Medicine* 374 (February 2016): 507–9.

223. See, for instance, David J. Waters, "Aging Research 2011: Exploring the Pet Dog Paradigm," *ILAR Journal* 52, no. 1 (January 1, 2011): 98, https://doi.org/10.1093/ilar.52.1.97.

224. Sára Sándor et al., "Man's Best Friend in Life and Death: Scientific Perspectives and Challenges of Dog Brain Banking," *GeroScience*, May 10, 2021, 1653, https://doi.org/10.1007/s11357-021-00373-7.

225. Sándor et al., 1654.

226. These numbers come from a report by the NRC, although use of dogs in research remains difficult to track on a worldwide scale. See Committee on Scientific and Humane Issues in the Use of Random Source Dogs and Cats in Research, *Scientific and Humane Issues in the Use of Random Source Dogs and Cats in Research* (Washington, DC: Institute for Laboratory Animal Research, Division on Earth and Life Studies, National Research Council of the National Academies, 2009), 72, https://www.ncbi.nlm.nih.gov/books/NBK32671/.

227. Brad Bolman, "In the Animal House: Rabies, Labor, and Salvage Dogs in Jim Crow Birmingham," *Historical Studies in the Natural Sciences* 52, no. 5 (November 2022): 600–601.

228. Lesley A. Sharp, *Animal Ethos: The Morality of Human-Animal Encounters in Experimental Lab Science* (Berkeley: University of California Press, 2018), 209. On the 3Rs, see Gail Davies et al., "Science, Culture, and Care in Laboratory Animal Research: Interdisciplinary Perspectives on the History and Future of the 3Rs," *Science, Technology, and Human Values* 43, no. 4 (July 1, 2018): 603–21, https://doi.org/10.1177/0162243918757034; Robert G. W. Kirk, "Recovering the Principles of Humane Experimental Technique: The 3Rs and the Human Essence of Animal Research," *Science, Technology, and Human Values* 43, no. 4 (July 1, 2018): 622–48, https://doi.org/10.1177/0162243917726579.

229. Robert E. Kohler, *Lords of the Fly: Drosophila Genetics and the Experimental Life* (Chicago: University of Chicago Press, 1994), 11.

230. Simon Werrett, *Thrifty Science: Making the Most of Materials in the History of Experiment* (Chicago: University of Chicago Press, 2019).

231. On the connection between tissue databases and other traces of past research, see Brad Bolman, "Strontium," in *An Anthropogenic Table of Elements*, ed. Timothy Neale, Thao Phan, and Courtney Addison (Toronto: University of Toronto Press, 2023).

232. For additional work on "citizen science," see Alan Irwin, *Citizen Science: A Study of People, Expertise and Sustainable Development* (New York: Routledge, 1995); Caren B. Cooper and Bruce V. Lewenstein, "Two Meanings of Citizen Science," in *The Rightful Place of Science: Citizen Science*, ed. Darlene Cavalier and Eric B. Kennedy (Tempe, AZ: Consortium of Science, Policy and Outcomes, 2016), 51–61; Rebecca Jordan et al., "Citizen

Science as a Distinct Field of Inquiry," *BioScience* 65, no. 2 (February 1, 2015): 208–11, https://doi.org/10.1093/biosci/biu217.

233. Amy Harmon, "Dogs Test Drug Aimed at Humans' Biggest Killer: Age," *New York Times*, May 16, 2016, https://www.nytimes.com/2016/05/17/us/aging-research-disease-dogs.html.

234. Waters, "Aging Research 2011," 97.

235. Mitch Leslie, "Massive Study of Dog Aging Likely to Lose Funding," *Science* 383, no. 6679 (2024): 137–38.

236. Matt Kaeberlein, "The Biology of Aging: Citizen Scientists and Their Pets as a Bridge between Research on Model Organisms and Human Subjects," *Veterinary Pathology* 53, no. 2 (March 1, 2016): 295, https://doi.org/10.1177/0300985815591082.

237. Chand Khanna and David M. Vail, "Targeting the Lung: Preclinical and Comparative Evaluation of Anticancer Aerosols in Dogs with Naturally Occurring Cancers," *Current Cancer Drug Trends* 3 (2003): 266.

238. "Are You Conducting Experiments on Captive Dogs?," Dog Aging Project, 2024, https://dogagingproject.zendesk.com/hc/en-us/articles/360057966873-Are-you-conducting-experiments-on-captive-dogs, accessed March 2024.

239. Kate L. Tsai, Leigh Anne Clark, and Keith E. Murphy, "Understanding Hereditary Diseases Using the Dog and Human as Companion Model Systems," *Mammalian Genome* 18, no. 6 (July 1, 2007): 444–51, https://doi.org/10.1007/s00335-007-9037-1.

240. Sam Jones, "Dog Cancer Research Advances Pursuit of Drugs for Humans and Canines," *Washington Post*, June 13, 2023, https://www.washingtonpost.com/wellness/2023/06/19/dog-cancer-trials-drugs-human-treatment/.

241. See, for instance, Michael K. Guy et al., "The Golden Retriever Lifetime Study: Establishing an Observational Cohort Study with Translational Relevance for Human Health," *Philosophical Transactions of the Royal Society B: Biological Sciences* 370, no. 1673 (July 19, 2015): 20140230, https://doi.org/10.1098/rstb.2014.0230.

242. A widely referenced version of this story is provided in Carl Bialik, "The Seven-Year Glitch: Looking for a Better Man-to-Dog Lifespan Ratio," *Wall Street Journal*, August 29, 2008, sec. Technology, https://www.wsj.com/articles/SB121997338176882143.

243. Marcus M. Mason and Albert M. Scheflen, "Senility in Dogs," *Cornell Veterinarian* 43 (1953): 10.

244. Knud Sand, "Experiments on the Endocrinology of the Sexual Glands," *Endocrinology: The Bulletin of the Association for the Study of Internal Secretions* 7 (1923): 285.

245. "Curious New Discoveries about Dogs," *San Francisco Examiner*, June 12, 1932, AW9.

246. Albert Lebeau, "L'âge du chien et celui de l'homme: Essai de statistique sur la mortalité canine," *Bulletin de l'Académie vétérinaire de France* 26 (April 1953): 230.

247. The numbers provided in news accounts of the Gaines scale match Lebeau's exactly. See, for instance, "Man-Dog Age Equivalents," *Veterinary News* 22, no. 6 (November 1958): 10. For the beagle chart, see Leo K. Bustad et al., "The Choice of the Beagle for Radiobiologic Studies," in *Radiobiology of Plutonium* (Salt Lake City: J.W., 1972), 209.

248. Tina Wang et al., "Quantitative Translation of Dog-to-Human Aging by Conserved Remodeling of the DNA Methylome," *Cell Systems* 11, no. 2 (August 26, 2020): 176–85.e6, https://doi.org/10.1016/j.cels.2020.06.006.

249. Scientists involved with the University of Vienna's Clever Dog Project have also experimentally demonstrated that canine personalities change over time, a breakthrough that many dog owners would consider far more predictable. See James Gorman, "Old

Dogs, New Research and the Secrets of Aging," *New York Times*, November 9, 2020, https://www.nytimes.com/2020/11/09/science/dogs-aging-behavior.html.

250. The project's logo could be said to feature three beagle-looking dogs, but beagles appear in almost none of the images now available on the website. The description of dogs as "citizen scientists" has disappeared since I initially analyzed it in Brad Bolman, "How Experiments Age: Gerontology, Beagles, and Species Projection at Davis," *Social Studies of Science* 48, no. 2 (2018): 252–53, https://doi.org/10.1177/0306312718759822. In 2024, the program referred to its style of research as "community science."

Conclusion

1. *One Minute Speeches*, 117th Congress (October 26, 2021), statement of Madison Cawthorn, Representative from North Carolina, https://www.c-span.org/congress/?chamber=house&date=2021-10-26.

2. An initial evaluation of these claims appeared in D'Angelo Gore, "Answering Questions about #BeagleGate," FactCheck.org, November 2, 2021, https://www.factcheck.org/2021/11/answering-questions-about-beaglegate/. Following the publication of Fauci's *On Call: A Doctor's Journey in Public Service* (New York: Viking, 2024), additional analysis of the beagle accusations confirmed earlier reasons for skepticism. See Martin Pengelly, "'Fauci Kills Puppies': Inside the Forgotten Fake 'Beaglegate' Scandal," *Daily Beast*, June 14, 2024, https://www.thedailybeast.com/fauci-kills-puppies-memoir-details-forgotten-fake-beaglegate-scandal. Glenn Kessler has nevertheless noted that Fauci's NIH was less transparent than it might have been about the studies. See Glenn Kessler, "Unpacking the Story of Fauci and Painful Experiments Involving Dogs," *Washington Post*, June 7, 2024, https://www.washingtonpost.com/politics/2024/06/07/unpacking-story-fauci-painful-experiments-involving-dogs/.

3. Lauren Boebert (@laurenboebert), "Beagle Lives Matter," Twitter (now X), October 25, 2021, https://x.com/laurenboebert/status/1452608951166672899.

4. Yasmeen Abutaleb and Beth Reinhard, "Fauci Swamped by Angry Calls over Beagle Experiments after Campaign That Included Misleading Image," *Washington Post*, January 7, 2022, https://www.washingtonpost.com/investigations/2021/11/19/fauci-beagle-white-coat-waste/.

5. April Rubin, "Harry and Meghan Adopt a Mistreated Beagle Mom," *New York Times*, August 24, 2022, sec. US, https://www.nytimes.com/2022/08/24/us/meghan-prince-harry-beagle-rescue-adoption.html. On the iconicity of beagles, see Lesley A. Sharp, *Animal Ethos: The Morality of Human-Animal Encounters in Experimental Lab Science* (Berkeley: University of California Press, 2018), 71.

6. This was reported the following month in "Envigo Pleads Guilty to Neglecting Dogs, Faces $35M Fine," *American Veterinary Medical Association News*, July 24, 2024, https://www.avma.org/news/envigo-pleads-guilty-neglecting-dogs-faces-35m-fine.

7. Kristen Bobst, "Astronaut Snoopy's 50-Plus Year History with NASA," *PBS SoCal*, July 17, 2019, https://www.pbssocal.org/science/snoopy-has-been-working-with-nasa-since-the-late-1950s. For one such study, see Kenneth C. Back and Anthony A. Thomas, "Aerospace Problems in Pharmacology and Toxicology," *Annual Review of Pharmacology* 10, no. 1 (1970): 395–412. The *Beagle* lander was named for Darwin's ship, *The Beagle*, one of many ships to be given that name by the British Royal Navy.

8. Jan Lindhe, Sven-Erik Hamp, and Harald Löe, "Plaque Induced Periodontal Disease in Beagle Dogs: A 4-Year Clinical, Roentgenographical and Histometrical Study," *Journal of Periodontal Research* 10 (1975): 243–55; Stanley R. Saxe et al., "Oral Debris, Calculus, and Periodontal Disease in the Beagle Dog," *Periodontics* 5, no. 5 (October 1967): 217–25.

9. Clive D. L. Wynne, "Dogs' (*Canis lupus familiaris*) Behavioral Adaptations to a Human-Dominated Niche: A Review and Novel Hypothesis," in *Advances in the Study of Behavior*, vol. 53 (Cambridge, MA: Academic, 2021), 139.

10. J. Newell Stannard, "Interview with Robley D. Evans," 1978, 6, J. Newell Stannard Papers, Betsey B. Creekmore Special Collections and University Archives, University of Tennessee, Knoxville.

11. Dylan Mulvin, *Proxies: The Cultural Work of Standing In* (Cambridge, MA: MIT Press, 2021), 4, emphasis in the original.

12. From this, one could develop an extensive typology of criteria for selecting an experimental organism, as for instance in Michael R. Dietrich et al., "How to Choose Your Research Organism," *Studies in History and Philosophy of Science Part C: Studies in History and Philosophy of Biological and Biomedical Sciences* 80 (April 1, 2020): 101227, https://doi.org/10/ggv7qx. My goal here was to show the operation of these criteria in practice and reveal the forces that make some more significant than others in specific contexts.

13. Arnold Arluke, "'We Build a Better Beagle': Fantastic Creatures in Lab Animal Ads," *Qualitative Sociology* 17, no. 2 (1994): 143.

14. Daniel P. Todes, "Pavlov's Physiology Factory," *Isis* 88, no. 2 (June 1, 1997): 220, https://doi.org/10.1086/383690.

15. An expanded version of this argument is offered, for instance, in Andrew Linzey and Clair Linzey, eds., *The Ethical Case against Animal Experiments* (Champaign: University of Illinois Press, 2018).

16. I draw here on work by Melanie Jeske, including her study of organ chips and the debates over animal alternatives, in Melanie Jeske, "Organs and Humans on Chips: Translation, Biomedical Models, and the Political Economy of Innovation" (PhD diss., University of California, San Francisco, 2022), 41.

17. Aviva Luria, "Creators of Virtual Heart Shine in Multimedia World," *IT Times*, August 1998, http://ittimes.ucdavis.edu/v6n10aug98/calf.html; Lois Kazakoff, "Dissection without Death," *San Francisco Chronicle*, October 21, 1997, C3.

18. Gina Kolata, "New Antibiotics Offer Freedom from Hospitals," *New York Times*, January 12, 1988, C3.

19. Quoted in "Tips for the Antibiotic Top," *Times*, January 23, 1987, 19.

20. "Quinolone: Family with a Future?," *Times*, March 3, 1987, 10.

21. Gregory S. Bisacchi, "Origins of the Quinolone Class of Antibacterials: An Expanded 'Discovery Story,'" *Journal of Medicinal Chemistry* 58 (2015): 4875.

22. Brian Ingham et al., "Arthropathy Induced by Antibacterial Fused *N*-Alkyl-4-Pyridone-3-Carboxylic Acids," *Toxicology Letters* 1 (1977): 22.

23. Alec W. Gough et al., "Quinolone Arthropathy—Acute Toxicity to Immature Articular Cartilage," *Toxicologic Pathology* 20, no. 3-1 (1992): 436.

24. John E. Burkhardt, Juan N. Walterspiel, and Urs B. Schaad, "Quinolone Arthropathy in Animals versus Children," *Clinical Infectious Diseases* 25 (1997): 1197.

25. Gough et al., "Quinolone Arthropathy," 436.

26. Gough et al., 437.

27. Roger M. Echols, "Introduction: Historical Perspective—Use of Ciprofloxacin in Children," *Pediatric Infectious Disease Journal* 16, no. 1 (January 1997): 89–90.

28. At the height of America's escalating war on terror, the drug also became the subject of minor controversy when shortages in the face of the anthrax crisis nearly prompted the US government to override Bayer's patent and purchase the drug from generic manufacturers. David B. Resnik and Kenneth A. De Ville, "Bioterrorism and Patent Rights: 'Compulsory Licensure' and the Case of Cipro," *American Journal of Bioethics* 2, no. 3 (June 1, 2002): 29, https://doi.org/10.1162/152651602760250084. Anne Pollock has shown how Cipro's use and distribution reflected anti-Black racism in medicine. See Anne Pollock, *Sickening: Anti-Black Racism and Health Disparities in the United States* (Minnesota: University of Minnesota Press, 2021), 28–29.

29. World Health Organization, *Model List of Essential Medicines*, 2023, https://www.who.int/publications/i/item/WHO-MHP-HPS-EML-2023.02.

30. Martin Chalumeau et al., "Fluoroquinolone Safety in Pediatric Patients: A Prospective, Multicenter, Comparative Cohort Study in France," *Pediatrics* 111, no. 6 (June 2003): e714–19.

31. Judith Sendzik, Hartmut Lode, and Ralf Stahlmann, "Quinolone-Induced Arthropathy: An Update Focusing on New Mechanistic and Clinical Data," *International Journal of Antimicrobial Agents* 33 (2009): 194–200.

32. News Staff, "Use Fluoroquinolones Only as Drug of Last Resort for Some Infections," American Academy of Family Physicians, July 27, 2016, https://www.aafp.org/news/health-of-the-public/20160727fluoroquinolones.html.

33. Jen Christensen, "Certain Antibiotics May Cause Aortic Aneurysm, FDA Warns," CNN, December 21, 2018, https://www.cnn.com/2018/12/20/health/antibiotics-aortic-aneurysm-cipro-levaquin-bn/index.html.

34. Deanna J. Buehrle, Marilyn M. Wagener, and Cornelius J. Clancy, "Outpatient Fluoroquinolone Prescription Fills in the United States, 2014 to 2020: Assessing the Impact of Food and Drug Administration Safety Warnings," *Antimicrobial Agents and Chemotherapy* 65, no. 7 (June 17, 2021): 10.1128/aac.00151-21, https://doi.org/10.1128/aac.00151-21.

35. See, for instance, Jonathan Binz, Christina K. Adler, and Tsz-Yin So, "The Risk of Musculoskeletal Adverse Events with Fluoroquinolones in Children: What Is the Verdict Now?," *Clinical Pediatrics* 55, no. 2 (February 1, 2016): 107–10, https://doi.org/10.1177/0009922815599959; Rana E. El Feghaly and Jennifer L. Goldman, "Perceived Harm May Be Helpful: Fear of Fluoroquinolone-Associated Adverse Events in Children," *Pediatrics* 147, no. 6 (June 1, 2021): e2021050321, https://doi.org/10.1542/peds.2021-050321.

36. Reported in T. J. Hayes, G. K. S. Roberts, and W. H. Halliwell, "An Idiopathic Febrile Necrotizing Arteritis Syndrome in the Dog: Beagle Pain Syndrome," *Toxicologic Pathology* 17, no. 1, part 2 (January 1, 1989): 129–37, https://doi.org/10.1177/019262338901700109.

37. For a more recent report, see A. Tipold and S. J. Schatzberg, "An Update on Steroid Responsive Meningitis-Arteritis," *Journal of Small Animal Practice* 51, no. 3 (March 2010): 150–54.

38. On the quote, and its possibly earlier origin in the work of Albert Jan Kluyver, see Herbert C. Friedmann, "From 'Butyribacterium' to '*E. coli*': An Essay on Unity in Biochemistry," *Perspectives in Biology and Medicine* 47, no. 1 (Winter 2004): 47–66.

39. Richard Kluger, "Oscar Auerbach Notes," n.d., 2, box 17, folder 186, Richard Kluger Papers (MS 1443), Manuscripts and Archives, Yale University Library. I spoke with Auerbach's son, Bruce, in 2020. One contemporaneous report noted a sign that read, "Veterans

Hospital—Auschwitz . . . Auerbach—Commandant." See Edward Higgins, "Dog Use Picketed: Group Protests VA Hospital Study," *Newark Evening News*, October 1, 1967, 28, Newark Public Library.

40. Janny Scott, "Animal Rights Controversy Hits Encyclopedia Pages," *AV Magazine*, May 1992, 2–3.

41. "Hundreds of Beagles Yelp in Brief Bid for Freedom," *Davis Daily Democrat*, April 10, 1980, clippings, Marvin D. Goldman / Laboratory for Energy-Related Health Research (LEHR) Collection, Archives and Special Collections, University of California, Davis.

42. "Animal Liberation Front (ALF); Domestic Security/Terrorism (DS/T): OO; Los Angeles" (Federal Bureau of Investigation, March 1990), FOIA Request. This and other documents related to the investigation were produced in response to a FOIA request submitted in 2024. A follow-up request was still being processed as this book went to press.

43. Jim Carlton, "$28,500 Offered for Safe Return of UCI Lab's Dogs," *Los Angeles Times*, February 2, 1988, 1.

44. Associated Press, "Animal Liberation Front Raids Dog-and-Ferret Breeding Farm," *Associated Press Newswires*, December 6, 2001, Factiva.

45. In April 2023, the group changed its name from Animal Rebellion to Animal Rising.

46. "Beagle Breeding Protests: Employee Barraged with Insults, Court Hears," *BBC News*, April 29, 2023, https://www.bbc.com/news/uk-england-cambridgeshire-65436250.

47. It is worth noting that the study focused principally on veterinary research and found that breed was not specified in 25.3 percent of publications. The relative proportion of beagles could, thus, be higher. The most popular breeds identified were the German shepherd, Labrador retriever, and golden retriever. Because these are extremely popular companion dogs and the survey included many studies of companion animals, the results make sense. See Evelyn Schulte and Sebastian P. Arlt, "What Kinds of Dogs Are Used in Clinical and Experimental Research?," *Animals* 12, no. 12 (January 2022): 1487, https://doi.org/10.3390/ani12121487. Circularly, the study cited an earlier version of my own research on beagles: Brad Bolman, "Dogs for Life: Beagles, Drugs, and Capital in the Twentieth Century," *Journal of the History of Biology* 55, no. 1 (2022), 147–79, https://doi.org/10.1007/s10739-021-09649-2.

48. Chris Thomas, "Rescued Research Beagles Get New 'Leash' on Life," *Public News Service*, June 9, 2011, https://www.publicnewsservice.org/2011-06-09/social-justice/rescued-research-beagles-get-new-leash-on-life/a20603-1.

49. Jesse McKinley, "Dog-Related Measures Flood Legislature in Final Days," *New York Times*, June 12, 2015, A22.

50. "Crowdsourcing Animal Research," *Science*, August 21, 2015. I sent the organization a copy of one of my early articles but, to my knowledge, never received a response.

51. The story ran under its original title, "Gruesome Dog Testing Outrage," on September 26. The article is available now as Donovan Slack, "VA Tightens Oversight after Outrage over Conducting Experiments on Dogs, Then Killing Them," *USA Today*, September 25, 2017, https://www.usatoday.com/story/news/politics/2017/09/25/va-tightens-oversight-medical-experiments-dogs-after-surgery-failures-canine-deaths/690684001/.

52. David J. Shulkin, "VA Secretary: Disabled Veterans Need Research on Dogs to Continue," *USA Today*, September 12, 2017, sec. Opinion.

53. Unlike many actors subject to PETA investigations, Huntingdon fought back aggressively, suing the organization under American antiracketeering statutes. See Gina

Kolata, "Tough Tactics in One Battle over Animals in the Lab," *New York Times*, March 24, 1998, sec. Science, https://www.nytimes.com/1998/03/24/science/tough-tactics-in-one-battle-over-animals-in-the-lab.html. Huntingdon eventually became Envigo as part of a multicompany merger in 2015.

54. Eric Garcia, "Brat Bill Would Limit VA Dog Experiments," *Roll Call*, July 13, 2017, https://rollcall.com/2017/07/13/brat-bill-would-limit-va-dog-experiments/.

55. Donovan Slack, "Fatal Experiments on Dogs Are Happening at the VA. Now, the Inspector General Is Investigating," *USA Today*, February 13, 2019, https://www.usatoday.com/story/news/politics/2019/02/13/fatal-va-dog-medical-experiments-investigation-lara-trump-veterans-affairs/2841951002/.

56. See, for instance, Ryan Noah Shapiro, "Bodies at War: National Security in American Controversies over Animal and Human Experimentation from WWI to the War on Terror" (PhD diss., Massachusetts Institute of Technology, 2018), 426.

57. "Experiments on Animals: A Moral Problem" (New York: National Catholic Society for Animal Welfare, Inc., 1967), Harrison A. Williams Jr. Papers, box 524, folder 28, Special Collections and University Archives, Rutgers University, New Brunswick, NJ.

58. David Grimm, "Conservatives, Liberals Team Up against Animal Research," *Science Insider*, November 21, 2016, https://www.sciencemag.org/news/2016/11/conservatives-liberals-team-against-animal-research.

59. Peter Aldhous, "This Activist Group Tapped into Partisan COVID Politics to Make Big Trouble for Anthony Fauci and the NIH," *Buzzfeed News*, May 4, 2022, https://www.buzzfeednews.com/article/peteraldhous/white-coat-waste-anthony-fauci-nih-ecohealth-alliance-lab.

60. "Congressional Animal Protection Caucus—Members," available in profile of US congressman Earl Blumenauer, https://blumenauer.house.gov/congressional-animal-protection-caucus-members, accessed July 2024.

61. On the animal rights and terrorism connections, see Shapiro, "Bodies at War," 581.

62. John Connor Cleveland, "Why Animal Welfare Is a Conservative Cause," *National Review*, April 19, 2016, https://www.nationalreview.com/2016/04/why-animal-welfare-conservative-cause/.

63. The "Dial for Dogs" campaign appeared on the White Coat Waste Project's website, whitecoatwaste.org, in 2019.

64. COST stands for Cost Openness and Spending Transparency. See "Contact Congress," White Coat Waste Project, https://tinyurl.com/5x5stm8n, accessed July 2024.

65. Nikki Schwab, "Beagles 'Lobby' against Taxpayer-Funded Animal Experiments," *US News and World Report*, July 9, 2014, Washington Whispers Blog, https://www.usnews.com/news/blogs/washington-whispers/2014/07/09/beagles-lobby-against-taxpayer-funded-animal-experiments.

66. Darren Samuelsohn, "Lara Trump's Controversial Pet Issue," *Politico*, May 29, 2017, https://www.politico.com/story/2017/05/29/lara-trump-pet-problem-238906.

67. Sharp, *Animal Ethos*, 73.

68. Glenn Greenwald and Leighton Akio Woodhouse, "Bred to Suffer: Inside the Barbaric US Industry of Dog Experimentation," *Intercept*, May 17, 2018, https://theintercept.com/2018/05/17/inside-the-barbaric-u-s-industry-of-dog-experimentation/.

69. Christine Stevens, "Statement of the AWI in AIBS Bulletin," *Animal Welfare Institute Information Report* 5, no. 3 (June 1956): 2, https://awionline.org/sites/default/files/uploads/documents/AWI-1956-IR.pdf (AIBS is the American Institute of Biological Sciences).

70. Rhonda Cornum and W. Ron DeHaven, "Preface," in *Necessity, Use, and Care of Laboratory Dogs at the US Department of Veterans Affairs*, Committee on Assessment of the Use and Care of Dogs in Biomedical Research, Consensus Study Report (Washington, DC: National Academies Press, 2020), ix, https://www.nap.edu/read/25772/chapter/1.

71. Committee on Assessment of the Use and Care of Dogs in Biomedical Research, *Necessity, Use, and Care of Laboratory Dogs at the US Department of Veterans Affairs*, 4.

72. See Jhinuk Basu Mullick, Chelsey S. Simmons, and Janak Gaire, "Animal Models to Study Emerging Technologies against SARS-CoV-2," *Cellular and Molecular Bioengineering* 13, no. 4 (August 1, 2020): 293–303, https://doi.org/10.1007/s12195-020-00638-9.

73. See Jane Stafford, "Mongrels Get Highest Prize of Dogdom: Whipple Award Conferred by General Kirk," *Syracuse Post Standard*, February 17, 1946.

74. Committee on Assessment of the Use and Care of Dogs in Biomedical Research, *Necessity, Use, and Care of Laboratory Dogs at the US Department of Veterans Affairs*, 4.

75. Melanie Haid, "The Most Popular Dog Breeds of 2022," *American Kennel Club* (blog), March 15, 2023, https://www.akc.org/expert-advice/dog-breeds/most-popular-dog-breeds-2022/.

76. In June 2024, researchers associated with Northwestern University reported on new analysis of preserved tissues and medical data from the UC Davis beagle project. They argued for continuing attention to these and other archival data sources. See Alexander D. Glasco et al., "Revisiting the Historic Strontium-90 Ingestion Beagle Study Conducted at the University of California–Davis: Opportunity in Archival Materials," *Radiation Research*, 2024, https://doi.org/10.1667/RADE-24-000022.1.

77. David Grimm, "Suffering in Silence: Caring for Research Animals Can Take a Severe Mental Toll; Is Anyone Listening?" *Science* 379, no. 6636 (March 2023), https://www.science.org/doi/epdf/10.1126/science.adh4770.

78. *USDA's Detector Dogs: Protecting American Agriculture*, Miscellaneous Publication 1539 (Riverdale Park, MD: US Department of Agriculture Animal and Plant Health Inspection Service, December 1996).

79. "Making a Difference (Beagle Brigades)," *NBC Evening News*, October 8, 2012.

80. Heather Paxson, "'Don't Pack a Pest': Parts, Wholes, and the Porosity of Food Borders," *Food, Culture and Society* 22, no. 5 (October 20, 2019): 661, https://doi.org/10.1080/15528014.2019.1638136; see also Heather Paxson, *Sensing Food Safety at the Border* (Sydney: Society for the Social Studies of Science, 2018).

81. Deborah Block, "Dogs Sniff Out COVID-19," *VOA News*, February 24, 2021, https://www.voanews.com/covid-19-pandemic/dogs-sniff-out-covid-19.

82. "The Candidates Speak: Quotes from the '64 Oratory," *New York Times*, November 1, 1964, E1.

83. Ian Haney López, *Dog Whistle Politics: How Coded Racial Appeals Have Reinvented Racism and Wrecked the Middle Class* (Oxford: Oxford University Press, 2015).

84. Shane Goldmacher, "How Alvin the Beagle Helped Usher In a Democratic Senate," *New York Times*, January 23, 2021, sec. US, https://www.nytimes.com/2021/01/23/us/politics/raphael-warnock-puppy.html.

85. The reference is to Steven Shapin and Simon Schaffer, *Leviathan and the Air-Pump: Hobbes, Boyle, and the Experimental Life*, 2nd ed. (Princeton, NJ: Princeton University Press, 2011), 332.

86. "Man's Best Friend Faces Strontium 90," *San Francisco Chronicle*, September 17, 1956, 26.

INDEX